Green Chemistry and Sustainable Technology

Series editors

Prof. Liang-Nian He
*State Key Laboratory of Elemento-Organic Chemistry, Nankai University,
Tianjin, China*

Prof. Robin D. Rogers
*Department of Chemistry, Center for Green Manufacturing,
The University of Alabama, Tuscaloosa, USA*

Prof. Dangsheng Su
*Shenyang National Laboratory for Materials Science, Institute of Metal Research,
Chinese Academy of Sciences, Shenyang, China
and
Department of Inorganic Chemistry, Fritz Haber Institute of the Max Planck
Society, Berlin, Germany*

Prof. Pietro Tundo
*Department of Environmental Sciences, Informatics and Statistics, Ca' Foscari
University of Venice, Venice, Italy*

Prof. Z. Conrad Zhang
*Dalian Institute of Chemical Physics, Chinese Academy of Sciences,
Dalian, China*

For further volumes:
http://www.springer.com/series/11661

Green Chemistry and Sustainable Technology

Aims and Scope

The series *Green Chemistry and Sustainable Technology* aims to present cutting-edge research and important advances in green chemistry, green chemical engineering and sustainable industrial technology. The scope of coverage includes (but is not limited to):

- Environmentally benign chemical synthesis and processes (green catalysis, green solvents and reagents, atom-economy synthetic methods etc.)
- Green chemicals and energy produced from renewable resources (biomass, carbon dioxide etc.)
- Novel materials and technologies for energy production and storage (biofuels and bioenergies, hydrogen, fuel cells, solar cells, lithium-ion batteries etc.)
- Green chemical engineering processes (process integration, materials diversity, energy saving, waste minimization, efficient separation processes etc.)
- Green technologies for environmental sustainability (carbon dioxide capture, waste and harmful chemicals treatment, pollution prevention, environmental redemption etc.)

The series *Green Chemistry and Sustainable Technology* is intended to provide an accessible reference resource for postgraduate students, academic researchers and industrial professionals who are interested in green chemistry and technologies for sustainable development.

Fangming Jin
Editor

Application of Hydrothermal Reactions to Biomass Conversion

Springer

Editor
Fangming Jin
School of Environmental Science
 and Engineering
Shanghai Jiao Tong University
Shanghai
China

ISSN 2196-6982   ISSN 2196-6990 (electronic)
ISBN 978-3-662-52409-1   ISBN 978-3-642-54458-3 (eBook)
DOI 10.1007/978-3-642-54458-3
Springer Heidelberg New York Dordrecht London

© Springer-Verlag Berlin Heidelberg 2014
Softcover reprint of the hardcover 1st edition 2014
This work is subject to copyright. All rights are reserved by the Publisher, whether the whole or part of the material is concerned, specifically the rights of translation, reprinting, reuse of illustrations, recitation, broadcasting, reproduction on microfilms or in any other physical way, and transmission or information storage and retrieval, electronic adaptation, computer software, or by similar or dissimilar methodology now known or hereafter developed. Exempted from this legal reservation are brief excerpts in connection with reviews or scholarly analysis or material supplied specifically for the purpose of being entered and executed on a computer system, for exclusive use by the purchaser of the work. Duplication of this publication or parts thereof is permitted only under the provisions of the Copyright Law of the Publisher's location, in its current version, and permission for use must always be obtained from Springer. Permissions for use may be obtained through RightsLink at the Copyright Clearance Center. Violations are liable to prosecution under the respective Copyright Law.
The use of general descriptive names, registered names, trademarks, service marks, etc. in this publication does not imply, even in the absence of a specific statement, that such names are exempt from the relevant protective laws and regulations and therefore free for general use.
While the advice and information in this book are believed to be true and accurate at the date of publication, neither the authors nor the editors nor the publisher can accept any legal responsibility for any errors or omissions that may be made. The publisher makes no warranty, express or implied, with respect to the material contained herein.

Springer is part of Springer Science+Business Media (www.springer.com)

Preface

The earth's sustainable development is threatened by energy exhaustion and rising atmospheric concentrations of carbon dioxide linked to global warming. One of the causes for energy crisis and increased atmospheric carbon dioxide could be the imbalance between the rapid consumption of fossil fuels in anthropogenic activities and the slow formation of fossil fuels. An efficient method for counteracting the imbalance in the carbon cycle should involve the rapid conversion of biomass and organic waste into fuels and chemicals. For this purpose, we can learn from the geologic formation of fossil fuels. It is known that hydrothermal reaction plays an important role in forming petroleum, natural gas, and coal from organic wastes, and thus can be recognized as another pathway in the carbon cycle.

Hydrothermal reaction is generally defined as a reaction occurring in the presence of an aqueous solvent at high temperature and high pressure. The application of hydrothermal reaction to the conversion of biomass, as a relatively new technology, is receiving increasing attention. It has been demonstrated that the hydrothermal conversion of biomass shows excellent potential for the rapid conversion of a wide variety of biomass into fuels and/or value-added products. It is because high-temperature water exhibits very different properties from ambient liquid water and is environmentally friendly due to the nature of the reaction medium, i.e., water. Thus, if the geologic formation of fossil fuels in nature could be combined with the hydrothermal methods being studied for biomass conversions, an efficient scheme could be realized to recycle carbon and produce fuels and/or chemicals.

This book compiles recent advances in hydrothermal conversion of biomass into chemicals and/or fuels and consists of 15 chapters. It introduces the properties of high-temperature water, the merits of hydrothermal conversion of biomass, and some novel hydrothermal conversion processes, such as hydrothermal production of value-added products (with an emphasis on the production of organic acids), hydrothermal gasification, hydrothermal liquefaction, and hydrothermal carbonization. A wide range of biomass and biomass waste is involved in this book, from carbohydrates, lignocelluloses, and glycerine, to bio-derived chemicals and sewage sludge.

This book will help readers to expand their knowledge of biomass conversion and the carbon cycle, and facilitate understanding of how the problems associated with biomass conversion, shortage of energy, and the environment, can be solved.

It is the editor's hope that materials compiled in this book will be useful in conveying a fundamental understanding of hydrothermal conversion of biomass in the carbon cycle so that a contribution can be made to achieving sustainable energy and environment.

Fangming Jin

Contents

Part I Characters of High Temperature Water and Hydrothermal Reactions

1 Water Under High Temperature and Pressure Conditions and Its Applications to Develop Green Technologies for Biomass Conversion 3
Fangming Jin, Yuanqing Wang, Xu Zeng, Zheng Shen and Guodong Yao

Part II Hydrothermal Conversion of Biomass into Chemicals

2 Hydrothermal Conversion of Cellulose into Organic Acids with a CuO Oxidant 31
Yuanqing Wang, Guodong Yao and Fangming Jin

3 Hydrothermal Conversion of Lignin and Its Model Compounds into Formic Acid and Acetic Acid 61
Xu Zeng, Guodong Yao, Yuanqing Wang and Fangming Jin

4 Production of Lactic Acid from Sugars by Homogeneous and Heterogeneous Catalysts 83
Ayumu Onda

5 Catalytic Conversion of Lignocellulosic Biomass to Value-Added Organic Acids in Aqueous Media 109
Hongfei Lin, Ji Su, Ying Liu and Lisha Yang

6 Catalytic Hydrothermal Conversion of Biomass-Derived Carbohydrates to High Value-Added Chemicals 139
Zhibao Huo, Lingli Xu, Xu Zeng, Guodong Yao and Fangming Jin

Part III Hydrothermal Conversion of Biomass into Fuels

7 Effective Utilization of Moso-Bamboo (*Phyllostachys heterocycla*) with Hot-Compressed Water........................... 155
Satoshi Kumagai and Tsuyoshi Hirajima

8 Hydrothermal Liquefaction of Biomass in Hot-Compressed Water, Alcohols, and Alcohol-Water Co-solvents for Biocrude Production............................. 171
Chunbao Charles Xu, Yuanyuan Shao, Zhongshun Yuan, Shuna Cheng, Shanghuang Feng, Laleh Nazari and Matthew Tymchyshyn

9 Hydrothermal Liquefaction of Biomass................. 189
Saqib Sohail Toor, Lasse Aistrup Rosendahl, Jessica Hoffmann, Thomas Helmer Pedersen, Rudi Pankratz Nielsen and Erik Gydesen Søgaard

10 Hydrothermal Gasification of Biomass for Hydrogen Production............................. 219
Jude A. Onwudili

Part IV Hydrothermal Conversion of Biomass into Other Useful Products

11 Review of Biomass Conversion in High Pressure High Temperature Water (HHW) Including Recent Experimental Results (Isomerization and Carbonization)................ 249
Masaru Watanabe, Taku M. Aida and Richard Lee Smith

12 Hydrothermal Carbonization of Lignocellulosic Biomass....... 275
Charles J. Coronella, Joan G. Lynam, M. Toufiq Reza and M. Helal Uddin

Part V Hydrothermal Conversion of Biomass Waste into Fuels

13 Organic Waste Gasification in Near- and Super-Critical Water............................. 315
Liejin Guo, Yunan Chen and Jiarong Yin

14	**Hydrothermal Treatment of Municipal Solid Waste for Producing Solid Fuel**...............................	355
	Kunio Yoshikawa and Pandji Prawisudha	
15	**Sewage Sludge Treatment by Hydrothermal Process for Producing Solid Fuel**...............................	385
	Kunio Yoshikawa and Pandji Prawisudha	

Contributors

Taku M. Aida Department of Environmental Study, Tohoku University, Sendai, Japan

Yunan Chen State Key Laboratory of Multiphase Flow in Power Engineering, International Research Center for Renewable Energy, Xi'an Jiaotong University, Xi'an, China

Shuna Cheng Institute for Chemical and Fuels from Alternative Resources, The University of Western Ontario, London, ON, Canada

Charles J. Coronella Chemical Engineering/170, University of Nevada, Reno, NV, USA

Shanghuang Feng Institute for Chemical and Fuels from Alternative Resources, The University of Western Ontario, London, ON, Canada

Liejin Guo State Key Laboratory of Multiphase Flow in Power Engineering, International Research Center for Renewable Energy, Xi'an Jiaotong University, Xi'an, China

Tsuyoshi Hirajima Faculty of Engineering, Kyushu University, Nishi-ku, Fukuoka, Japan

Jessica Hoffmann Department of Energy Technology, Aalborg University, Aalborg Ø, Denmark

Zhibao Huo School of Environmental Science and Engineering, Shanghai Jiao Tong University, Shanghai, China

Fangming Jin School of Environmental Science and Engineering, Shanghai Jiao Tong University, Shanghai, China

Satoshi Kumagai Research and Education Center of Carbon Resource, Kyushu University, Nishi-ku, Fukuoka, Japan; Organization for Cooperation with Industry and Regional Community, Honjyo, Saga, Japan

Hongfei Lin Department of Chemical and Materials Engineering, University of Nevada, Reno, NV, USA

Ying Liu Department of Chemical and Materials Engineering, University of Nevada, Reno, NV, USA

Joan G. Lynam Chemical Engineering/170, University of Nevada, Reno, NV, USA

Laleh Nazari Institute for Chemical and Fuels from Alternative Resources, The University of Western Ontario, London, ON, Canada

Rudi Pankratz Nielsen Department of Biotechnology, Chemistry and Environmental Engineering, Section of Chemical Engineering, Aalborg University, Esbjerg, Denmark

Ayumu Onda Research Laboratory of Hydrothermal Chemistry, Faculty of Science, Kochi University, Kochi, Japan

Jude A. Onwudili School of Process, Environmental and Materials Engineering, Energy Research Institute, The University of Leeds, Leeds, UK

Thomas Helmer Pedersen Department of Energy Technology, Aalborg University, Aalborg Ø, Denmark

Pandji Prawisudha Department of Mechanical Engineering, Bandung Institute of Technology, Bandung, Indonesia

M. Toufiq Reza Chemical Engineering/170, University of Nevada, Reno, NV, USA

Lasse Aistrup Rosendahl Department of Energy Technology, Aalborg University, Aalborg Ø, Denmark

Yuanyuan Shao Institute for Chemical and Fuels from Alternative Resources, The University of Western Ontario, London, ON, Canada

Zheng Shen National Engineering Research Center for Facilities Agriculture, Institute of Modern Agricultural Science and Engineering, Tongji University, Shanghai, China

Richard Lee Smith Research Center of Supercritical Fluid Technology, Tohoku University, Sendai, Japan; Department of Environmental Study, Tohoku University, Sendai, Japan

Ji Su Department of Chemical and Materials Engineering, University of Nevada, Reno, NV, USA

Erik Gydesen Søgaard Department of Biotechnology, Chemistry and Environmental Engineering, Section of Chemical Engineering, Aalborg University, Esbjerg, Denmark

Saqib Sohail Toor Department of Energy Technology, Aalborg University, Aalborg Ø, Denmark

Matthew Tymchyshyn Institute for Chemical and Fuels from Alternative Resources, The University of Western Ontario, London, ON, Canada

M. Helal Uddin Chemical Engineering/170, University of Nevada, Reno, NV, USA

Yuanqing Wang RIKEN Research Cluster for Innovation Nakamura Laboratory, Saitama, Japan

Masaru Watanabe Research Center of Supercritical Fluid Technology, Tohoku University, Sendai, Japan; Department of Environmental Study, Tohoku University, Sendai, Japan

Chunbao Charles Xu Institute for Chemical and Fuels from Alternative Resources, The University of Western Ontario, London, ON, Canada

Lingli Xu School of Environmental Science and Engineering, Shanghai Jiao Tong University, Shanghai, China

Lisha Yang Department of Chemical and Materials Engineering, University of Nevada, Reno, NV, USA

Guodong Yao School of Environmental Science and Engineering, Shanghai Jiao Tong University, Shanghai, China

Jiarong Yin International Research Center for Renewable Energy, State Key Laboratory of Multiphase Flow in Power Engineering, Xi'an Jiaotong University, Xi'an, China

Kunio Yoshikawa Department of Environmental Science and Technology, Tokyo Institute of Technology, Tokyo, Japan

Zhongshun Yuan Institute for Chemical and Fuels from Alternative Resources, The University of Western Ontario, London, ON, Canada

Xu Zeng School of Environmental Science and Engineering, Shanghai Jiao Tong University, Shanghai, China

Part I
Characters of High Temperature Water and Hydrothermal Reactions

Chapter 1
Water Under High Temperature and Pressure Conditions and Its Applications to Develop Green Technologies for Biomass Conversion

Fangming Jin, Yuanqing Wang, Xu Zeng, Zheng Shen and Guodong Yao

Abstract This chapter introduces the chemical and physical properties of water under high temperature and pressure, such as ion product, density, dielectric constant and hydrogen bonding, and the applications of these properties on biomass conversion. These properties that are adjustable by changing the reaction temperature and pressure or adding additives are central to the reactivity of the biomass feedstock to break the C–C or C–O bonds. For example, glucose will follow different reaction pathways under acidic or alkali environment which is related to the ion product of water. Presently, hundreds of strategies utilizing these properties to transform biomass into target products intentionally or unintentionally are proposed. In this chapter, the hydrothermal processes applied in the conversion of biomass including cellulose, hemicelluloses, lignin and glycerin into commodity chemicals such as organic acids are mainly reviewed. In addition, the production of CO_2 as a byproduct from biomass conversion is sometimes inevitable. To achieve 100 % carbon yield, the process of reduction of CO_2 is often neglected but required. In the last section, the one pot reaction of glycerin conversion and CO_2 reduction is reviewed based on the hydrogen bonding property.

F. Jin (✉) · X. Zeng · G. Yao
School of Environmental Science and Engineering, Shanghai Jiao Tong University, 800 Dongchuan RD, Shanghai 200240, China
e-mail: fmjin@sjtu.edu.cn

Y. Wang
RIKEN Research Cluster for Innovation Nakamura Laboratory, 2-1 Hirosawa, Wako, Saitama 351-0198, Japan

Z. Shen
National Engineering Research Center for Facilities Agriculture, Institute of Modern Agricultural Science and Engineering, Tongji University, Shanghai 200092, China

1.1 Introduction

The terminology used in the literature for water under high temperature and pressure conditions (WHTP) is quite diverse. For instance, hot compressed water (HCW) is used to denote water above 200 °C and at sufficiently high pressure [1]. High temperature water (HTW) is also defined as liquid water above 200 °C [2]. Based on the critical point of water (T_c = 373 °C, P_c = 22.1 MPa), water can be divided into sub-critical water (below its critical point) and super-critical water (above its critical point). The lower limit of temperature of subcritical water can be 100 °C in the liquid state [3]. The terminology "near-critical water" is also often employed [4]. Aqueous phase processing (APP) is employed in the liquid water at 200–260 °C and 10–50 bar to produce H_2, CO, and light alkanes from sugar-derived feed [5]. More broadly and popularly, the terminology "hydrothermal", originally from geology, is used in the literatures to denote the reaction medium of high temperature and pressure water. According to their main product, it can be divided into hydrothermal carbonization (usually conducted at 100–200 °C) [6], hydrothermal liquefaction (often at 200–350 °C) [7], hydrothermal gasification (often at 350–750 °C) [8], Thus in this chapter, the terminology of hydrothermal will be mostly adopted to denominate water above 100 °C and 0.1 MPa including the sub and super-critical water.

The distribution of products from hydrothermal biomass conversion, such as gas, liquid or solid, is largely dependent on the properties of water at different states. Two competing reaction mechanisms are present: an ionic or polar reaction mechanism typical of liquid phase chemistry at low temperature and a free radical reaction mechanism typical of gas phase reactions at high temperature [9, 10]. The latter radical reactions are preferred to lead gas formation [11]. In addition, molecular reaction, which is different from ionic and radical reactions, is molecular rearrangement enhanced by coordination with water and proceeds around the critical region of water [12].

Therefore, in the following sections, we will introduce the representative properties of WHTP such as ion product, density, dielectric constant and hydrogen bonding, and discuss the effect of these properties on biomass conversion.

1.2 Ion Product

The ion product (K_w), also called self-ionization constant, is defined as the product of the concentrations of H^+ and OH^- in the water in units of mol^2/kg^2. When increasing the temperature, the ion product of water increases from $K_w = 10^{-14}$ mol^2/kg^2 at room temperature to approximately 10^{-11} mol^2/kg^2 at around 300 °C at constant pressure (250 bar) [2]. Above the critical temperature, the ion product decreases sharply with increasing temperature [2]. In the ranges when water has a bigger K_w number, water may show enhancement of acid and base catalyzed reactions due to the high concentration of H^+ and OH^- ions [7]. Furthermore, it is expected to get higher yield of target chemicals by adding minimal amounts of either acid or base catalysts. Antal et al. proposed that the ionic reaction are favored at $K_w > 10^{-14}$, and free radical reactions are favored at $K_w < 10^{-14}$ [13]. In this section, five classes of reactions, often taken place in the conversion of biomass, are discussed with one typical example to show the influence of ion product of water in the acid or base catalyzed reaction.

1.2.1 Hydrolysis

As shown in Fig. 1.1, hydrolysis is one of the major and usually initial reactions happened in conversion of biomass in which glycosidic bonds between sugar units are cleaved to form simple sugars such as glucose and partially hydrolyzed oligomers. Hydrolysis can happen both in acid and base catalyzed reactions, while the former reaction condition (acidic) is more often adopted because base catalysis lead to more side reactions [14, 15]. The hydrolysis of cellulose to glucose is a widely investigated reaction in biomass conversion because cellulose is the major component of plant biomass and the product glucose is a very important intermediate [16]. Under hydrothermal conditions, cellulose reacts with water and is hydrolyzed into glucose or other monomers proceeding through C–O–C bond cleavage and accompanied by further degradation. Three possible reaction paths of cellobiose hydrolysis are demonstrated including acid, base and water catalyzed ways [15]. Acid hydrolysis proceeds through the formation of a conjugated acid followed by the glycosidic bond cleavage and leads to the two glucose units. In the base pathway, the OH^- attacks at the anomeric carbon atom, renders the cleavage of the O bridge and again yields the two glucose units. The water catalyzed reaction is characterized by H_2O adsorption. Then water and the glycosidic bond split simultaneously and form two glucoses again. Sasaki et al. [17, 18] conducted cellulose decomposition experiments with a flow reactor type reactor from 290 to 400 °C at 25 MPa. Higher hydrolysis product yields (around 75 %) were obtained in supercritical water (SCW) than in subcritical water. The reason was attributed to the difference of reaction rate in the formation and degradation of oligomer or glucose. At a low temperature region, the glucose or oligomer conversion rate was much faster than the hydrolysis rate of cellulose. However, around the critical point, the hydrolysis rate jumped to more than an order of magnitude higher level and became faster than the glucose or oligomer decomposition rate. The direct observation by diamond anvil cell showed that the cellulose disappeared with a more than two orders of magnitude faster rate at 300–320 °C than that estimated [18]. This phenomenon indicated that the presence of a homogeneous hydrolysis atmosphere caused by the dissolution of cellulose or hydrolyzed oligomers around the critical temperature and thus resulted in the high cellulose hydrolysis rate. The additional acid catalysts including homogenous and heterogenous catalysts would

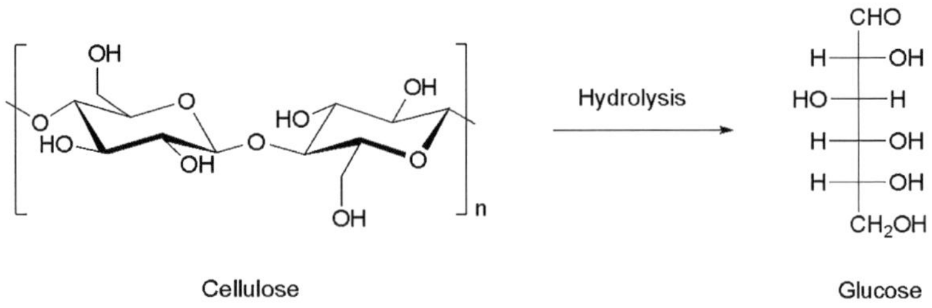

Fig. 1.1 Hydrolysis of cellulose

also enhance the yield of glucose which was around 50–80 % [16]. The base catalyst might cause more side reactions [15] but could inhibit the formation of char which was very crucial in the continuous flow reactor to prevent plug [19].

1.2.2 Isomerization

As shown in Fig. 1.2, the isomerization between glucose and fructose is very common and has been considered as one key step in biomass conversion. The difference of their reactivity and selectivity for target materials makes the tunable transformation to specific one (usually from glucose to fructose) highly desirable [20]. This reaction is typically catalyzed by the base catalyst, named as Lobry de Bruyn-Alberda van Ekensterin transformation (LBAE). The mechanism proceeds by deprotonation of alpha carbonyl carbon of glucose by base, resulting in the formation of a series of enolate intermediates. The overall process involves hydrogen transfer from C2 to C1 and from O2 to O1 of an alpha hydroxy aldehyde to form the corresponding ketone. Kabyemela et al. [21] found the isomerization from fructose to glucose is negligible compared with its reversion under hydrothermal conditions because glucose and fructose have same product distribution except for 1,6-anhydroglucose which is not observed in the decomposition of fructose. Recently, Davis et al. [22] reported another Lewis acid catalyzed pathway of isomerization via intramolecular hydride transfer for glucose–fructose. In addition to glucose–fructose isomerization, there is another important isomerization between glyceraldehydes and dihydroxyacetone under hydrothermal conditions [23].

1.2.3 Dehydration

Dehydration reactions of biomass comprise an important class of reactions in the area of sugar chemistry. As shown in Fig. 1.3, fructose can be dehydrated into hydroxymethylfurfural (HMF) with loss of three water molecules by acid

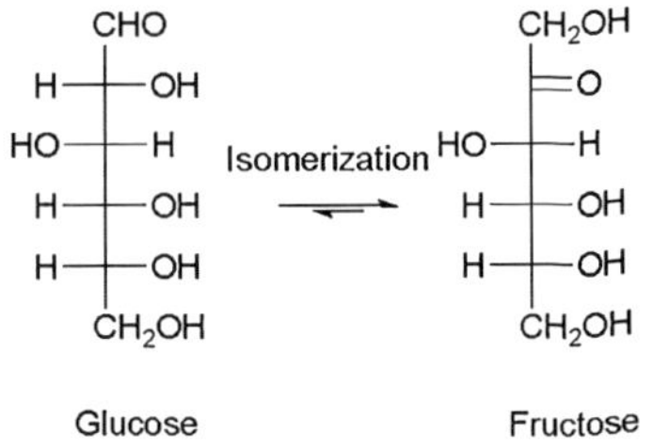

Fig. 1.2 Isomerization between glucose and fructose

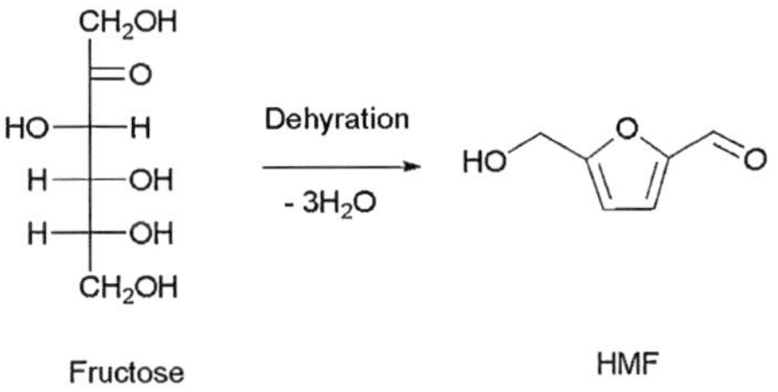

Fig. 1.3 Dehydration of fructose into HMF

catalyzed reaction. Antal et al. [24] proposed that HMF is produced from fructose via cyclic intermediates. Recent studies confirmed that the HMF formation was from the acid-catalyzed dehydration of C6-sugars in the furanose form [25, 26]. Hence, fructose which contains 21.5 % of furanose tautomers in aqueous solution can be converted to HMF easier than glucose which contains only 1 % of furanose tautomers in aqueous solutions. The rehydration of HMF with two molecules of water would produce levulinic acid and formic acid [27]. Levulinic acid can be further converted into g-valerolactone (GVL) via hydrogenation with hydrogen [28], which can be converted to liquid alkenes in the molecular weight range appropriate for transportation fuel [29].

Yoshida et al. obtained the best yield of HMF (65 %) from fructose achieved at a temperature of 513 K for a residence time of 120 s [30]. Since glucose is more common than fructose in biomass conversion, researchers usually adopt a two-step strategy to produce HMF from glucose: (1) isomerization of glucose into fructose catalyzed by base and (2) dehydration of fructose into HMF by acid [31]. Since water under high temperatures and pressures can play the roles of both acid and base catalysts, high yield of HMF can be obtained under hydrothermal conditions in one step. Jin et al. [32] reported the total highest yields of HMF and levulinic acid from glucose were about 50 %, which occurred at 523 K for 5 min with H_3PO_4 as a catalyst and the highest yield of levulinic acid was about 55 % at 523 K for 5 min with HCl as a catalyst. For the three mineral acids (HCl, H_2SO_4 and H_3PO_4), it was found that not only the pH, but also the nature of the acids, had great influence on the decomposition pathway [30]. The order for the production of HMF using the three acids was in the sequence of $H_3PO_4 > H_2SO_4 >$ HCl [32]. On the contrary, the order for production of levulinic acid followed HCl $> H_2SO_4 > H_3PO_4$ [32].

There are some drawbacks in the acid catalyzed formation of HMF from fructose or glucose. Kinetics studies [33–35] showed that humins formation from glucose and HMF cannot be neglected. The activation energy of its formation from glucose and HMF were estimated at 51 and 142 kJ/mol, respectively, while dehydration of glucose to HMF and rehydration of HMF to levulinic acid were 160 and 95 kJ/mol, respectively [35]. To minimize the formation of humins and enhance the selectivity towards HMF, a biphasic solution with water and organic phase was adopted that would continuously extract HMF as it is produced [36–39]. Dumesic et al. reported a 61 % yield of HMF from glucose using a biphasic reactor of water/tetrahydrofuran with $AlCl_3 \cdot 6H_2O$ catalyst at 160 °C [37].

1.2.4 Retro Aldol Reaction

Many researchers [17, 21, 23, 40, 41] have examined intermediate products for the hydrothermal degradation of glucose and cellulose at a reaction temperature of near 300 °C. As shown in Fig. 1.4, through these studies, it was revealed that some compounds containing three carbon atoms, such as glyceraldehyde, dihydroxyacetone and pyruvaldehyde, were formed by the base catalytic role of HTW. Furthermore, there was isomerization occurring between glyceraldehyde and dihydroxyacetone followed by their subsequent dehydration to pyruvaldehyde [23]. The ketone (fructose) can undergo reverse aldol reaction by C3–C4 bond cleavage to form glyceraldehydes. These C3 carbon compounds were considered as the precursors of lactic acid from transformation of pyruvaldehyde [40]. On the other hand, the intermediates glycoaldehyede and erythrose were transformed from glucose by retro aldol reaction [17, 21]. In Organic Chemistry, retro aldol reaction can usually be catalyzed by either an acid or a base. Experimental data suggested, however, that retro aldol reaction under hydrothermal conditions was base-catalyzed [2]. Sasaki et al. [42] reported that the retro aldol reaction selectively proceeded at higher temperatures (above 673 K) and lower pressure (below 25 MPa). At a low temperature, the retro aldol reaction was preferred in alkali environment [43].

These formed intermediates from C2–C3 or C3–C4 bond cleavage by reverse aldol reaction from hexoses can then be fast transformed into mainly lactic acid and other low molecular acid in which glyceraldehyde can produce a higher yield of lactic acid [41]. Lactic acid is a key chemicals as a building block for biodegradable lactic acid polymers with limited environmental impact. Jin et al. [44, 45] showed that the addition of base catalyst [NaOH and $Ca(OH)_2$] can increase the yield of lactic acid. The highest yield of lactic acid from glucose was 27 % with 2.5 M NaOH and 20 % with 0.32 M $Ca(OH)_2$ at 300 °C for 60 s [44]. A very recent study by Labidi et al. [46] also found that the highest yield lactic acid of 45 % from corn cobs was obtained using 0.7 M $Ca(OH)_2$ at 300 °C for 30 min. The reason that base catalyst increased the yield of lactic acid can be attributed to the enhancement of reaction pathway for lactic acid production discussed above.

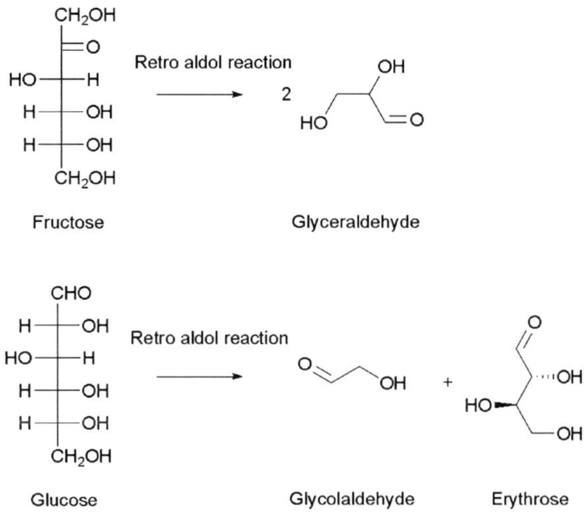

Fig. 1.4 Retro aldol reaction of fructose and glucose

Another reason may be that the lactate formed actually by alkaline solution prevents it from decomposition [47]. Compared with NaOH at lower alkaline concentration [44], Ca(OH)$_2$ promoted more effectively the production of lactic acid than NaOH at the same OH$^-$ concentration. It is probably because Ca^{2+} was more capable than Na$^+$ in forming complexes with two oxygen atoms in the hexoses. When the concentration of Ca(OH)$_2$ increased from 0.32 to 0.4 M, it did not lead to increase in lactic acid yield; while the optimum OH$^-$ concentration for NaOH was 2.5 M. This difference can be attributed to the fact that the saturated solubility of NaOH is higher than that of Ca(OH)$_2$.

1.2.5 Decarboxylation and Decarbonylation

$$[\text{HCOOH} \rightarrow \text{H}_2 + \text{CO}_2 (\text{decarboxylation})] \quad (1.1)$$

$$[\text{HCOOH} \rightarrow \text{H}_2\text{O} + \text{CO} (\text{decarbonylation})] \quad (1.2)$$

The reactions of formic acid play a key roles in the chemistry of hydrothermal reaction partly because it was the simplest acid and product of many acid/base catalyzed or oxidation reactions, and partly because itself or formate is considered to be the intermediate of water gas shift reaction and reduction of carbon dioxide [48]. The understanding of its reactivity especially coupled with the properties of water will facilitate the researches on energy production and environment protection. As shown in Eqs. 1.1 and 1.2, for the decomposition of formic acid, there existed two competitive pathways: decarboxylation and decarbonylation. Early experimental results showed that in the gas phase the decarbonylation dominated, but in the liquid phase the decarboxylation dominated [49, 50]. Savage et al. [49]

conducted the formic acid decomposition experiments from 320 to 500 °C, at pressures from 18.3 to 30.4 MPa, and at 1.4–80 s reaction times. Conversion rates ranging from 38 to 100 % were obtained with the major products of CO_2 and H_2. In their experiments, the decarbonylation product CO was also detected and the yields were always at least an order of magnitude lower than the yields of decarboxylation. The reason why decarboxylation dominated in the liquid phase can be explained by the presence of water as a homogeneous catalyst that can catalyze decarboxylation more than decarbonylation by a theoretical calculation [51]. The kinetic data also supported the assumption of a homogenous reaction based on the consistency with the reaction rate law that was first order in formic acid [49]. However, Nakahara et al. [52] indicated that the reactor wall might show catalytic role in the formic acid decomposition that is a heterogeneous reaction according to an NMR investigation. Compared with its acidic environment, Jin et al. [47] found the addition of alkali could prevent the formic acid decomposition even with the presence of oxidant H_2O_2 at 250 °C for 60 s.

1.3 Water Density

Water density can be varied greatly with temperature and pressure under hydrothermal conditions. Water density decreases with the increase in temperature at constant pressure. For example, water density decreases from about 800 kg/m^3 like liquid phase to about 150 kg/m^3 like gas phase without phase change as the temperature increases from 300 to 450 °C. Meanwhile, water density controlled by temperature and pressure can be related to ion product by Eq. 1.3 using a fitting method proposed by Marshall et al. [53].

$$[\log K_w = A + \frac{B}{T} + \frac{C}{T^2} + \frac{D}{T^3} + (E + \frac{F}{T} + \frac{G}{T^2}) \log \rho] \quad (1.3)$$

where T is temperature in Kelvin, ρ is density in g/cm^3, and A–G are fitting parameters. This result indicated that the chemistry of biomass conversion can also be controlled by water density. However, it is not to say that water density affects the reaction mechanism only by changing ion product of water. Water density changes can reflect the changes of water in molecular level such as solvation effect, hydrogen bonding, polarity, dielectric strength, molecular diffusivity and viscosity that will influence the chemistry inside [54]. In super-critical water, the reaction mechanism varies from a reaction atmosphere that favors radical reaction to one that favors ionic reactions dictated by the water density [55]. Experimental data showed that reactions seemed to proceed via ionic pathways in the high density water while radical reactions seemed to be the main reaction pathways in the less dense super-critical water [9]. Westacott et al. [56, 57] investigated *tert*-butyl chloride dissociation in super-critical water by computational methods and showed that water density affect the competition between ionic and radical reaction mechanisms. The ionic heterolytic dissociation was preferred over the

1 Water Under High Temperature and Pressure Conditions

radical homolytic dissociation when water density was larger than 0.03 g/cm^3 [56, 57]. In this section, different reaction mechanisms via ionic or radical pathways affected by water density were introduced from different feedstock.

1.3.1 Cellulose

Aida et al. [58, 59] examined the reactions of D-glucose and D-fructose under high temperature (up to 400 °C) and pressure (up to 100 MPa) water. The benzilic acid rearrangement reaction from pyruvaldehyde to lactic was enhanced by the water density [59]. And the dehydration reaction to 5-HMF and the hydrolysis of 5-HMF were both enhanced by the increase in water density at 400 °C [58]. One explanation for the enhancement of water density was that water can lower the activation energy by forming "water bridge" in the transition state [51] and thus the increased water concentration would be advantageous for this effect. Another explanation for the density effect on the dehydration and hydrolysis reactions may due to the change in of ion product of water like mentioned above.

1.3.2 Glycerin

Glycerin, as a byproduct of biodiesel formation, can be a promising feedstock for chemicals and fuel production with hydrothermal treatment. The main product of glycerin degradation under hydrothermal conditions could be acrolein, acetaldehyde, formaldehyde, propionaldehyde, allyl alcohol, methanol, ethanol, lactic acid, carbon monoxide, carbon dioxide and hydrogen based on reaction conditions [9, 60–62]. Buhler et al. [9] conducted glycerin degradation experiments in near and super-critical water in the temperature range of 622–748 K, at pressures of 25, 35, or 45 MPa. They found a great change in the product composition with pressure (density) indicating the presence of different reaction mechanisms (ionic and radical reaction). The relative yield of acetaldehyde and formaldehyde increased with the increase in pressure (increase in water density) while that of methanol and allyl alcohol decreased. They suggested that acetaldehyde and formaldehyde were formed by ionic reactions and the latter by radical reactions. The relative stable yield of acrolein with pressure showed that both mechanisms were present.

1.3.3 Lignin

Lignin, the second most abundant polymeric aromatic organic substance in wood biomass after cellulose, has been considered as an important alternative source of chemical compounds. Wahyudiono et al. [63–65] performed lignin and its model compound decomposition experiments under hydrothermal conditions. The yields of catechol, phenol and o-cresol from guaiacol increased with increasing water density from 0.17 to 0.60 g/cm^3 at 653–673 K and at 25–40 MPa [63]. The results suggested that the increase in water density could enhance the hydrolysis of guaiacol into its derivatives and the dehydration of alcohols [63]. Sato et al. [66] reported that the yield of gas products from the gasification of alkylphenols can be increased with the increase in water density from 0 to 0.3 g/cm^3 in SCW at 673 K. Osada et al. [67] compared the gas yield from lignin and 4-propylphenol which is a model of low-molecular weight compounds from lignin at 673 K with water density from 0.1 to 0.5 g/cm^3. The results indicated that the step (decomposition of lignin to low-molecular weight compounds) in the gasification was enhanced by increasing the water density, and the rate of gasification of 4-propylphenol was not affected by water density.

1.3.4 C_1 Compounds

Formaldehyde is very reactive under hydrothermal conditions. Because two formaldehyde molecules can produce methanol and formic acid by Cannizzaro reaction, and one formaldehyde can decompose into carbon monoxide and hydrogen. The product of carbon monoxide with H_2O can further produce carbon dioxide and hydrogen by water gas shift reaction. The produced formic acid will undergo competitive pathways to produce CO_2 or CO like mentioned above. Osada et al. [68] discussed the water density (0.17–0.50 g/cm^3) dependence of formaldehyde reaction in super-critical water with batch experiments. It was found that the Cannizzaro reaction mechanism was the preferred reaction pathway for formaldehyde with the product of methanol at higher water densities. At lower water densities, monomolecular decomposition became the main reaction pathway with higher yield of carbon monoxide. The water density dependence on rate constant of formic acid disappearance at super-critical water was studied by Yu et al. [49]. It was found that the rate constant of formic acid disappearance increased with increasing of the water concentration from 5 to 10 mol/L, then decreased, and then increased again when water concentration was greater than 25 mol/L. As the water concentration increased from 1.8 to 5.7 mol/L in super-critical water at 500 °C, the OH radical was proposed to increase which promoted the oxidation of methanol [69].

1.4 Dielectric Constant

The dielectric constant is the ratio of the permittivity of a substance to the permittivity of free space. The dielectric constant of water under ambient condition is 78.5. Water under this condition could be used as good solvent for the polar

Fig. 1.5 Dielectric constant of water as a function of temperature. *Dashed line* 25 MPa; *solid line* 50 MPa; *dotted line* 100 MPa. Reprinted from Ref. [70], Copyright 2013, with permission from Elsevier

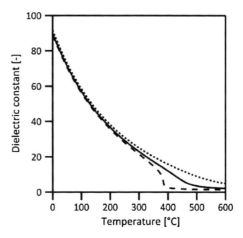

Fig. 1.6 Dielectric constant of water as a function of pressure at constant temperatures (273, 298, 323 and 348 K) (*fine lines* Bradley's equation [72]; *thick lines* adjusted values extracted from the *International Association for the Properties of Water and Steam* [73]). Reprinted with the permission from Ref. [72]. Copyright 1979 American Chemical Society

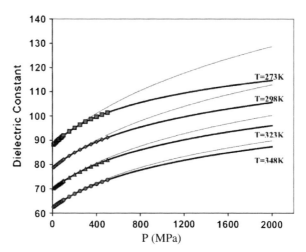

materials. However, it cannot be used to dissolve hydrocarbon and gas. Dielectric constant of water as a function of temperature can be seen in Fig. 1.5 [70]. As shown in Fig. 1.5, the dielectric constant of water reduces sharply with the increase of temperature of water. HTW in sub-critical and super-critical condition behaves like many organic solvents which can dissolve organic compounds completely forming a single fluid phase. The advantages of a single supercritical phase reaction medium are that higher concentrations of reactants can often be attained and no interphase mass transport processes which will hinder the reaction rates were indispensable.

As a consequence of this lack of data, attempts to estimate the properties of aqueous species at high temperature and/or high pressure rely on the estimated or extrapolated dielectric constant values [71]. The dielectric constant dependence on the pressure, proposed by Bradley and Pitzer [72], can be seen in Fig. 1.6. Bradley used an equation suggested by Tait in 1880 for volumetric data. As shown in figure1.6, at constant temperature, the dielectric constant values increased linearly with the increase of pressure. It should be noted that the original Bradley's equation does not reproduce adequately the data available from the *International Association for the Properties of Water and Steam* [73] used in Fig. 1.6 for P above 400 MPa, particularly at and above 323 K. However, the trends are similar, which can approximately represent the change of dielectric constant with different pressure.

1.4.1 Dielectric Constant of High Temperature Water

Park et al. compared the dielectric constant (ε) of superheated water at different temperature and pressure, as shown in Table 1.1 [74]. The dielectric constant values of water decreased with the increase of temperature from 44 at 150°C to 2 at 350 °C. These values are between those of organic solvent ethanol ($\varepsilon = 24$ at 25 °C) and methanol ($\varepsilon = 33$ at 25 °C). This indicates that superheated water can be used as an organic solvent. Moreover, superheated water is readily available, non-toxic, reusable and very low in cost as well as environmentally friendly. Therefore, superheated water can be used as an alternative cleaning technology, instead of using organic solvents or toxic and strong aqueous liquid media. For the extraction of dioxins [75], pesticides [76], PCBs [77], and PAHs [78]. Lagadec et al. reported that the optimum subcritical water extraction was at 275 °C in 35 min for all low and high molecular weight PAHs from contaminated Manufactured Gas Plant (MGP) soil [76]. Moreover, it can also be used to determine a superior instant analytical technique (using GC oven as heater) by using organic solvent [78]. However, a complete extraction technology with shorter extraction time at a temperature range (from 100 to 300 °C) using subcritical water for industrial application has not been determined; therefore, an additional study is necessary [74].

The dielectric constant of SCW is very special, because the dielectric constant under this condition is much lower, and the number of hydrogen bonds is much smaller and their strength is much weaker. Supercritical water above 374 °C and 221 bar shows water is greatly diminished-frequently less than reduced local molecular ordering and less effective hydrogen bonding as characterized by its lower dielectric constant (about 1 to 3) [79]. As a result, SCW behaves like many organic solvents so that organic compounds have complete miscibility with SCW. Moreover, gases are also miscible in SCW, thus a SCW reaction environment provides an opportunity to conduct the chemical reactions in a single fluid phase that would otherwise occur in a multiphase system under conventional conditions [80]. Therefore, SCW exhibits considerable characters of solvent, which can dissolve nonpolar materials and gas, and the characters of easy diffusion and motion [81]. The dielectric constant of SCW corresponds to the value of polar solvent under ambient condition. The dielectric constant of ambient water varies continuously over a much larger range in the supercritical state. This variation offers the possibility of using pressure and temperature to influence the properties of the reaction medium. Therefore, it is possible for the formation of C–C bond

1 Water Under High Temperature and Pressure Conditions

Table 1.1 Dielectric constant (ε) of subcritical water and common organic solvent

ε (at subcritical water °C)	ε of common organic solvent at 25 °C
44 (150)	1.9 (n-hexane)
35 (200)	21 (acetone)
27 (250)	24 (ethanol)
20 (300)	33 (methanol)
2 (350)	39 (acetonitrile)

Table 1.2 Dielectric constant and density of water at some supercritical conditions

Temperature (°C)	Pressure (MPa)	Density (g/cm^3)	Dielectric constant
400	25	0.17	2.4
400	30	0.35	5.9
500	25	0.09	1.5
500	30	0.12	1.7
350	25	0.63	14.85

with organ metallic catalytic reactions which always needs organic solvent. Gomez-Briceno et al. compared the dielectric constant of water at different supercritical conditions, 400 and 500 °C and two pressures values, 25 and 30 MPa, as shown in Table 1.2 [82]. The data showed that the dielectric constant decreased significantly with the decrease of temperature. However, the influence was very small.

Water with large dielectric constant will exhibited strong effect with microstructure of water, and eventually influence the reaction [1]. The large dielectric constant means that substances whose molecules contain ionic bonds tend to dissociate in water yielding solutions containing ions. This occurs because water as a solvent opposes the electrostatic attraction between positive and negative ions that would prevent ionic substances from dissolving. These separated ions become surrounded by the oppositely charged ends of the water dipoles and become hydrated. This ordering tends to be counteracted by the random thermal motions of the molecules. Water molecules are always associated with each other through as many as four hydrogen bonds and this ordering of the structure of water greatly resists the random thermal motions. Indeed it is this hydrogen bonding which is responsible for its large dielectric constant.

1.4.2 Effects of Dielectric Constant on the Application of High Temperature Water

In this section, different reaction mechanisms affected by dielectric constant of HTW are introduced.

(1) Hydrolysis of organic compounds

Townsend studied the relationship between the hydrolysis rate constants and the dielectric constant, and the results showed that the hydrolysis rate constants correlated well with the dielectric constant of water [83]. Marrone and Tester also have studied the hydrolysis of methylene chloride to form formaldehyde and HCl [84, 85]. Their research results showed that the dielectric constant of the reaction medium influenced the rate of hydrolysis significantly. The reaction slowed down as the temperature increased along with the decrease of dielectric constant of water. With higher dielectric constant, the intermediates were stabilized much better and the hydrolysis reaction was accelerated. Marrone has developed a quantitative kinetics model and showed that more accurately with experimental data [86], as shown in Fig. 1.7. From Fig. 1.7, it can be seen that in the subcritical region where water is still quite polar, the rate constant is small; however, in the range of temperatures just beyond the critical point where dielectric constant drops by an order of magnitude, the rate constant increases dramatically. These researches showed that in the process of hydrolysis, the reaction rate could be regulated by the dielectric constant via the change of temperature.

(2) Hydrothermal conversion of carbohydrate biomass

The hydrothermal process is one of the most promising processes for the conversion of carbohydrate biomass into chemicals, because HTW has unique properties as a reaction medium, such as a lower dielectric constant, fewer and weaker hydrogen bonds [2, 4]. By the variation of the relative dielectric constant with temperature and pressure, reaction rates can be controlled. There have been extensive researches on the conversion of biomass into chemicals under hydrothermal conditions. Sasaki et al. studied the hydrothermal conversion of guaiacol in sub- and SCW [87]. Results showed that the reaction rate constant was different with the change of temperature, which was related to the dielectric constant, as shown in Fig. 1.8.

Ragauskas et al. reviewed the application of high temperature water [88]; for example, near-critical water (200–300 °C) exhibited a reduction in dielectric constant (20 to 30) relative to ambient water, and the ability of HTW to dissolve both nonpolar organic molecules and inorganic salts was comparable to that of the popular organic solvent acetone. Fangming Jin performed a series of studies on the hydrothermal conversion of biomass and CO_2 due to the unique characters of HTW including the dielectric constant [7, 89–91]. These results showed that the lower dielectric constant, caused by the high temperature, affected the reaction significantly, which induced the effective conversion of biomass and CO_2 gas under hydrothermal conditions. Franck et al. studied the cellulose conversion with solid acid catalyst in supercritical state, which showed that the dissolution of nonpolar organic macromolecules such as cellulose was accelerated with low dielectric constant [92].

Fig. 1.7 Water dielectric constant as a function of temperature. Reprinted with the permission from Ref. [86]. Copyright 1998 American Chemical Society

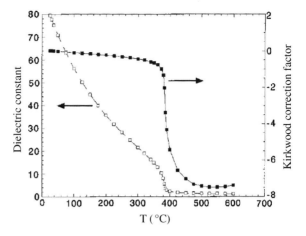

Fig. 1.8 Rate constants of guaiacol conversion in subcritical water (at 483–563 K, 8 MPa: *squares*) and supercritical water (at 653–673 K, 30 MPa) Reprinted from Ref. [87], with permission from Springer Science+Business Media

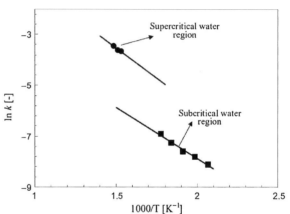

(3) Degradative extraction

Morimoto et al. has studies the miscibility of SCW with asphaltene at 400–450 °C and 20–35 MPa [93]. Relationship between extraction yield and pressure, dielectric constant can be seen in Fig. 1.9. With increasing the pressure at 440 °C, the degradative extraction yield of AS using SCW reached a maximum at around 30 MPa. The extraction behavior was thought to be controlled mainly by the water properties represented by the dielectric constant and Hansen solubility parameter. supercritical water extraction at >400 °C and >25 MPa has been used in several types of heavy crude, including oil sand bitumen [94], vacuum residue [95], asphalt [96], heavy oil [97], and coal tar [98]. Wang et al. reviewed the conventional Soxhlet extraction and the new alternative methods used for the extraction of nutraceuticals from plants [99]. The microwave-assisted extraction depends on the dielectric susceptibility of solvent and matrix, better recoveries can be obtained by moistening samples with a substance that possesses a relatively high dielectric constant such as water.

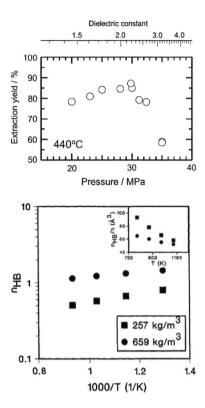

Fig. 1.9 Relationship between extraction yield and pressure (the extra scale above: dielectric constant of pure water at each) Reprinted from Ref. [93], Copyright 2012, with permission from Elsevier

Fig. 1.10 Number of hydrogen bonds per water molecules. Reprinted with the permission from Ref. [100]. Copyright 1996 American Chemical Society

1.5 Hydrogen Bonding

WHTP exhibits properties that are very different from those of ambient liquid water, but hydrogen bonding is the source of many unique properties of liquid water.

It can be shown in Fig. 1.10 that with increasing temperature and decreasing density, the hydrogen bonding in water becomes weaker and less persistent [100]. For example, water at 673 K and 0.5 g/cm^3 retains 30–45 % of the hydrogen bonding that exists at ambient conditions, whereas water at 773 K and 0.1 g/cm^3 retains 10–14 % [101]. The hydrogen bonding network in ambient liquid water exists in the form of infinite percolating large clusters of hydrogen-bonded water molecules, but the hydrogen bonding network in WHTP exists in the form of small clusters of hydrogen-bonded water molecules [100, 102–104]. In general, the average cluster size of hydrogen-bonded water molecules decreases with increasing temperature and decreasing density. For instance, most of the clusters at 773–1073 K and 0.12–0.66 g/cm^3 consist of five water molecules or less, although existing a small number of clusters that are as large as about 20 water molecules [100, 103, 104]. These results shows that the less hydrogen bonding results in much less order in WHTP than ambient liquid water, and then individual water

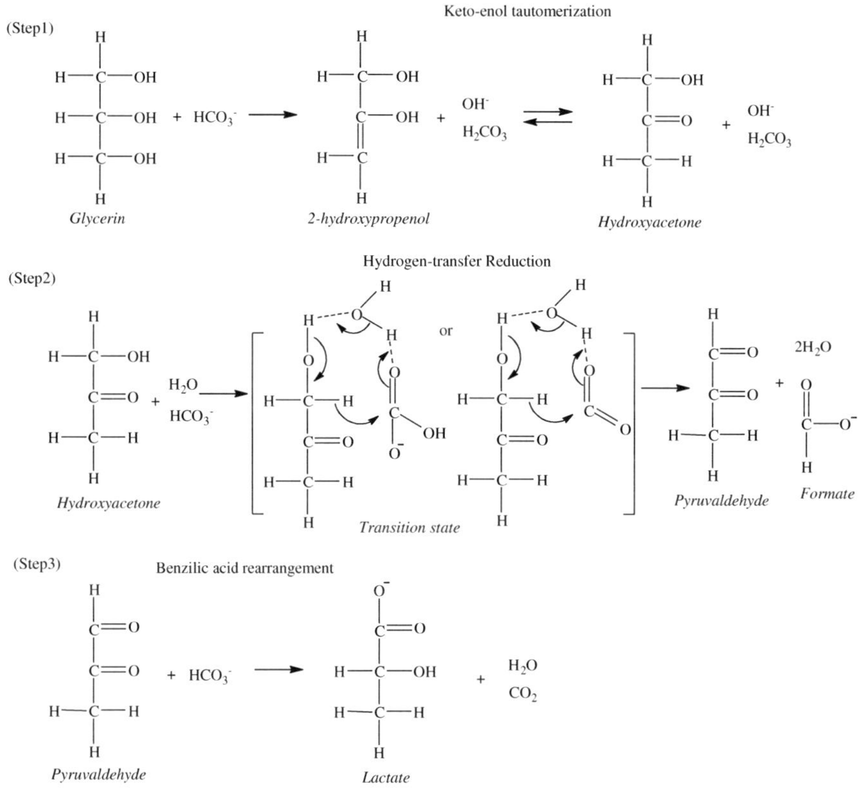

Fig. 1.11 The proposed pathway of the hydrogen-transfer reduction of $NaHCO_3$ with glycerine [118] Reproduced by permission of The Royal Society of Chemistry

molecules can participate in elementary reaction steps as a hydrogen source or catalyst during hydrothermal conversion of biomass into high-valued chemicals.

There has been much previous research about that water molecules can supply hydrogen atoms to participate in reactions during the steam reforming of glucose [105, 106] and biomass [107, 108], the pyrolysis of alkyldiammonium dinitrate [109], and the oxidation of methylene chloride [84], lactic acid [110], and carbon monoxide [111–113], hydrogenation of dibenzothiophene [114] and heavy oils [115], co-liquefaction of coal and cellulose [116], and alcohol-mediated reduction of CO_2 and $NaHCO_3$ into formate [117, 118]. They produced the hydrogen in situ by partially oxidizing the organic compounds to generate carbon monoxide, which then underwent the water-gas shift reaction ($CO + H_2O \leftrightarrow CO_2 + H_2$). The authors proposed that the reactive intermediate generated by the water-gas shift reaction was the actual hydrogenation agent, not the hydrogen molecule itself. As shown in Fig. 1.11, our recent study found that CO_2 or $NaHCO_3$ could be transformed into formate by alcohol-mediated reduction under hydrothermal alkaline conditions [117, 118].

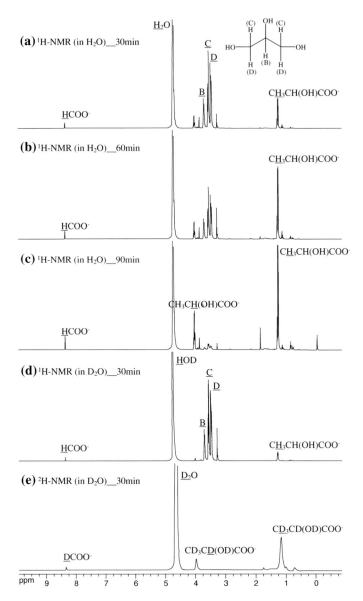

Fig. 1.12 ^{1}H-NMR spectra for the solution after the hydrothermal reaction of 0.33 M glycerol at 300 °C with 1.25 M NaOH in H_2O for **a** 30 min, **b** 60 min, **c** 90 min, and **d** ^{1}H-NMR and **e** ^{2}H-NMR spectra with 1.25 M NaOD in D_2O for 30 min [120]. Reproduced by permission of The Royal Society of Chemistry

Hydrogen deuterium exchange data also provide evidence for hydrogen supply by water. Deuterium can be incorporated into the products of hydrocarbon pyrolyses in supercritical D_2O [119, 120]. More recently, to discover the reaction

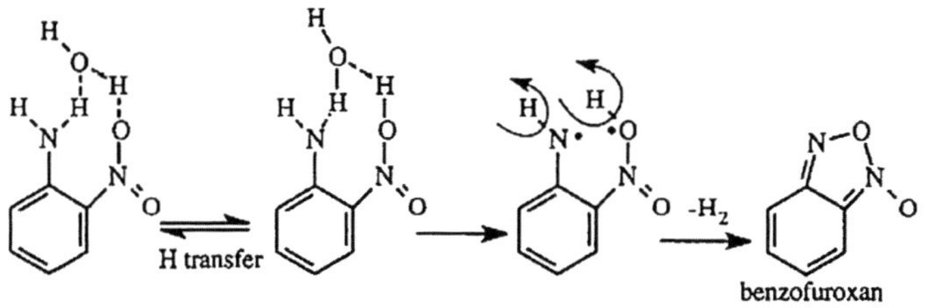

Fig. 1.13 Water catalysis for the intramolecular hydrogen-transfer during the conversion of nitroaniline to benzofurozan. Reprinted from Ref. [131], Copyright 1995, with permission from Elsevier

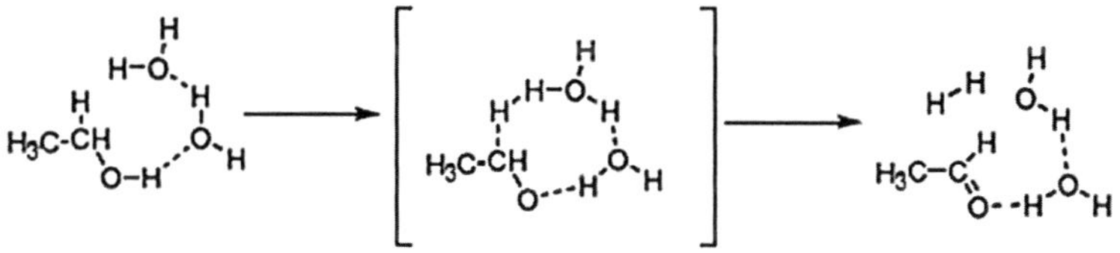

Fig. 1.14 Proposed transition state consisting of an ethanol molecule and two water molecules in SCW without catalyst. Reprinted from Ref. [128], Copyright 2003, with permission from Elsevier

mechanism for the production of hydrogen and lactic acid from glycerol under alkaline hydrothermal conditions, we identified the different intermediates involved during reactions by investigating the water solvent isotope effect with ^{1}H-NMR, ^{2}H-NMR, LC-MS and GC-MS analyses as shown in Fig. 1.12 [120]. The results from solvent isotope studies showed that (1) almost all of the H on the β-C of lactic acid was exchanged by D_2O, which suggested that the hydroxyl (−OH) group on the 2-C of glycerol was first transformed into a carbonyl (C=O) group and then was converted back into a −OH group to form lactic acid; (2) a large amount of D was found in the produced hydrogen gas, which shows that the water molecules acted as a reactant; and (3) D % in the produced hydrogen gas was far more than 50 %, which straightforwardly showed that acetol was formed in the first place as the most probable intermediate by undergoing a dehydration reaction rather than a dehydrogenation reaction.

The natural abundance of hydronium and hydroxide ions suggests that some acid and base-catalyzed reaction may proceed in HTW in the absence of an added catalyst [40, 121–133]. Alcohol dehydration is nominally catalyzed by either acid or base in the presence of added catalysts. In WHTP, however, experimental data suggest that the dominant mechanism is acid catalysis and the dehydration reactivity depends on the structure of the alcohol [121–127].

Experimental data suggest that water molecules can also catalyze a reaction by directly participating in the transition state and reducing its energy. This form of catalysis is important for reactions involving some types of intramolecular hydrogen transfer. For example, Klein et al. proposed a type of water catalysis for the intramolecular hydrogen-transfer step during the conversion of nitroaniline to benzofurozan as shown in Fig. 1.13 [131], and decarboxylation of acetic acid derivatives in WHTP [132].

Arita et al. reported that hydrogen can be generated by an ethanol oxidation reaction catalyzed by water molecules and that half of the produced hydrogen could come from the water in accordance with the proposed reaction mechanism in Fig. 1.14 [128]. Moreover, Takahashi et al. suggested that water molecules played significant catalytic roles in ethanol oxidation reactions based on ab initio density functional theory calculation [133].

References

1. Kruse A, Dinjus E (2007) Hot compressed water as reaction medium and reactant—Properties and synthesis reactions. J Supercrit Fluids 39(3):362–380
2. Akiya N, Savage PE (2002) Roles of water for chemical reactions in High-Temperature Water. Chem Rev 102(8):2725–2750
3. Pourali O, Asghari FS, Yoshida H (2009) Sub-critical water treatment of rice bran to produce valuable materials. Food Chem 115(1):1–7
4. Watanabe M, Sato T, Inomata H, Smith RL, Arai K, Kruse A, Dinjus E (2004) Chemical reactions of C-1 compounds in near-critical and supercritical water. Chem Rev 104(12):5803–5821
5. Huber GW, Iborra S, Corma A (2006) Synthesis of transportation fuels from biomass: Chemistry, catalysts, and engineering. Chem Rev 106(9):4044–4098
6. Titirici MM, White RJ, Falco C, Sevilla M (2012) Black perspectives for a green future: hydrothermal carbons for environment protection and energy storage. Energy Environ Sci 5(5):6796–6822
7. Jin FM, Enomoto H (2011) Rapid and highly selective conversion of biomass into value-added products in hydrothermal conditions: chemistry of acid/base-catalysed and oxidation reactions. Energy Environ Sci 4(2):382–397
8. Kruse A, Bernolle P, Dahmen N, Dinjus E, Maniam P (2010) Hydrothermal gasification of biomass: consecutive reactions to long-living intermediates. Energy Environ Sci 3(1):136–143
9. Buhler W, Dinjus E, Ederer HJ, Kruse A, Mas C (2002) Ionic reactions and pyrolysis of glycerol as competing reaction pathways in near- and supercritical water. J Supercrit Fluids 22(1):37–53
10. Kruse A, Dinjus E (2007) Hot compressed water as reaction medium and reactant—2 degradation reactions. J Supercrit Fluids 41(3):361–379
11. Kruse A (2008) Supercritical water gasification. Biofuels Biopro Biorefin 2(5):415–437
12. Sato T, Sekiguchi G, Adschiri T, Arai K (2002) Ortho-selective alkylation of phenol with 2-propanol without catalyst in supercritical water. Ind Eng Chem Res 41(13):3064–3070
13. Antal MJ, Brittain A, Dealmeida C, Ramayya S, Roy JC (1987) Heterolysis and homolysis in supercritical water. ACS Symp Ser 329:77–86
14. Chheda JN, Huber GW, Dumesic JA (2007) Liquid-phase catalytic processing of biomass-derived oxygenated hydrocarbons to fuels and chemicals. Angewandte Chemie (International Edition) 46(38):7164–7183
15. Bobleter O (2005) Hydrothermal degradation and fractionation of saccharides and polysaccharides. In: Dumitriu S (ed) Polysaccharides: structural diversity and functional versatility, 2nd edn. Marcel Dekker, New York, pp 893–937

16. Huang Y-B, Fu Y (2013) Hydrolysis of cellulose to glucose by solid acid catalysts. Green Chem 15(5):1095–1111
17. Sasaki M, Kabyemela B, Malaluan R, Hirose S, Takeda N, Adschiri T, Arai K (1998) Cellulose hydrolysis in subcritical and supercritical water. J Supercrit Fluids 13(1–3):261–268
18. Sasaki M, Fang Z, Fukushima Y, Adschiri T, Arai K (2000) Dissolution and hydrolysis of cellulose in subcritical and supercritical water. Ind Eng Chem Res 39(8):2883–2890
19. Brunner G (2009) Near critical and supercritical water. Part I. Hydrolytic and hydrothermal processes. J Supercrit Fluids 47(3):373–381
20. Takagaki A, Nishimura S, Ebitani K (2012) Catalytic transformations of biomass-derived materials into value-added chemicals. Catal Surv Asia 16(3):164–182
21. Kabyemela BM, Adschiri T, Malaluan RM, Arai K (1999) Glucose and fructose decomposition in subcritical and supercritical water: detailed reaction pathway, mechanisms, and kinetics. Ind Eng Chem Res 38(8):2888–2895
22. Wang Y, Kovacik R, Meyer B, Kotsis K, Stodt D, Staemmler V, Qiu H, Traeger F, Langenberg D, Muhler M, Woell C (2007) CO_2 activation by ZnO through the formation of an unusual tridentate surface carbonate. Angewandte Chemie (International Edition) 46(29):5624–5627
23. Kabyemela BM, Adschiri T, Malaluan R, Arai K (1997) Degradation kinetics of dihydroxyacetone and glyceraldehyde in subcritical and supercritical water. Ind Eng Chem Res 36(6):2025–2030
24. Antal MJ Jr, Mok WSL, Richards GN (1990) Mechanism of formation of 5-(hydroxymethyl)-2-furaldehyde from d-fructose and sucrose. Carbohydr Res 199(1):91–109
25. Amarasekara AS, Williams LD, Ebede CC (2008) Mechanism of the dehydration of d-fructose to 5-hydroxymethylfurfural in dimethyl sulfoxide at 150 degrees C: an NMR study. Carbohydr Res 343(18):3021–3024
26. Guan J, Cao Q, Guo X, Mu X (2011) The mechanism of glucose conversion to 5-hydroxymethylfurfural catalyzed by metal chlorides in ionic liquid: a theoretical study. Comput Theor Chem 963(2–3):453–462
27. Weingarten R, Conner WC, Huber GW (2012) Production of levulinic acid from cellulose by hydrothermal decomposition combined with aqueous phase dehydration with a solid acid catalyst. Energy Environ Sci 5(6):7559–7574
28. Wettstein SG, Alonso DM, Chong YX, Dumesic JA (2012) Production of levulinic acid and gamma-valerolactone (GVL) from cellulose using GVL as a solvent in biphasic systems. Energy Environ Sci 5(8):8199–8203
29. Bond JQ, Alonso DM, Wang D, West RM, Dumesic JA (2010) Integrated catalytic conversion of gamma-valerolactone to liquid alkenes for transportation fuels. Science 327(5969):1110–1114
30. Asghari FS, Yoshida H (2006) Acid-catalyzed production of 5-hydroxymethyl furfural from D-fructose in subcritical water. Ind Eng Chem Res 45(7):2163–2173
31. Srokol Z, Bouche AG, van Estrik A, Strik RCJ, Maschmeyer T, Peters JA (2004) Hydrothermal upgrading of biomass to biofuel; studies on some monosaccharide model compounds. Carbohydr Res 339(10):1717–1726
32. Takeuchi Y, Jin FM, Tohji K, Enomoto H (2008) Acid catalytic hydrothermal conversion of carbohydrate biomass into useful substances. J Mater Sci 43(7):2472–2475
33. Shen J, Wyman CE (2012) Hydrochloric acid-catalyzed levulinic acid formation from cellulose: data and kinetic model to maximize yields. AIChE J 58(1):236–246
34. Cinlar B, Wang TF, Shanks BH (2013) Kinetics of monosaccharide conversion in the presence of homogeneous Bronsted acids. Appl Catal A:Gen 450:237–242
35. Weingarten R, Cho J, Xing R, Conner WC, Huber GW (2012) Kinetics and reaction engineering of levulinic acid production from aqueous glucose solutions. Chemsuschem 5(7):1280–1290

36. Pagan-Torres YJ, Wang TF, Gallo JMR, Shanks BH, Dumesic JA (2012) Production of 5-hydroxymethylfurfural from glucose using a combination of Lewis and Bronsted acid catalysts in water in a biphasic reactor with an alkylphenol solvent. Acs Catal 2(6):930–934
37. Yang Y, Hu CW, Abu-Omar MM (2012) Conversion of carbohydrates and lignocellulosic biomass into 5-hydroxymethylfurfural using $AlCl_3$ center dot $6H_2O$ catalyst in a biphasic solvent system. Green Chem 14(2):509–513
38. Wang TF, Pagan-Torres YJ, Combs EJ, Dumesic JA, Shanks BH (2012) Water-compatible lewis acid-catalyzed conversion of carbohydrates to 5-hydroxymethylfurfural in a biphasic solvent system. Top Catal 55(7–10):657–662
39. Roman-Leshkov Y, Chheda JN, Dumesic JA (2006) Phase modifiers promote efficient production of hydroxymethylfurfural from fructose. Science 312(5782):1933–1937
40. Jin FM, Zhou ZY, Enomoto H, Moriya T, Higashijima H (2004) Conversion mechanism of cellulosic biomass to lactic acid in subcritical water and acid-base catalytic effect of subcritical water. Chem Lett 33(2):126–127
41. Kishida H, Jin FM, Yan XY, Moriya T, Enomoto H (2006) Formation of lactic acid from glycolaldehyde by alkaline hydrothermal reaction. Carbohydr Res 341(15):2619–2623
42. Sasaki M, Goto K, Tajima K, Adschiri T, Arai K (2002) Rapid and selective retro-aldol condensation of glucose to glycolaldehyde in supercritical water. Green Chem 4(3):285–287
43. Yang BY, Montgomery R (1996) Alkaline degradation of glucose: effect of initial concentration of reactants. Carbohydr Res 280(1):27–45
44. Yan X, Jin F, Tohji K, Kishita A, Enomoto H (2010) Hydrothermal conversion of carbohydrate biomass to lactic acid. AIChE J 56(10):2727–2733
45. Yan XY, Jin FM, Tohji K, Moriya T, Enomoto H (2007) Production of lactic acid from glucose by alkaline hydrothermal reaction. J Mate Sci 42(24):9995–9999
46. Sánchez C, Egüés I, García A, Llano-Ponte R, Labidi J (2012) Lactic acid production by alkaline hydrothermal treatment of corn cobs. Chem Eng J 181–182:655–660
47. Jin FM, Yun J, Li GM, Kishita A, Tohji K, Enomoto H (2008) Hydrothermal conversion of carbohydrate biomass into formic acid at mild temperatures. Green Chem 10(6):612–615
48. Wang W, Wang SP, Ma XB, Gong JL (2011) Recent advances in catalytic hydrogenation of carbon dioxide. Chem Soc Rev 40(7):3703–3727
49. Yu JL, Savage PE (1998) Decomposition of formic acid under hydrothermal conditions. Ind Eng Chem Res 37(1):2–10
50. Saito K, Kakumoto T, Kuroda H, Torii S, Imamura A (1984) Thermal unimolecular decomposition of formic acid. J Chem Phys 80(10):4989–4997
51. Akiya N, Savage PE (1998) Role of water in formic acid decomposition. AIChE J 44(2):405–415
52. Wakai C, Yoshida K, Tsujino Y, Matubayasi N, Nakahara M (2004) Effect of concentration, acid, temperature, and metal on competitive reaction pathways for decarbonylation and decarboxylation of formic acid in hot water. Chem Lett 33(5):572–573
53. Marshall W, Franck E (1981) Ion product of water substance, 0–1000 °C, 1–10,000 bars new international formulation and its background. J Phys Chem Ref Data 10(2):295–304
54. Peterson AA, Vogel F, Lachance RP, Froling M, Antal MJ, Tester JW (2008) Thermochemical biofuel production in hydrothermal media: A review of sub- and supercritical water technologies. Energy Environ Sci 1(1):32–65
55. Katritzky AR, Nichols DA, Siskin M, Murugan R, Balasubramanian M (2001) Reactions in high-temperature aqueous media. Chem Rev 101(4):837–892
56. Westacott RE, Johnston KP, Rossky PJ (2001) Stability of ionic and radical molecular dissociation pathways for reaction in supercritical water. J Phys Chem B 105(28):6611–6619
57. Westacott RE, Johnston KP, Rossky PJ (2001) Simulation of an SN1 reaction in supercritical water. J Am Chem Soc 123(5):1006–1007
58. Aida TM, Sato Y, Watanabe M, Tajima K, Nonaka T, Hattori H, Arai K (2007) Dehydration of d-glucose in high temperature water at pressures up to 80 MPa. J Supercrit Fluids 40(3):381–388

59. Aida TM, Tajima K, Watanabe M, Saito Y, Kuroda K, Nonaka T, Hattori H, Smith RL, Arai K (2007) Reactions of D-fructose in water at temperatures up to 400 °C and pressures up to 100 MPa. J Supercrit Fluids 42(1):110–119
60. Chakinala AG, Brilman DWF, van Swaaij WPM, Kersten SRA (2010) Catalytic and non-catalytic supercritical water gasification of microalgae and glycerol. Ind Eng Chem Res 49(3):1113–1122
61. Kishida H, Jin FM, Zhou ZY, Moriya T, Enomoto H (2005) Conversion of glycerin into lactic acid by alkaline hydrothermal reaction. Chem Lett 34(11):1560–1561
62. Qadariyah L, Mahfud Sumarno, Machmudah S, Wahyudiono, Sasaki M, Goto M (2011) Degradation of glycerol using hydrothermal process. Bioresour Technol 102(19):9267–9271
63. Wahyudiono, Kanetake T, Sasaki M, Goto M (2007) Decomposition of a lignin model compound under hydrothermal conditions. Chem Eng Technol 30(8):1113–1122
64. Wahyudiono, Sasaki M, Goto M (2008) Recovery of phenolic compounds through the decomposition of lignin in near and supercritical water. Chem Eng Process 47(9–10):1609–1619
65. Wahyudiono, Sasaki M, Goto M (2009) Conversion of biomass model compound under hydrothermal conditions using batch reactor. Fuel 88(9):1656–1664
66. Sato T, Osada M, Watanabe M, Shirai M, Arai K (2003) Gasification of alkylphenols with supported noble metal catalysts in supercritical water. Ind Eng Chem Res 42(19):4277–4282
67. Osada M, Sato O, Watanabe M, Arai K, Shirai M (2006) Water density effect on lignin gasification over supported noble metal catalysts in supercritical water. Energy Fuels 20(3):930–935
68. Osada M, Watanabe M, Sue K, Adschiri T, Arai K (2004) Water density dependence of formaldehyde reaction in supercritical water. J Supercrit Fluids 28(2–3):219–224
69. Henrikson JT, Grice CR, Savage PE (2006) Effect of water density on methanol oxidation kinetics in supercritical water. J Phys Chem A 110(10):3627–3632
70. Akizuki M, Fujii T, Hayashi R, Oshima Y Effects of water on reactions for waste treatment, organic synthesis, and bio-refinery in sub- and supercritical water. J Biosci Bioeng. (in press)
71. Floriano WB, Nascimento MAC (2004) Dielectric constant and density of water as a function of pressure at constant temperature. Braz J Phys 34:38–41
72. Bradley DJ, Pitzer KS (1979) Thermodynamics of electrolytes. 12. Dielectric properties of water and Debye-Hueckel parameters to 350 °C and 1 kbar. J Phys Chem 83(12):1599–1603
73. CRC Handbook of chemistry and physics (1993) 74 edn. CRC Press
74. Islam MN, Jo YT, Park JH (2012) Remediation of PAHs contaminated soil by extraction using subcritical water. J Ind Eng Chem 18(5):1689–1693
75. Hashimoto S, Watanabe K, Nose K, Morita M (2004) Remediation of soil contaminated with dioxins by subcritical water extraction. Chemosphere 54(1):89–96
76. Lagadec AJM, Miller DJ, Lilke AV, Hawthorne SB (2000) Pilot-scale subcritical water remediation of polycyclic aromatic hydrocarbon- and pesticide-contaminated soil. Environ Sci Technol 34(8):1542–1548
77. Yang Y, Bowadt S, Hawthorne SB, Miller DJ (1995) Subcritical water extraction of polychlorinated-biphenyls from soil and sediment. Anal Chem 67(24):4571–4576
78. Hawthorne SB, Yang Y, Miller DJ (1994) Extraction of organic pollutants from environmental solids with sub- and supercritical water. Anal Chem 66(18):2912–2920
79. Smith KA, Griffith P, Harris JG, Herzog HJ, Howard JB, Latanision R, Peters WA (1995) Supercritical water oxidation: principles and prospects. Paper presented at the proceedings of the international water conference, Pittsburgh
80. Savage PE (1999) Organic chemical reactions in supercritical water. Chem Rev 99(2):603–621
81. Eckert CA, Knutson BL, Debenedetti PG (1996) Supercritical fluids as solvents for chemical and materials processing. Nature 383(6598):313–318

82. Gomez-Briceno D, Blazquez F, Saez-Maderuelo A (2013) Oxidation of austenitic and ferritic/martensitic alloys in supercritical water. J Supercrit Fluids 78:103–113
83. Townsend SH, Abraham MA, Huppert GL, Klein MT, Paspek SC (1988) Solvent effects during reactions in supercritical water. Ind Eng Chem Res 27(1):143–149
84. Marrone PA, Gschwend PM, Swallow KC, Peters WA, Tester JW (1998) Product distribution and reaction pathways for methylene chloride hydrolysis and oxidation under hydrothermal conditions. J Supercrit Fluids 12(3):239–254
85. Tester JW, Marrone PA, DiPippo MM, Sako K, Reagan MT, Arias T, Peters WA (1998) Chemical reactions and phase equilibria of model halocarbons and salts in sub- and supercritical water (200–300 bar, 100–600 °C). J Supercrit Fluids 13(1–3):225–240
86. Marrone PA, Arias TA, Peters WA, Tester JW (1998) Solvation effects on kinetics of methylene chloride reactions in sub- and supercritical water: theory, experiment, and ab initio calculations. J Phys Chem A 102(35):7013–7028
87. Wahyudiono, Sasaki M, Goto M (2011) Thermal decomposition of guaiacol in sub- and supercritical water and its kinetic analysis. J Mater Cycles Waste Manage 13(1):68–79
88. Ragauskas AJ, Williams CK, Davison BH, Britovsek G, Cairney J, Eckert CA, Frederick WJ, Hallett JP, Leak DJ, Liotta CL, Mielenz JR, Murphy R, Templer R, Tschaplinski T (2006) The path forward for biofuels and biomaterials. Science 311(5760):484–489
89. Jin F, Gao Y, Jin Y, Zhang Y, Cao J, Wei Z, Smith RL Jr (2011) High-yield reduction of carbon dioxide into formic acid by zero-valent metal/metal oxide redox cycles. Energy Environ Sci 4(3):881–884
90. Jin FM, Zhong H, Cao JL, Cao JX, Kawasaki K, Kishita A, Matsumoto T, Tohji K, Enomoto H (2010) Oxidation of unsaturated carboxylic acids under hydrothermal conditions. Bioresour Technol 101(19):7624–7634
91. Jin FM, Zhou ZY, Moriya T, Kishida H, Higashijima H, Enomoto H (2005) Controlling hydrothermal reaction pathways to improve acetic acid production from carbohydrate biomass. Environ Sci Technol 39(6):1893–1902
92. Rataboul F, Essayem N (2010) Cellulose reactivity in supercritical methanol in the presence of solid acid catalysts: direct synthesis of methyl-levulinate. Ind Eng Chem Res 50(2):799–805
93. Morimoto M, Sato S, Takanohashi T (2012) Effect of water properties on the degradative extraction of asphaltene using supercritical water. J Supercrit Fluids 68:113–116
94. Kishita A, Takahashi S, Kamimura H, Miki M, Moriya T, Enomoto H (2002) Hydrothermal visbreaking of bitumen in supercritical water with alkali. J Jpn Petrol Inst 45(6):361–367
95. Cheng Z-M, Ding Y, Zhao L-Q, Yuan P-Q, Yuan W-K (2009)Effects of supercritical water in vacuum residue upgrading. Energy Fuels 23(6):3178–3183
96. Sato T, Adschiri T, Arai K, Rempel GL, Ng FTT (2003) Upgrading of asphalt with and without partial oxidation in supercritical water. Fuel 82(10):1231–1239
97. Kokubo S, Nishida K, Hayashi A, Takahashi H, Yokota O, Inage S-i (2008) Effective demetalization and suppression of coke formation using supercritical water technology for heavy oil upgrading. J Jpn Petrol Inst 51(5):309–314
98. Han L, Zhang R, Bi J (2009) Experimental investigation of high-temperature coal tar upgrading in supercritical water. Fuel Process Technol 90(2):292–300
99. Wang L, Weller CL (2006) Recent advances in extraction of nutraceuticals from plants. Trends Food Sci Technol 17(6):300–312
100. Mizan TI, Savage PE, Ziff RM (1996) Temperature dependence of hydrogen bonding in supercritical water. J Phys Chem 100(1):403–408
101. Hoffmann MM, Conradi MS (1997) Are there hydrogen bonds in supercritical water? J Am Chem Soc 119(16):3811–3817
102. Jedlovszky P, Brodholt JP, Bruni F, Ricci MA, Soper AK, Vallauri R (1998) Analysis of the hydrogen-bonded structure of water from ambient to supercritical conditions. J Chem Phys 108(20):8528–8540
103. Kalinichev AG, Churakov SV (1999) Size and topology of molecular clusters in supercritical water: a molecular dynamics simulation. Chem Phys Lett 302(5–6):411–417

104. Mountain RD (1999) Voids and clusters in expanded water. J Chem Phys 110(4):2109–2115
105. Holgate HR, Meyer JC, Tester JW (1995) Glucose hydrolysis and oxidation in supercritical water. AIChE J 41(3):637–648
106. Yu DH, Aihara M, Antal MJ (1993) Hydrogen-production by steam reforming glucose in supercritical water. Energy Fuels 7(5):574–577
107. Antal MJ, Allen SG, Schulman D, Xu XD, Divilio RJ (2000) Biomass gasification in supercritical water. Ind Eng Chem Res 39(11):4040–4053
108. Xu XD, Antal MJ (1998) Gasification of sewage sludge and other biomass for hydrogen production in supercritical water. Environ Prog 17(4):215–220
109. Maiella PG, Brill TB (1996) Spectroscopy of hydrothermal reactions. 3. The water-gas reaction, "hot spots", and formation of volatile salts of NCO- from aqueous [$NH_3(CH_2)_nNH_3$]NO_3 (n = 2,3) at 720 K and 276 bar by T-jump/FT-IR spectroscopy. Appl Spectrosc 50(7):829–835
110. Li L, Portela JR, Vallejo D, Gloyna EF (1999) Oxidation and Hydrolysis of Lactic Acid in Near-Critical Water. Ind Eng Chem Res 38(7):2599–2606
111. Helling RK, Tester JW (1987) Oxidation kinetics of carbon monoxide in supercritical water. Energy Fuels 1(5):417–423
112. Holgate HR, Tester JW (1994) Oxidation of hydrogen and carbon-monoxide in subcritical and supercritical water-reaction-kinetics, pathways, and water-density effects.1. experimental results. J Phys Chem 98(3):800–809
113. Holgate HR, Webley PA, Tester JW, Helling RK (1992) Carbon monoxide oxidation in supercritical water the effects of heat-transfer and the water gas shift reaction on observed kinetics. Energy Fuels 6(5):586–597
114. Adschiri T, Shibata R, Sato T, Watanabe M, Arai K (1998) Catalytic hydrodesulfurization of dibenzothiophene through partial oxidation and a water-gas shift reaction in supercritical water. Ind Eng Chem Res 37(7):2634–2638
115. Arai K, Adschiri T, Watanabe M (2000) Hydrogenation of hydrocarbons through partial oxidation in supercritical water. Ind Eng Chem Res 39(12):4697–4701
116. Matsumura Y, Nonaka H, Yokura H, Tsutsumi A, Yoshida K (1999) Co-liquefaction of coal and cellulose in supercritical water. Fuel 78(9):1049–1056
117. Shen Z, Zhang YL, Jin FM (2011) From $NaHCO_3$ into formate and from isopropanol into acetone: hydrogen-transfer reduction of $NaHCO_3$ with isopropanol in high-temperature water. Green Chem 13(4):820–823
118. Shen Z, Zhang YL, Jin FM (2012) The alcohol-mediated reduction of CO_2 and $NaHCO_3$ into formate: a hydrogen transfer reduction of $NaHCO_3$ with glycerine under alkaline hydrothermal conditions. Rsc Adv 2(3):797–801
119. Kruse A, Ebert KH (1996) Chemical reactions in supercritical water. 1. Pyrolysis of tert-butylbenzene. Ber Bunsen Phys Chem 100(1):80–83
120. Zhang YL, Shen Z, Zhou XF, Zhang M, Jin FM (2012) Solvent isotope effect and mechanism for the production of hydrogen and lactic acid from glycerol under hydrothermal alkaline conditions. Green Chem 14(12):3285–3288
121. Ramayya S, Brittain A, Dealmeida C, Mok W, Antal MJ (1987) Acid-catalyzed dehydration of alcohols in supercritical water. Fuel 66(10):1364–1371
122. Narayan R, Antal MJ (1989) Kinetic elucidation of the acid-catalyzed mechanism of 1-propanol dehydration in supercritical water. ACS Symp Ser 406:226–241
123. Xu XD, Dealmeida CP, Antal MJ (1991) Mechanism and kinetics of the acid-catalyzed formation of ethene and diethyl-ether from ethanol in supercritical water. Ind Eng Chem Res 30(7):1478–1485
124. Antal MJ, Leesomboon T, Mok WS, Richards GN (1991) Kinetic studies of the reactions of ketoses and aldoses in water at high-temperature.3. mechanism of formation of 2-furaldehyde from d-xylose. Carbohydr Res 217:71–85
125. Xu XD, Antal MJ (1994) Kinetics and mechanism of isobutene formation from t-butanol in hot liquid water. AIChE J 40(9):1524–1534

126. Xu XD, Antal MJ, Anderson DGM (1997) Mechanism and temperature-dependent kinetics of the dehydration of tert-butyl alcohol in hot compressed liquid water. Ind Eng Chem Res 36(1):23–41
127. Antal MJ, Carlsson M, Xu X, Anderson DGM (1998) Mechanism and kinetics of the acid-catalyzed dehydration of 1- and 2-propanol in hot compressed liquid water. Ind Eng Chem Res 37(10):3820–3829
128. Arita T, Nakahara K, Nagami K, Kajimoto O (2003) Hydrogen generation from ethanol in supercritical water without catalyst. Tetrahedron Lett 44(5):1083–1086
129. Shen Z, Jin FM, Zhang YL, Wu B, Cao JL (2009) Hydrogen transfer reduction of ketones using formic acid as a hydrogen donor under hydrothermal conditions. J Zhejiang Univ-Sc A 10(11):1631–1635
130. Shen Z, Zhang YL, Jin FM, Zhou XF, Kishita A, Tohji K (2010) Hydrogen-transfer reduction of ketones into corresponding alcohols using formic acid as a hydrogen donor without a metal catalyst in high-temperature water. Ind Eng Chem Res 49(13):6255–6259
131. Wang XG, Gron LU, Klein MT, Brill TB (1995) The influence of high-temperature water on the reaction pathways of nitroanilines. J Supercrit Fluids 8(3):236–249
132. Belsky AJ, Maiella PG, Brill TB (1999) Spectroscopy of hydrothermal reactions. 13. Kinetics and mechanisms of decarboxylation of acetic acid derivatives at 100–260 degrees C under 275 bar. J Phys Chem A 103(21):4253–4260
133. Takahashi H, Hisaoka S, Nitta T (2002) Ethanol oxidation reactions catalyzed by water molecules: $CH_3CH_2OH + nH_2O = CH_3CHO + H_2 + nH_2O$ (n = 0, 1, 2). Chem Phys Lett 363(1–2):80–86
134. Liu J, Zeng X, Cheng M, Yun J, Li Q, Jing Z, Jin F (2012) Reduction of formic acid to methanol under hydrothermal conditions in the presence of Cu and Zn. Bioresour Technol 114:658–662

Part II
Hydrothermal Conversion of Biomass into Chemicals

Chapter 2
Hydrothermal Conversion of Cellulose into Organic Acids with a CuO Oxidant

Yuanqing Wang, Guodong Yao and Fangming Jin

Abstract In this chapter, we review some recent progress on the acid/base-catalyzed hydrothermal conversion and oxidation of cellulose into organic acids mainly in our research group. A novel one-pot production of organic acids and metal copper from cellulose and CuO under alkaline hydrothermal conditions is introduced based on our former research. The mechanism of formation of organic acids and metal copper is discussed. A principal reaction pathway from cellulose to organic acids and their reactions are also discussed. The results show that from cellulose to organic acids, the production processes are mainly composed of four stages of reactions. The reaction conditions were also optimized for production of organic acids and copper. These results show that a selective production of organic acids including lactic acid, glycolic acid, acetic acid, and formic acid can be achieved by varying reaction temperature and time and ratio of CuO and NaOH addition.

2.1 Introduction

To decrease the dependence of the human race on fossil resources and relieve the global warming caused by anthropogenic CO_2 emission, renewable and green resources of fuels and chemicals should be found. Cellulosic biomass emerges as a promising source for many reasons. First, cellulosic biomass is sustainable as long as light, water, CO_2, and some other necessary conditions are provided. Secondly,

---

Y. Wang
RIKEN Research Cluster for Innovation Nakamura Laboratory, 2-1 Hirosawa, Saitama, Wako 351-0198, Japan

G. Yao · F. Jin (✉)
School of Environmental Science and Engineering, Shanghai Jiao Tong University, 800 Dongchuan Road, Shanghai 200240, China
e-mail: fmjin@sjtu.edu.cn

cellulosic biomass is nonedible in comparison with starch or corn. So, it will not conflict with the need of human beings for food. Some solid wastes mainly composed of cellulose such as paper, sawdust, and municipal wastes can also be used. Thirdly, biomass is commonly recognized as carbon neutral that has no net release of CO_2.

Many effective chemical methods for transformation of cellulose into valuable chemicals have been developed and reviewed [1, 2]. Generally speaking, gasification, pyrolysis, liquefaction, and solidification are the main approaches for transformation of cellulosic biomass leading to different products. For instance, CO, CO_2, and H_2 are normally formed from gasification when using oxygen as the gasifying agent which can then be synthesized to liquid alkanes and alcohols by Fischer-Tropsch process [3]. Oxygen containing low molecular weight organics is formed through liquefaction technology such as 5-hydroxymethyl-2-furaldehyde (HMF) [4]. Carbons can also be formed to be employed as energy and storage materials by solidification [5].

Recently, hydrothermal technology applied in transformation of biomass to gas, liquid, and carbon products has attracted attention for its high efficient and environment-friendly advantages [5–10]. High temperature water has a lot of unique features. For example, water around 250–300 °C has high ion product (k_w), which can act both as acid and base catalyst [6]. The low dielectric constant of high temperature water can greatly increase solubility of nonpolar feedstock. Hydrothermal reaction can quickly hydrolyze insoluble cellulose into water-soluble saccharides and desirable chemicals afterwards [10, 11]. Besides, direct use of wet biomass is available without the drying process, which can decrease the cost a lot. Readers are suggested to find classic reviews [5–10] on hydrothermal technology. In this chapter, selective conversion of cellulose into desired products (HMF and organic acids) is discussed through acid/base-catalyzed and oxidation reactions and recent progress in organic acid production by CuO oxidant under hydrothermal conditions in our group is reviewed.

2.1.1 Conversion of Cellulose by Acid/Base Catalyst

2.1.1.1 Acid Catalyst

Figure 2.1 shows a principal reaction pathway of cellulose by hydrothermal conversion. It is well known that cellulose can decompose into HMF by acid-catalyzed reaction and lactic acid by base-catalyzed reaction without adding any catalyst under hydrothermal conditions as high-temperature water can act both as acid and base catalyst [6, 12]. However, the yields of some specific value-added products, such as HMF and lactic acid, are low without adding catalyst. Therefore, it is expected that the selective production of HMF and lactic acid can be enhanced by adding acid or base catalyst under hydrothermal conditions.

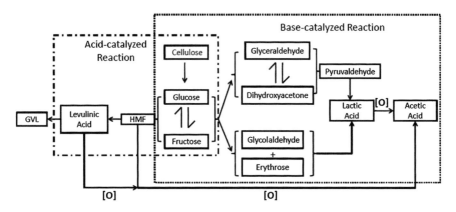

Fig. 2.1 A principal reaction pathway of cellulose by hydrothermal conversion

As shown in Fig. 2.1, HMF is produced from the acid-catalyzed dehydration of C6-sugars (i.e., hexoses) in the furanose form [13, 14]. Hence, fructose which contains 21.5 % of furanose tautomers in aqueous solution can be converted into HMF easier than glucose which contains only 1 % of furanose tautomers in aqueous solutions. However, acid catalyst can improve the yield of HMF from glucose by increasing isomerization of glucose into fructose followed by Brønsted acid-catalyzed dehydration of fructose to HMF [4]. Mineral acids (HCl, H_2SO_4 and H_3PO_4) [15] and acid metal salts [16] are usually adopted acid. The rehydration of HMF with two molecules of water would produce levulinic acid and formic acid [17]. Levulinic acid can be further converted into g-valerolactone (GVL) via hydrogenation with hydrogen [18], which can be converted to liquid alkenes in the transportation fuel range [19]. Kinetics studies [20–22] show that humins formation from glucose and HMF cannot be neglected. The activation energy of its formation from glucose and HMF was estimated at 51 and 142 kJ/mol, respectively, while dehydration of glucose to HMF and rehydration of HMF to levulinic acid was 160 and 95 kJ/mol, respectively [22].

Currently, numerous researches [16, 23–25] on production of HMF from cellulosic biomass have been concentrated on the temperature range of 100–200 °C. To minimize the formation of humins and enhance selectivity toward HMF, a biphasic solution with water and organic phase was adopted that would continuously extract HMF as it is produced [16, 23–25]. Dumesic et al. reported a 61 % yield of HMF from glucose using a biphasic reactor of water/tetrahydrofuran with $AlCl_3 \cdot 6H_2O$ catalyst at 160 °C [23]. However, hydrothermal reactions can afford a fast transformation of biomass within a few minutes [26] compared with the case in lower temperature range that normally needs 30 min to hours [16]. Yoshida et al. obtained the best yield of HMF (65 %) from fructose achieved at a temperature of 513 K for a residence time of 120 s [26]. Our study [15] found the highest yield of levulinic acid is about 55 % at 523 K for 5 min with HCl as a catalyst, and the total highest yields of HMF and levulinic acid are about 50 %, which occurred at 523 K for 5 min with H_3PO_4 as a catalyst. For the three mineral acids (HCl, H_2SO_4, and H_3PO_4), it was found that not only the pH, but also the nature of the acids, had great influence on the decomposition pathway [26]. At lower pH, a rehydration of HMF to levulinic and formic acids was favored, whereas at higher pH, polymerization reactions was favored [26]. The order for the production of HMF using the three acids is in the sequence of $H_3PO_4 > H_2SO_4 > HCl$. By contrast, the order for production of levulinic acid follows $HCl > H_2SO_4 > H_3PO_4$ [15].

2.1.1.2 Base Catalyst

Lactic acid has received attention as a building block for biodegradable lactic acid polymers with limited environmental impact. Currently, the fermentation of starch is the main method for producing lactic acid. Bioconversion (bacterial fermentation), however, is not available directly to cellulose and lignocelluloses. In general, pretreatment is needed, and also a large amount of residue is acquired for further treatment. Besides, the fermentation is a complex and sensitive process requiring 2–8 days to complete the reaction, of which the pH and temperature must be carefully monitored. In contrast, hydrothermal reactions have been shown to convert cellulose and lignocelluloses into lactic acid directly and effectively.

Researchers [27–31] have examined intermediate products for hydrothermal degradation of glucose and cellulose at a reaction temperature of nearly 300 °C. As shown in Fig. 2.1, through these studies it was revealed that fructose and some compounds containing three carbon atoms (C3 carbon compounds), such as glyceraldehyde, dihydroxyacetone, and pyruvaldehyde, are formed by the base catalytic role of high-temperature water. Furthermore, there is isomerization occurring between glyceraldehyde and dihydroxyacetone followed by their subsequent dehydration to pyruvaldehyde [30]. These C3 carbon compounds are considered as the precursors of lactic acid from transformation of pyruvaldehyde [27]. On the other hand, the intermediates glycolaldehyde and erythrose transformed from glucose [28, 29] can also produce lactic acid [31]. Although these intermediates from C2–C3 or C3–C4 bond cleavage by reverse aldol condensation from hexoses can all produce lactic acid, glyceraldehyde can produce a higher yield of lactic acid [31]. Thus, it is suggested that if the selective bond cleavage can be achieved between these two bond cleavages, the yield of lactic acid can be higher.

Our recent studies [32, 33] show that the addition of base catalyst (NaOH and $Ca(OH)_2$) can increase the yield of lactic acid. The highest yield of lactic acid from glucose was 27 % with 2.5 M NaOH and 20 % with 0.32 M $Ca(OH)_2$ at 300 °C for 60 s [32]. A recent study by Labidi et al. [34] also found that the highest yield lactic acid of 45 % from corn cobs was obtained using 0.7 M $Ca(OH)_2$ at 300 °C for 30 min. The reason that base catalyst increases the yield of lactic acid can be attributed to the enhancement of reaction pathway for lactic acid production discussed above. Another reason may be that the lactate formed actually by alkaline solution prevents it from decomposition [35]. For comparison with NaOH and Ca(OH)$_2$ [32], at lower alkaline concentration, Ca(OH)$_2$ promotes more effectively for production of lactic acid than NaOH in the same OH$^-$ concentration. This is probably because Ca^{2+} is more capable than Na$^+$ for forming complexes with two oxygen atoms in the hexoses. When Ca(OH)$_2$ increased higher from 0.32 to 0.4 M, it did not lead to increase in lactic acid yield, while the optimum OH$^-$ concentration for NaOH was 2.5 M. This difference can be attributed to the fact that the saturated solubility of NaOH is higher than that of Ca(OH)$_2$.

2.1.2 Conversion of Cellulose by Acid/Base Catalyst Coupling with Novel Oxidant

2.1.2.1 H$_2$O$_2$ Oxidant

Hydrogen peroxide, H$_2$O$_2$ is an attractive oxidant for liquid-phase reactions in an economically, technically, and environmentally satisfying manner [36]. In hydrothermal reactions, it has been adopted to convert biomass into stable oxidation products such as acetic acid [37–39]. Acetic acid is also an important raw material in the industry primarily for producing vinyl acetate monomer (VAM) and acetic anhydride, and a solvent for purified terephthalic acid (PTA) production [40]. More than 60 % of the world acetic acid production employs the methanol carbonylation methods with drawbacks of catalyst solubility limitations and the loss of expensive Rh metal due to precipitation in the separation sections [40]. Thus, the development of an environmentally friendly and highly effective method for converting biomass into acetic acid is strongly desired. Daimon et al. [37] studied the production of low molecular weight carboxylic acids by the hydrothermal treatment of representative organic waste and compounds with H$_2$O$_2$. They found that acetic acid was dominant among the several carboxylic acids obtained including formic, propanoic, succinic, and lactic acids. About 29 mg/g yield of acetic acid from glucose was obtained. The reactions of H$_2$O$_2$ in the supercritical water were also investigated experimentally and theoretically [41–43]. It was found that the rate of dissociation of H$_2$O$_2$ in supercritical water is density-dependent and faster than its high pressure limit rate in the gas phase.

A strategy for improving acetic acid yield consisting of two steps was proposed by our group [12, 44]. As shown in Fig. 2.2, the two-step process consists of both a hydrothermal reaction without a supply of oxygen (the first-step reaction) and an oxidation reaction (the second-step reaction). In Fig. 2.2a, the first step is to accelerate the formation of HMF, 2-FA, and lactic acid because these furans and lactic acid can produce a large amount of acetic acid by their oxidation. In Fig. 2.2b, the first step is a reaction in the presence of an added acid catalyst that accelerates the formation of levulinic acid. The second step is to further convert the furans and lactic acid or levulinic acid produced in the first step into acetic acid

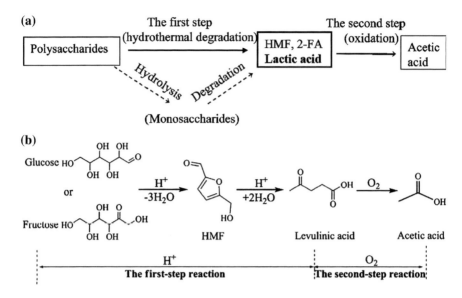

Fig. 2.2 Two proposed two-step processes for enhancing acetic acid yield. **a** Reprinted with permission from Ref. [12]. Copyright 2005 American Chemical Society. **b** Reprinted with permission from Ref. [44]. Copyright 2012 American Chemical Society

by oxidation with newly supplied H_2O_2. It was shown that the acetic acid yield was greatly increased by the two-step process. The yield of acetic acid increased to 23 % from glucose and 26 % from fructose at 523 K for 300 s with HCl and H_2O_2 [44].

On the other hand, if the method of hydrothermal treatment by adding base catalyst and H_2O_2 together was adopted, high yield of formic acid can be obtained [35]. Formic acid is an important organic chemical because it can be seen as an energy material that produces hydrogen as well as its other chemical application [45]. Formic acid is probably formed mainly from the direct oxidation of glucose by α scission [35]. The addition of alkali was considered as an effective way to prevent oxidative decomposition of formic acid [35]. The highest yield of formic acid through hydrothermal oxidation of glucose in the absence of alkali was about 24 %. In the case of adding alkali, the yield of formic acid increased to about 75 % from glucose [35].

2.1.2.2 CuO Oxidant

Recently, we have found that some metal oxides, such as CuO, could be reduced to its metal form with carbohydrates [46–48], which suggest that these metal oxides show oxidative potential for carbohydrates. That is, metal oxide might have the effect of promoting the production of organic acids from carbohydrates. With these in mind, a new process for selective production of acetic acid and lactic acid using

CuO as oxidant from glucose was proposed [49]. CuO has an obvious promotion effect on the production of acetic acid and lactic acid from glucose in the presence of NaOH [49]. Furthermore, we extended the feedstock from glucose to cellulose, and proposed a process for production of organic acids and refined copper in one-pot reaction from cellulosic biomass. An overall reaction scheme from cellulose to organic acids with a CuO oxidant under alkaline hydrothermal conditions was formulated based on these results. In the following, the progress on the production of organic acids and copper from glucose and cellulose with CuO in our group was reviewed including some recent unpublished results.

2.2 Conversion of Cellulose with CuO

In this section, there are three types of reactors used. Figure 2.3 shows a schematic diagram of the reactors used. As shown in Fig. 2.3, two types of batch reactor and a continuous flow type reactor were adopted. The detailed experimental methods can be found in our previous reports [12, 49]. The biggest differences are their heating time and flow type. As shown in Fig. 2.3a, the time required to heat up the oven from room temperature to 300 °C was 23 min. It needs another 38 min for the reactor to reach 300 °C and become stable. For the smaller tube reactor (see Fig. 2.3b), the time required to raise the temperature of reaction medium from 20 to 300 °C was about 15 s, which was faster than the former one. Therefore, the real reaction time performed in the former reactor was longer than that in the latter because of the heating time difference, although their nominal reaction times were the same. As shown in Fig. 2.3c, compared with the batch reactors, the continuous flow type reactor can be conducted at low temperatures and very short residence times (less than 1 min).

2.2.1 Conversion of Glucose

2.2.1.1 Effect of CuO on Products

Experiments with glucose were conducted in the presence and absence of CuO at 300 °C for 1 min using batch reactor 1. 300 °C was chosen because our previous research [10] on conversion of cellulosic biomass into chemicals showed that 300 °C was the optimum reaction temperature under hydrothermal conditions. As shown in Fig. 2.4, glycolic acid, lactic acid, formic acid, and acetic acid have been detected in the presence of CuO. As illustrated in Fig. 2.5, in the case without CuO (entry 1), the yields of these organic acids were all low, below 1.0 %. The conversion of glucose was above 99 %, thus, the yield almost equals the selectivity. In the case of adding CuO, the yield of acetic acid increased to 6.6 % and no formation of lactic acid was observed. This result suggests that CuO had a promotion

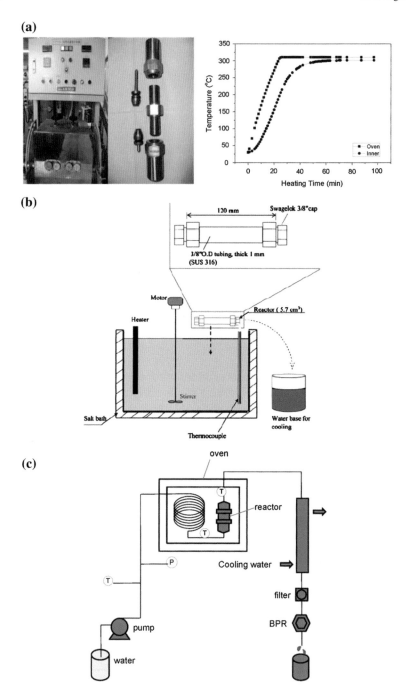

Fig. 2.3 Schematic diagram of reactors used. **a** Batch rector 1: photo and its heating time profile, reproduced from Ref. [49] by permission of John Wiley & Sons Ltd; **b** Batch reactor 2, reprinted with permission from Ref. [12]. Copyright 2013 American Chemical Society; **c** Diagram of continuous flow reactor, reproduced from Ref. [49] by permission of John Wiley & Sons Ltd

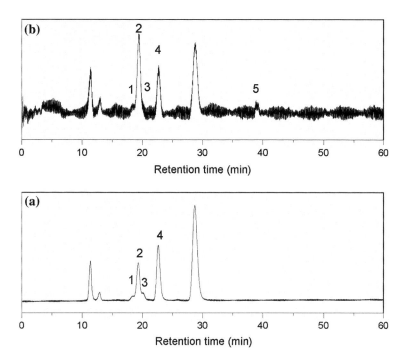

Fig. 2.4 a HPLC–UV chromatogram obtained from 0.0525 g D(+)-glucose, 2 mmol CuO, and 1 M NaOH after 1 min of reaction at 300 °C. 1: glycolic acid, 2: lactic acid, 3: formic acid, and 4: acetic acid. **b** HPLC-RI chromatogram obtained from 0.0525 g D(+)-glucose, 2 mmol CuO, and 1 M NaOH after 1 min of reaction at 300°C. 1: glycolic acid, 2: lactic acid, 3: formic acid, 4: acetic acid, and 5: 5-HMF. Reactor used: batch reactor 1. Reproduced from Ref. [49] by permission of John Wiley & Sons Ltd

effect on the acetic acid yield. In addition, as mentioned above, some researchers [33, 50] have proved that alkali had an obvious catalysis effect on the production of lactic acid from carbohydrate biomass under hydrothermal conditions. Thus, an experiment with glucose was conducted in the presence of CuO and NaOH to examine the role of OH^- in the production of organic acids in the presence of CuO. The yield of lactic acid and acetic acid increased to 25 and 23 %, respectively, suggesting that the role of CuO for the production of organic acids can be improved in the presence of OH^-. As shown in Fig. 2.5, in the case of adding CuO alone, the percentage of TOC in the liquid samples to the total input carbons based on carbon number was 23 %, even lower than that without CuO and NaOH (42 %). This percentage increased to 60 % when NaOH was added in the presence of CuO (see entry 3), which suggests that OH^- can protect glucose from carbonization into solids [51]. These observations further suggest that OH^- was crucial in the conversion of glucose into organic acids. Therefore, the combination of CuO and NaOH was adopted in the following for selective production of lactic acid and acetic acid.

Fig. 2.5 Effect of CuO on the yields of organic acid from glucose at 300 °C for 1 min (Entry 1: without addition; Entry 2: 2 mmol CuO; and Entry 3: 2 mmol CuO and 1 M NaOH. Reactor used: batch reactor 1) ([a] the percentage of TOC in the liquid samples to the total input carbons based on carbon number after reactions). Reproduced from Ref. [49] by permission of John Wiley & Sons Ltd

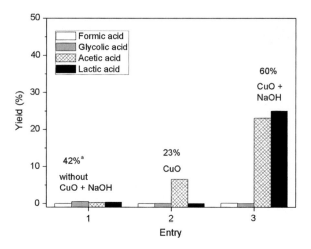

2.2.1.2 Effect of the Ratio of CuO:NaOH

The optimization ratio of CuO to NaOH for obtaining high yields of lactic acid and acetic acid was studied. As shown in Fig. 2.6, when fixing NaOH concentration to 1 M, the yield of acetic acid increased from 15 to 31 % with the increase of the amount of CuO from 1 to 4 mmol, which further confirmed that CuO had a selective production for acetic acid. However, the yield of lactic acid decreased from 39 to 7.2 % with the increase of the amount of CuO from 1 to 4 mmol. When lowering the NaOH concentration to 0.1 M with 2 mmol CuO, both yields of lactic acid and acetic acid decreased. This result indicates that a low concentration of alkali was disadvantageous for the production of lactic acid and acetic acid. Thus, the ratio of 2 mmol CuO to 1 M NaOH was adopted in the following discussion for attaining both high yields of lactic acid and acetic acid.

2.2.1.3 Effect of Reaction Temperature and Time

Figure 2.7a shows the effect of reaction temperature at 250 and 300 °C for 1 min on yields of lactic acid and acetic acid. The yield of lactic acid decreased from 47 to 25 % and the yield of acetic acid increased from 4.4 to 23 % with the increase in reaction temperature from 250 to 300 °C. The yields of formic acid and glycolic acid were very low at both temperatures. Figure 2.7b shows the effect of reaction time on yields of organic acids. Similarly, with increase in reaction time the yield of acetic acid increased, while the yield of lactic acid increased. A high yield of acetic acid (32 %) was obtained for 60 min. The yields of formic acid and glycolic acid were very low, below 1.0 %. It is suggested that the increase in acetic acid and the decrease in lactic acid with the increase in reaction temperature and time are probably because the formation of acetic acid is through the oxidative decomposition of lactic acid in the presence of CuO.

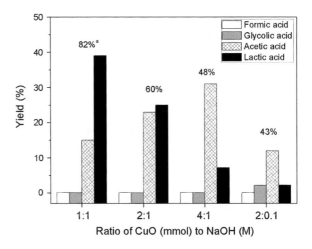

Fig. 2.6 Effect of the ratio of CuO to NaOH (mmol to M) on the yields of organic acid from glucose at 300 °C for 1 min. Reactor used: batch reactor 1 ([a] the percentage of TOC in the liquid samples to the total input carbons based on carbon number after reactions). Reproduced from Ref. [49] by permission of John Wiley & Sons Ltd

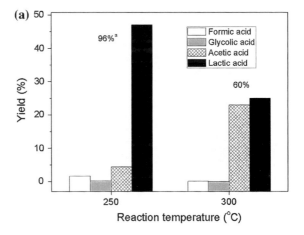

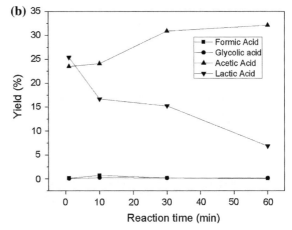

Fig. 2.7 a Effect of reaction temperature on the yields of organic acid (reaction conditions: 0.0525 g glucose, 2 mmol CuO, 1 M NaOH, and 1 min of reaction time; [a] the percentage of TOC in the liquid samples to the total input carbons based on carbon number after reactions); **b** effect of reaction time on the yields of organic acid (reaction conditions: 0.0525 g glucose, 2 mmol CuO, 1 M NaOH, and 300 °C). Reactor used: batch reactor 1. Reproduced from Ref. [49] by permission of John Wiley & Sons Ltd

As discussed above, a longer reaction time and higher reaction temperature was not advantageous for a high yield of lactic acid in the presence of CuO. That is, a higher yield of lactic acid may be obtained in a short reaction time. However, a short reaction time is not available with the used batch reactor system due to the long heat-up time. Thus, to obtain a higher yield of lactic acid, experiments were conducted in a continuous flow reactor, which can be conducted at low temperatures and short residence times. The temperature range was set from 188 to 284 °C and the residence time was set at 0–1 min under a stable pressure (10 MPa). As expected, a much higher yield of lactic acid was obtained in the continuous flow reactor compared to that in the batch reactor (see Fig. 2.8). The highest yield of lactic acid (59 %) was obtained at 188 °C with a short reaction time of 0.15 min. To the best of our knowledge, it is the highest yield of lactic acid reported so far obtained from glucose under hydrothermal conditions.

Similarly to the results obtained in the bath reactor, there is a tendency of decrease in yield of lactic acid with the increase in the residence time. For acetic acid, the yields were below 5 %, which were much lower than the results obtained in the batch reactor at all temperatures. These results suggest that in the initial time of reactions, glucose first transformed into lactic acid, which was then maybe converted into acetic acid, and the conversion of glucose into lactic acid was fast and conversion of lactic acid into acetic acid was relatively slow. These suggest that selective production of lactic acid and acetic acid can be achieved by controlling reaction time and temperature.

2.2.2 Reactions of Intermediates and Organic Acids with CuO

2.2.2.1 Reactions of Intermediates and Organic Acids

It has been reviewed before that aldose, aldehydes, and ketones, such as fructose, erthrose, glyceraldehyde, glycolaldehyde, pyruvaldehyde, and hydroxyacetone, are first formed as intermediates from glucose and then transformed into lactic acid under alkaline hydrothermal conditions. Therefore, fructose, erythrose, glyceraldhyde, glycolaldehyde, and pyruvaldehyde were chosen as starting materials to examine their availability to produce organic acids with CuO. In the presence of CuO (without NaOH) at 300 °C for 1 min, a yield of 2.7 % from fructose, 13.3 % from glyceraldehyde, 1.7 % from glcoaldehyde, and 18.5 % from pyruvaldehyde for production of acetic acid was observed, respectively [49]. For all these experiments, no peaks of starting materials and lactic acid were observed maybe because their concentrations were below the detection limit [49]. Due to the uncertainty in the purity of erythrose, the quantification is not available [49]. These results show that those selected intermediates were able to produce acetic acid with CuO by acid/base-catalyzed reactions. According to our recent studies

Fig. 2.8 Yields of organic acid with different residence times using the continuous flow reactor (1.75 wt % glucose feed solution; 1 M NaOH; and 1.5 g CuO). Reproduced from Ref. [49] by permission of John Wiley & Sons Ltd

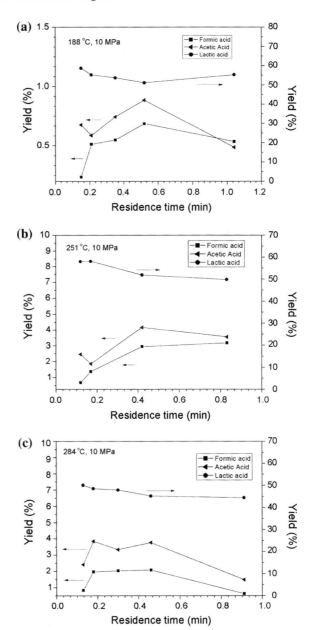

performed in the batch reactor 2 at 300 °C for 30 s, in the presence of CuO and NaOH, the intermediates of glyceraldyhde and glycolaldehyde were transformed into lactic acid mainly with glycolic acid, acetic acid, and formic acid (unpublished results), suggesting lactic acid was still initially mainly formed in the

Fig. 2.9 HPLC–UV chromatogram obtained from 0.35 M lactic acid, 2 mmol CuO, and 1 M NaOH at 300 °C for 1 min (1: lactic acid, 2: acetic acid, and 3: propionic acid). Reactor used: batch reactor 1. Reproduced from Ref. [49] by permission of John Wiley & Sons Ltd

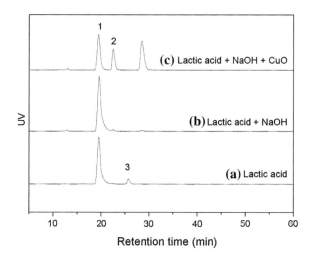

transformation of these intermediates in the addition of CuO under alkaline hydrothermal conditions.

Later, to examine the conversion of lactic acid into acetic acid, experiments with lactic acid as a starting material in the presence and absence of CuO were performed. As shown in Fig. 2.9a, lactic acid slightly decomposed and no peak of acetic acid was observed without CuO and NaOH. From Fig. 2.9b, in the case of adding 1 M NaOH without CuO, the peak of acetic acid was found. The yield and selectivity of acetic acid were 2.3 and 17 %, respectively. When adding 2 mmol CuO in the presence of 1 M NaOH, the yield and selectivity of acetic acid increased to 9.8 and 29 %, respectively. These results suggest that production of acetic acid from lactic acid can be improved with CuO. Figure 2.10 shows the change in the decomposition of lactic acid and the yield of acetic acid with CuO and NaOH by varying reaction time from 1 to 60 min. The remaining lactic acid decreased from 66.7 to 41.6 % and the yield of acetic acid increased from 9.8 to 18.1 % in accordance with the increase in reaction time from 1 to 60 min. The selectivity of acetic acid became stable with the increase in reaction time. Therefore, in the presence of CuO and NaOH, the high yield of acetic acid obtained in Sect. 2.1 was mainly attributed to the decomposition of lactic acid because lactic acid was initially formed with low yield of acetic acid at the same reaction time (see Fig. 2.8) and then lactic acid can be transformed into acetic acid.

For the reactions of other acids, such as glycolic acid, acetic, and formic acid, according to our recent study performed in the batch reactor 2 (unpublished results), they are relatively stable in the presence of NaOH whose conversions were lower than 10 % at 300 °C for 180 s. The presence of CuO (in addition of CuO and NaOH) can enhance the conversion of these acids except acetic acid compared with the case only in addition of NaOH. The conversion of these acids decreased in the following order: formic acid > glycolic acid > lactic acid > acetic acid in the

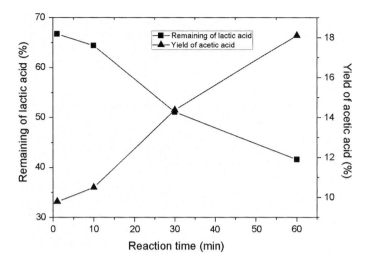

Fig. 2.10 Effect of reaction time on conversion of lactic acid and yield of acetic acid (0.35 M lactic acid, 2 mmol CuO, and 1 M NaOH at 300 °C). Reactor used: batch reactor 1. Reproduced from Ref. [49] by permission of John Wiley & Sons Ltd

presence of CuO. Acetic acid, glycolic acid, and lactic acid can all produce formic acid and formic acid can be further reacted to produce gas [52]. The fact that the yield of formic acid and glycolic acid from glucose performed in the batch reactor 1 was relatively low was attributed to the easy decomposition of these two acids in the presence of CuO and long heating time of batch reactor 1.

2.2.2.2 Role of CuO and Possible Mechanism

A possible mechanism of conversion of glucose into lactic acid and acetic acid in the presence of CuO was proposed and is presented in Fig. 2.11. At the beginning of reaction, a strong base (NaOH) under hydrothermal conditions may enhance the solubility of CuO to form hydroxo complex [53]. Subsequently, dissociated Cu(II) ions from the hydroxo complex may coordinate with hydroxyl oxygen atoms of glucose to form a comparatively stable coordination compound. The short distance between oxygen and Cu atoms in the coordination compound is favorable for electron transfer from the oxygen atom to Cu(II) ion, resulting in the reduction of Cu(II). Simultaneously, lactic acid is formed as the main product from the transformation of complex and glyceraldehyde [10]. Then, the formed lactic acid is further oxidized into acetic acid via a similar transformation by formation of Cu complex with the release of CO_2 as shown in Fig. 2.11.

The detailed discussion on reduction of CuO into Cu_2O and Cu by glucose and cellulose is given in Sects. 2.3.1 and 2.3.2.

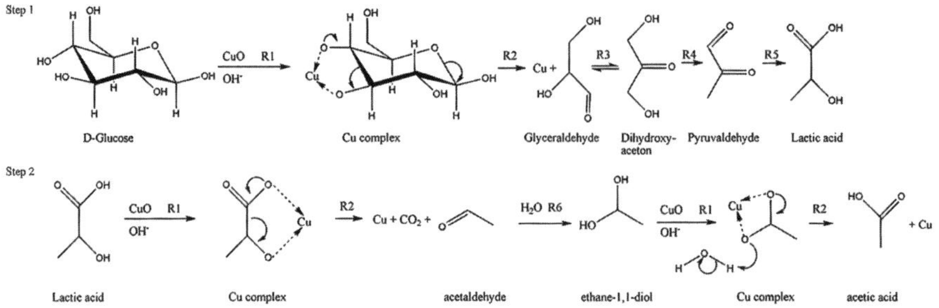

Fig. 2.11 Reaction pathway of production of lactic and acetic acid from D-glucose with CuO under alkaline hydrothermal conditions (R1: coordination; R2: redox; R3: Lobry de Bruyn-Alberda van Ekenstein transformation (LBAE); R4: elimination; R5: benzilic acid rearrangement; and R6: hydration). Reproduced from Ref. [49] by permission of John Wiley & Sons Ltd

2.2.3 Conversion of Cellulose with CuO

2.2.3.1 Effect of Reaction Temperature and Time

The feedstock in this section is extended to cellulose rather than its monomer glucose to investigate the reaction pathway of cellulosic biomass in this reaction system and optimize reaction conditions. Experiments of cellulose with CuO and NaOH under different reaction temperatures and time were conducted and the yield of organic acids and total organic carbon (TOC) are given in Fig. 2.12. As shown in Fig. 2.12, the total organic acids yield first increased and then decreased after 10 min at 275 °C, while it decreased after 5 min in the case of 300 °C and decreased after 1 min at 325 °C along with a similar trend of TOC yield. The higher the reaction temperature is, the sooner the yield of total organic acids will reach its highest value. The following decrease can be attributed to the reactions of carboxylic acids by decarboxylation or decarbonylation that lead to decrease in carbon yield.

For specific organic acids, it can be seen from Fig. 2.12 that mainly four kinds of organic acids including lactic acid, glycolic acid, acetic acid, and formic acid have been detected to be the same as the result obtained from glucose. Acetic acid increased with the increase in reaction time at all temperatures, which can be ascribed to the fact that it is very stable and can be obtained from lactic acid with CuO. As for relatively decomposable acids, such as formic acid, glycolic acid, and lactic acid, a trend of first increase and then decrease or direct decrease in the yield was observed, showing the competitive reactions between the formation and decomposition of these acids. These results suggest us that a selective production of these organic acids can be achieved by varying reaction temperature and time. It can be also found that an initial short reaction time, such as 5 min at 300 °C, mainly accounts for the formation of organic acids from cellulose. After that, the decomposition of organic acids dominates the whole process.

Fig. 2.12 Effect of reaction temperature and time on the yields of organic acids. Reaction conditions: 0.035 g cellulose, 2.0 mmol CuO, 1.0 mol/L NaOH, 2 mL water. Reactor used: batch reactor 2. Unpublished results

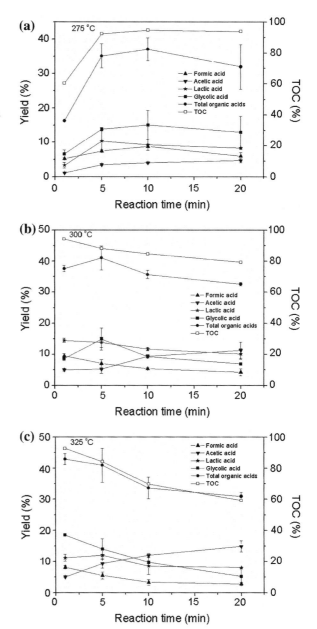

2.2.3.2 Reaction Scheme

On the basis of these results, we therefore propose a principal reaction pathway (Fig. 2.13) showing that cellulose is first hydrolyzed to glucose and fructose, then degraded into two carbons or three carbons aldehydes and ketones, and these aldehydes and ketones are subsequently transformed to carboxylic acids which can be further degraded from higher carbon numbers acid to lower carbon numbers acid. It can be concluded that the overall reaction pathways consist of four reaction stages: (i) stage from polysaccharide to monosaccharide; (ii) stage from monosaccharide to aldehydes and ketones; (iii) stage from aldehydes and ketones to carboxylic acids; (iv) stage of reactions of carboxylic acids.

2.3 Reduction of CuO with Cellulose

In the studies of conversion of cellulose with CuO, Cu_2O and Cu were found in solid products. This indicated that CuO can be reduced directly by cellulose. It is well known that Cu existing in nature is mainly in the form of ores containing oxygen or sulfur. Direct reduction of Cu oxides to Cu with cellulose as reductant is significant for green Cu smelting process. Compared to traditional CuO reducing agents in smelting, such as carbon, heavy oil, petroleum gas, and ammonia [54–56], cellulose is an environment-friendly reductant. This is because cellulose is one of the most abundant carbon neutral and renewable resources. Moreover, cellulose can reduce CuO efficiently at lower hydrothermal temperatures (<250 °C). Meanwhile, cellulose was converted into value-added chemicals (lactic acid, acetic acid, et al.). Thus, cellulose is a promising green reducing agent in Cu smelting.

In this section, we outline some recent advances on the conversion of CuO to Cu by using cellulose under alkaline hydrothermal conditions. Experiments in this study were carried out in a Teflon-lined batch reactor with an inner volume of 20 mL.

2.3.1 Examination of the Reduction of CuO to Cu with Cellulose

Figure 2.14 shows the XRD patterns of solid samples obtained after hydrothermal reactions at 250 and 180 °C with and without 0.50 mol/L NaOH. CuO can be partly reduced to Cu_2O and Cu, even at a lower temperature of 180 °C (see Fig. 2.14b), and only a Cu peak was observed when the temperature was increased to 250 °C. These results indicate that CuO can be completely reduced to highly pure Cu at 250 °C. Comparing the XRD patterns in the presence and absence of NaOH at the temperature of 180 °C (see Fig. 2.14b, c), it is clear that only a small quantity of Cu_2O and no Cu was produced, which suggest that alkali was favorable for the reduction of CuO to Cu_2O and Cu.

The particle size distributions for the original CuO and Cu obtained after the reaction at 250 °C for 1.5 h without any further processing were measured using a

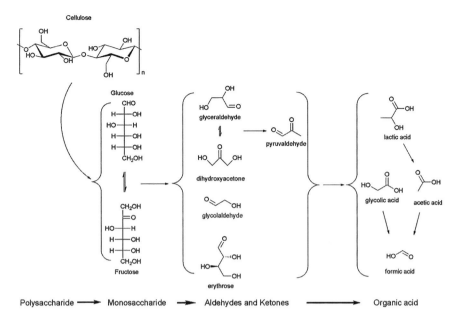

Fig. 2.13 Proposed reaction pathway from cellulose to organic acids

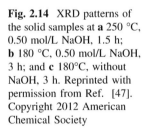

Fig. 2.14 XRD patterns of the solid samples at **a** 250 °C, 0.50 mol/L NaOH, 1.5 h; **b** 180 °C, 0.50 mol/L NaOH, 3 h; and **c** 180°C, without NaOH, 3 h. Reprinted with permission from Ref. [47]. Copyright 2012 American Chemical Society

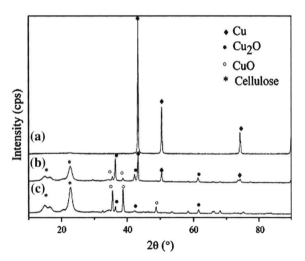

laser particle size analyzer. As shown in Table 2.1, the average particle sizes of the initial CuO and the Cu obtained were 6.19 μm (standard deviation (SD) = 3.50 μm) and 1.46 μm (SD = 1.03 μm), respectively. The range of the Cu particle size distribution was narrower than that of CuO, indicating that the particle size of Cu obtained after the reaction was more uniform.

Table 2.1 Distribution of particle size of the purchased CuO with 200 mesh and the obtained Cu with cellulose after the reaction at 250 °C with 0.50 mol/L NaOH after 1.5 h[a]

	av diam (μm)	D10 (μm)	D50 (μm)	D90 (μm)
Purchased CuO (200 mesh)	6.19	2.53	5.77	10.33
Obtained Cu	2.41	0.76	1.64	5.01

[a] D10, D50, and D90 mean that 10, 50, and 90 % of the powder particles are smaller than this value, respectively

Reprinted with permission from Ref. [47]. Copyright 2012 American Chemical Society

To determine the effect of CuO on the conversion of cellulose, the liquid products obtained after the hydrothermal degradation of cellulose with and without CuO were analyzed by HPLC. As shown in Fig. 2.15, lactic acid, formic acid, acetic acid, pyruvic acid, and some other low molecular weight carboxylic acids were detected in the absence and presence of CuO. However, in the presence of CuO, the amount of acetic acid increased from 2.0 to 4.0 g/L, the yield of formic acid decreased by 85 %, and the amount of lactic acid increased slightly. The increase of acetic acid may be attributed to the oxidative decomposition of lactic acid in the presence of CuO, as discussed above. Therefore, the production of lactic acid from cellulose was promoted in the presence of CuO. Moreover, the decrease of formic acid may be due to further oxidative decomposition to CO_2 [57].

2.3.2 *Effects of Reaction Conditions on the Conversion of CuO into Cu*

The effects of the reaction temperature, reaction time, and concentration of alkali on the reduction of CuO to Cu were investigated to determine the optimum conditions for the conversion of CuO into Cu. The yield of Cu_2O or Cu was defined as the ratio of the Cu_2O or Cu to the solid sample obtained based on the mass by XRD analysis. The Rietveld method is a standard technique for quantitative phase analysis. The Rietveld calculations in this work were performed by the software TOPAS 4.2 from Bruker AXS GmbH, Germany. This software is based on the fundamental parameter approach (FPA), which considers the geometric and unit-specific parameters [58].

The influence of reaction temperature on the reduction of CuO with cellulose was determined by varying the temperature from 180 to 250 °C after 2 h (Fig. 2.16). The yield of Cu greatly increased with the increase in reaction temperature. When the temperature increased to 250 °C, the CuO was completely reduced to Cu after 1.5 h. However, at temperature of 180 °C, the yield of Cu was only approximately 71 %, even for the longer reaction time of 8 h. As shown in the XRD patterns (Fig. 2.14b), there was unreacted cellulose remaining in the solid sample, which indicates that, at lower temperature, not all of the cellulose took

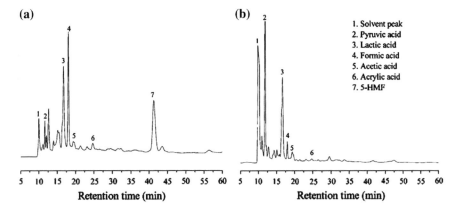

Fig. 2.15 HPLC chromatograms of liquid samples **a** without CuO and **b** with CuO at 250 °C, 1.5 h, 0.50 mol/L NaOH. Reprinted with permission from Ref. [47]. Copyright 2012 American Chemical Society

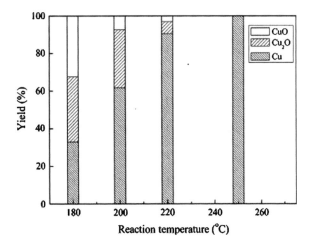

Fig. 2.16 Yields of Cu and Cu_2O with the variation of reaction temperature at 0.50 mol/L NaOH, 2 h. Reprinted with permission from Ref. [47]. Copyright 2012 American Chemical Society

part in the reduction reaction, leading to less conversion of CuO. These results suggest that the degradation of cellulose played an important role in the reduction of CuO to Cu. Moreover, in the view of thermodynamics of the reaction, ΔG_{red} (the Gibbs free energy for the reduction of oxides) slowly decreases as the temperature increases [59]. Therefore, the improvement in the reduction of CuO to Cu at a higher temperature, that is, above 200 °C, could be attributed to the decrease in ΔG_{red}.

As shown in Fig. 2.17, the yield of Cu increased from 61 % after 2 h to 93 % after 4 h at 200 °C, and no further obvious increase in the yield of Cu was observed after 6 h. Similar increase tendency was observed at 180 °C.

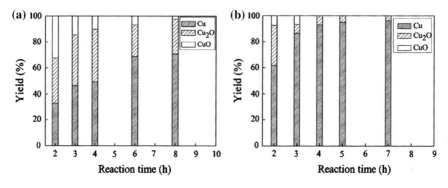

Fig. 2.17 Yields of Cu and Cu$_2$O with reaction time at temperatures of **a** 180 and **b** 200 °C, respectively, at 0.50 mol/L NaOH. Reprinted with permission from Ref. [47]. Copyright 2012 American Chemical Society

A large amount of Cu$_2$O was produced at a lower temperature after a short reaction time (Figs. 2.16 and 2.17), which suggests that the formation of Cu from CuO is a multistep reaction that proceeds via the formation of Cu$_2$O. The first step in the reduction of CuO to Cu$_2$O is exothermic with a calculated $\Delta_r H_m$ of −1306.94 kJ/mol, and the second step for reduction of Cu$_2$O to Cu is endothermic with a calculated $\Delta_r H_m$ of 703.07 kJ/mol. Therefore, CuO can easily be converted to Cu$_2$O at a lower temperature after a short reaction time, whereas a higher temperature is necessary to convert Cu$_2$O into Cu.

Experiments were conducted at 180 °C and 200 °C to investigate the effect of alkali by varying the concentration of NaOH from 0 to 1.0 mol/L. As shown in Fig. 2.18, a higher NaOH concentration was favorable for the reduction of CuO. The yield of Cu increased greatly with the increase in NaOH concentration and exceeded 90 % with 0.50 mol/L NaOH at 200 °C after 3 h. The notable increase in the conversion of CuO with the increase in NaOH concentration probably occurred for three reasons. First, alkali can promote the hydrolysis of cellulose. It has been reported that, NaOH can readily break the hydrogen bonds in cellulose and promote the degradation of cellulose in favor of C–C bond cleavage under hydrothermal conditions [60, 61]. Second, NaOH promoted the conversion of glucose into lactic acid due to its alkali catalytic activity which shows reducing function for CuO as discussed later [33, 62]. Moreover, according to the Pourbaix diagram in electrochemistry, the E_h (reduction potential) decreases as pH increases [63, 64]. Therefore, the increase of Cu with the increase of alkali concentration is probably due to the reduction of E_h by OH$^-$ ions. To maintain the safety of the experiments and the durability of the reactor, the optimum concentration of NaOH was set at 0.50 mol/L.

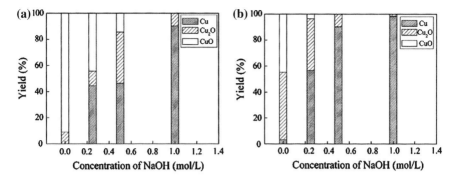

Fig. 2.18 Effect of concentration of NaOH on yields of Cu and Cu_2O at temperatures of a 180 and b 200 °C, respectively, 3 h. Reprinted with permission from Ref. [47]. Copyright 2012 American Chemical Society

2.3.3 Proposed Mechanism of Reduction of CuO to Cu with Cellulose

Generally, cellulose is a nonreducing sugar and is insoluble in water at a temperature; however, cellulose can be easily hydrolyzed to oligosaccharides and monosaccharides in HTW even without base or acid, due to inherent acid and base catalytic roles of HTW [65, 66]. Therefore, cellulose may first be hydrolyzed into glucose under the alkaline hydrothermal conditions [38], which subsequently is used for the reduction of CuO. Moreover, it is known that hydrogen or syngas can be produced from the hydrothermal gasification of biomass under subcritical water conditions [67–70]. Therefore, there are two possible explanations for the reduction of CuO to Cu in the presence of cellulose. One explanation is that CuO is reduced by the glucose formed by the hydrolysis of cellulose. The other explanation is that CuO is reduced by a reducing gas, such as CO and H_2 [71, 72], formed by the decomposition of cellulose. To test the latter hypothesis, gas samples were collected and examined by GC-TCD. Nearly no reducing gas, such as H_2 and CO, was produced, and approximately 58 % (v/v) of the gas collected was CO_2. Therefore, it was reasonable to propose that CuO was reduced by glucose formed by the hydrolysis of cellulose. The possible mechanism of reduction of CuO by glucose under hydrothermal conditions has been discussed in Sect. 2.2.2.2.

As mentioned above, the increase of acetic acid obtained in the presence of CuO is probably due to the oxidation of lactic acid. To test this assumption, further experiments with lactic acid as a reductant in the presence of CuO at 250 °C after 3 h were carried out under acidic, neutral, and alkaline conditions by adjusting the pH with NaOH. As shown in Fig. 2.19, which depicts the XRD patterns of the solid products after reactions with lactic acid at pH 12.0, 6.0, and 3.0, the peak of Cu was higher at a lower pH, and the peak of CuO was not observed at pH 3.0. These results indicate that acidic conditions are favorable for the reduction of CuO

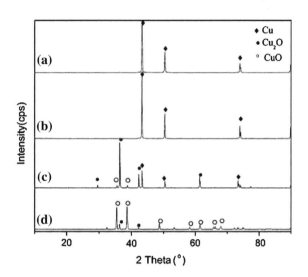

Fig. 2.19 XRD patterns of solid samples with lactic acid at temperature of 250 °C, 3 h, and **a** Cu = 6 mmol, pH 3.0; **b** CuO = 6 mmol, pH 3.0; **c** CuO = 6 mmol, pH 6.0; and **d** CuO = 6 mmol, pH 12.0. Reprinted with permission from Ref. [47]. Copyright 2012 American Chemical Society

to Cu with lactic acid as reductant. Figure 2.19a is the XRD pattern of solid sample after the blank experiment with Cu and lactic acid, which indicated that there were no reactions with Cu and lactic acid. Figure 2.20 shows the HPLC chromatograms of the liquid products obtained under the same conditions. These chromatograms show that the pyruvic acid and acetic acid were present in the liquid samples. Comparing the HPLC chromatograms at different pH values, the lactic acid peak was smaller and the acetic acid peak became higher with a decrease in the pH, thereby indicating that the increase of acetic acid could be attributed to the oxidation of lactic acid [73]. These results also demonstrate that acidic conditions are favorable for the decomposition of lactic acid, perhaps because organic acids are difficult to degrade under alkaline conditions, as demonstrated by our previous report [35]. A comparison of the HPLC chromatograms of the liquid samples obtained under the same conditions with and without CuO (Fig. 2.20c, d) reveals that the peak area of the acetic acid increased in the presence of CuO, indicating that CuO can promote the decomposition of lactic acid and improve the yield of acetic acid. Furthermore, a series of experiments with acetic acid (0.50 mol/L) and CuO at 250 °C were conducted to investigate the reducing capacity of acetic acid for CuO. As shown in Fig. 2.21, the reduction of CuO was not observed at pH 6.0 and 12.0, respectively, and a small amount of CuO was reduced to Cu_2O at pH 3.0. In contrast with experiments with lactic acid, the reducing ability of acetic acid is much lower in the reduction of CuO under hydrothermal conditions. Therefore, CuO can be reduced not only by cellulose but also by the products of cellulose decomposition, such as lactic acid.

Based on the reduction of CuO to Cu by cellulose as reducing agent under mild hydrothermal conditions, a facile and green process for producing Cu from CuO was proposed. As shown in Fig. 2.22, highly pure Cu can be obtained easily.

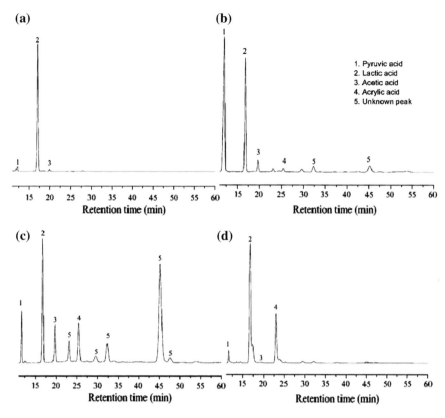

Fig. 2.20 HPLC chromatograms of liquid samples in the presence of CuO and lactic acid with different pH values, **a** pH = 12.0, **b** pH = 6.0, **c** pH = 3.0, **d** pH = 3.0 (without CuO), at 250 °C, 3 h. Reprinted with permission from Ref. [47]. Copyright 2012 American Chemical Society

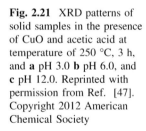

Fig. 2.21 XRD patterns of solid samples in the presence of CuO and acetic acid at temperature of 250 °C, 3 h, and **a** pH 3.0 **b** pH 6.0, and **c** pH 12.0. Reprinted with permission from Ref. [47]. Copyright 2012 American Chemical Society

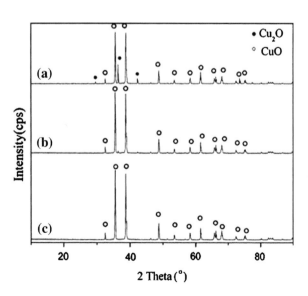

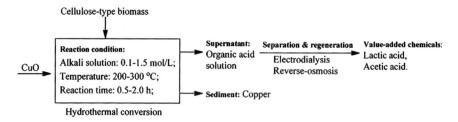

Fig. 2.22 Proposed flow sheet for converting CuO into Cu. Reprinted with permission from Ref. [47]. Copyright 2012 American Chemical Society

At the same time, some value-added chemicals, such as lactic acid and acetic acid, were also produced by suitable management [49]. Work along this line is now in progress.

References

1. Corma A, Iborra S, Velty A (2007) Chemical routes for the transformation of biomass into chemicals. Chem Rev 107(6):2411–2502
2. Gallezot P (2012) Conversion of biomass to selected chemical products. Chem Soc Rev 41(4):1538–1558
3. Tanksale A, Beltramini JN, Lu GM (2010) A review of catalytic hydrogen production processes from biomass. Renew Sustain Energy Rev 14(1):166–182
4. Gallo JMR, Alonso DM, Mellmer MA, Dumesic JA (2013) Production and upgrading of 5-hydroxymethylfurfural using heterogeneous catalysts and biomass-derived solvents. Green Chem 15(1):85–90
5. Titirici MM, White RJ, Falco C, Sevilla M (2012) Black perspectives for a green future: hydrothermal carbons for environment protection and energy storage. Energy Environ Sci 5(5):6796–6822
6. Akiya N, Savage PE (2002) Roles of water for chemical reactions in high-temperature water. Chem Rev 102(8):2725–2750
7. Watanabe M, Sato T, Inomata H, Smith RL, Arai K, Kruse A, Dinjus E (2004) Chemical reactions of C-1 compounds in near-critical and supercritical water. Chem Rev 104(12):5803–5821
8. Peterson AA, Vogel F, Lachance RP, Froling M, Antal MJ, Tester JW (2008) Thermochemical biofuel production in hydrothermal media: A review of sub- and supercritical water technologies. Energy Environ Sci 1(1):32–65
9. Kruse A (2009) Hydrothermal biomass gasification. J Supercrit Fluids 47(3):391–399
10. Jin FM, Enomoto H (2011) Rapid and highly selective conversion of biomass into value-added products in hydrothermal conditions: chemistry of acid/base-catalysed and oxidation reactions. Energy Environ Sci 4(2):382–397
11. Alonso DM, Bond JQ, Dumesic JA (2010) Catalytic conversion of biomass to biofuels. Green Chem 12(9):1493–1513
12. Jin FM, Zhou ZY, Moriya T, Kishida H, Higashijima H, Enomoto H (2005) Controlling hydrothermal reaction pathways to improve acetic acid production from carbohydrate biomass. Environ Sci Technol 39(6):1893–1902

13. Amarasekara AS, Williams LD, Ebede CC (2008) Mechanism of the dehydration of d-fructose to 5-hydroxymethylfurfural in dimethyl sulfoxide at 150 degrees C: an NMR study. Carbohydr Res 343(18):3021–3024
14. Guan J, Cao Q, Guo X, Mu X (2011) The mechanism of glucose conversion to 5-hydroxymethylfurfural catalyzed by metal chlorides in ionic liquid: A theoretical study. Comput Theor Chem 963(2–3):453–462
15. Takeuchi Y, Jin FM, Tohji K, Enomoto H (2008) Acid catalytic hydrothermal conversion of carbohydrate biomass into useful substances. J Mater Sci 43(7):2472–2475
16. Pagan-Torres YJ, Wang TF, Gallo JMR, Shanks BH, Dumesic JA (2012) Production of 5-Hydroxymethylfurfural from glucose using a combination of Lewis and Bronsted acid catalysts in water in a biphasic reactor with an alkylphenol solvent. ACS Catal 2(6):930–934
17. Weingarten R, Conner WC, Huber GW (2012) Production of levulinic acid from cellulose by hydrothermal decomposition combined with aqueous phase dehydration with a solid acid catalyst. Energy Environ Sci 5(6):7559–7574
18. Wettstein SG, Alonso DM, Chong YX, Dumesic JA (2012) Production of levulinic acid and gamma-valerolactone (GVL) from cellulose using GVL as a solvent in biphasic systems. Energy Environ Sci 5(8):8199–8203
19. Bond JQ, Alonso DM, Wang D, West RM, Dumesic JA (2010) Integrated Catalytic Conversion of gamma-Valerolactone to Liquid Alkenes for Transportation Fuels. Science 327(5969):1110–1114
20. Shen J, Wyman CE (2012) Hydrochloric acid-catalyzed levulinic acid formation from cellulose: data and kinetic model to maximize yields. AIChE J 58(1):236–246
21. Cinlar B, Wang TF, Shanks BH (2013) Kinetics of monosaccharide conversion in the presence of homogeneous Bronsted acids. Appl Catal A 450:237–242
22. Weingarten R, Cho J, Xing R, Conner WC, Huber GW (2012) Kinetics and reaction engineering of levulinic acid production from aqueous glucose solutions. ChemSusChem 5(7):1280–1290
23. Yang Y, Hu CW, Abu-Omar MM (2012) Conversion of carbohydrates and lignocellulosic biomass into 5-hydroxymethylfurfural using $AlCl_3 \cdot 6H_2O$ catalyst in a biphasic solvent system. Green Chem 14(2):509–513
24. Wang TF, Pagan-Torres YJ, Combs EJ, Dumesic JA, Shanks BH (2012) Water-compatible Lewis acid-catalyzed conversion of carbohydrates to 5-hydroxymethylfurfural in a biphasic solvent system. Top Catal 55(7–10):657–662
25. Roman-Leshkov Y, Chheda JN, Dumesic JA (2006) Phase modifiers promote efficient production of hydroxymethylfurfural from fructose. Science 312(5782):1933–1937
26. Asghari FS, Yoshida H (2006) Acid-catalyzed production of 5-hydroxymethyl furfural from D-fructose in subcritical water. Ind Eng Chem Res 45(7):2163–2173
27. Jin FM, Zhou ZY, Enomoto H, Moriya T, Higashijima H (2004) Conversion mechanism of cellulosic biomass to lactic acid in subcritical water and acid-base catalytic effect of subcritical water. Chem Lett 33(2):126–127
28. Kabyemela BM, Adschiri T, Malaluan RM, Arai K (1999) Glucose and fructose decomposition in subcritical and supercritical water: Detailed reaction pathway, mechanisms, and kinetics. Ind Eng Chem Res 38(8):2888–2895
29. Sasaki M, Kabyemela B, Malaluan R, Hirose S, Takeda N, Adschiri T, Arai K (1998) Cellulose hydrolysis in subcritical and supercritical water. J Supercrit Fluids 13(1–3):261–268
30. Kabyemela BM, Adschiri T, Malaluan R, Arai K (1997) Degradation kinetics of dihydroxyacetone and glyceraldehyde in subcritical and supercritical water. Ind Eng Chem Res 36(6):2025–2030
31. Kishida H, Jin FM, Yan XY, Moriya T, Enomoto H (2006) Formation of lactic acid from glycolaldehyde by alkaline hydrothermal reaction. Carbohydr Res 341(15):2619–2623
32. Yan X, Jin F, Tohji K, Kishita A, Enomoto H (2010) Hydrothermal conversion of carbohydrate biomass to lactic acid. AIChE J 56(10):2727–2733
33. Yan XY, Jin FM, Tohji K, Moriya T, Enomoto H (2007) Production of lactic acid from glucose by alkaline hydrothermal reaction. J Mater Sci 42(24):9995–9999

34. Sánchez C, Egüés I, García A, Llano-Ponte R, Labidi J (2012) Lactic acid production by alkaline hydrothermal treatment of corn cobs. Chem Eng J 181–182:655–660
35. Jin FM, Yun J, Li GM, Kishita A, Tohji K, Enomoto H (2008) Hydrothermal conversion of carbohydrate biomass into formic acid at mild temperatures. Green Chem 10(6):612–615
36. Noyori R, Aoki M, Sato K (2003) Green oxidation with aqueous hydrogen peroxide. Chem Commun 16:1977–1986
37. Quitain AT, Faisal M, Kang K, Daimon H, Fujie K (2002) Low-molecular-weight carboxylic acids produced from hydrothermal treatment of organic wastes. J Hazard Mater 93(2):209–220
38. Goto M, Obuchi R, Hiroshi T, Sakaki T, Shibata M (2004) Hydrothermal conversion of municipal organic waste into resources. Bioresour Technol 93(3):279–284
39. Shen Z, Zhou J, Zhou X, Zhang Y (2011) The production of acetic acid from microalgae under hydrothermal conditions. Appl Energy 88(10):3444–3447
40. Yoneda N, Kusano S, Yasui M, Pujado P, Wilcher S (2001) Recent advances in processes and catalysts for the production of acetic acid. Appl Catal A 221(1–2):253–265
41. Croiset E, Rice SF, Hanush RG (1997) Hydrogen peroxide decomposition in supercritical water. AIChE J 43(9):2343–2352
42. Akiya N, Savage PE (2000) Effect of water density on hydrogen peroxide dissociation in supercritical water. 1. Reaction equilibrium. J Phys Chem A 104(19):4433–4440
43. Akiya N, Savage PE (2000) Effect of water density on hydrogen peroxide dissociation in supercritical water. 2. Reaction kinetics. J Phys Chem A 104(19):4441–4448
44. Fang Y, Zeng X, Yan P, Jing Z, Jin F (2012) An acidic two-step hydrothermal process to enhance acetic acid production from carbohydrate biomass. Ind Eng Chem Res 51(12):4759–4763
45. Grasemann M, Laurenczy G (2012) Formic acid as a hydrogen source—recent developments and future trends. Energy Environ Sci 5(8):8171–8181
46. Yao GD, Huo ZB, Jin FM (2011) Direct reduction of copper oxide into copper under hydrothermal conditions. Res Chem Intermed 37(2):351–358
47. Li Q, Yao G, Zeng X, Jing Z, Huo Z, Jin F (2012) Facile and green production of Cu from CuO using cellulose under hydrothermal conditions. Ind Eng Chem Res 51(7):3129–3136
48. Yao GD, Zeng X, Li QJ, Wang YQ, Jing ZZ, Jin FM (2012) Direct and highly efficient reduction of NiO into Ni with cellulose under hydrothermal conditions. Ind Eng Chem Res 51(23):7853–7858
49. Wang Y, Jin F, Sasaki M, Wang F, Wahyudiono, Jing Z, Goto M (2013) Selective conversion of glucose into lactic acid and acetic acid with copper oxide under hydrothermal conditions. AIChE J 59(6):2096–2104
50. Zhang S, Jin F, Hu J, Huo Z (2011) Improvement of lactic acid production from cellulose with the addition of Zn/Ni/C under alkaline hydrothermal conditions. Bioresour Technol 102(2):1998–2003
51. Onwudili JA, Williams PT (2009) Role of sodium hydroxide in the production of hydrogen gas from the hydrothermal gasification of biomass. Int J Hydrogen Energy 34(14):5645–5656
52. Akiya N, Savage PE (1998) Role of water in formic acid decomposition. AIChE J 44(2):405–415
53. Fievet F, Fievet-Vincent F, Lagier J-P, Dumont B, Figlarz M (1993) Controlled nucleation and growth of micrometre-size copper particles prepared by the polyol process. J Mater Chem 3(6):627–632
54. Luo R (1987) Overall equilibrium diagrams for hydrometallurgical systems: copper–ammonia—water system. Hydrometallurgy 17(2):177–199
55. Oishi T, Yaguchi M, Koyama K, Tanaka M, Lee JC (2008) Hydrometallurgical process for the recycling of copper using anodic oxidation of cuprous ammine complexes and flow-through electrolysis. Electrochim Acta 53(5):2585–2592
56. Moskalyk RR, Alfantazi AM (2003) Review of copper pyrometallurgical practice: today and tomorrow. Miner Eng 16(10):893–919

57. Yu JL, Savage PE (1998) Decomposition of formic acid under hydrothermal conditions. Ind Eng Chem Res 37(1):2–10
58. Cheary RW, Coelho A (1992) A fundamental parameters approach to X-ray line-profile fitting. J Appl Crystallogr 25:109–121
59. Larcher D, Patrice R (2000) Preparation of metallic powders and alloys in polyol media: a thermodynamic approach. J Solid State Chem 154(2):405–411
60. Bobleter O (1994) Hydrothermal degradation of polymers derived from plants. Prog Polym Sci 19(5):797–841
61. Deng T, Sun J, Liu H (2010) Cellulose conversion to polyols on supported Ru catalysts in aqueous basic solution. Sci China-Chem 53(7):1476–1480
62. Yan X, Jini F, Kishita A, Enomoto H, Tohji K (2008) Formation of lactic acid from cellulosic biomass by alkaline hydrothermal reaction. In: Tohji K, Tsuchiya N, Jeyadevan B, Water Dyn 987:50–53
63. van der Weijden RD, Mahabir J, Abbadi A, Reuter MA (2002) Copper recovery from copper(II) sulfate solutions by reduction with carbohydrates. Hydrometallurgy 64(2): 131–146
64. Jeyadevan B, Joseyphus RJ, Kodama D, Matsumoto T, Sato Y, Tohji K (2007) Role of polyol in the synthesis of Fe particles. J Magn Magn Mater 310(2):2393–2395
65. Sakanishi K, Ikeyama N, Sakaki T, Shibata M, Miki T (1999) Comparison of the hydrothermal decomposition reactivities of chitin and cellulose. Ind Eng Chem Res 38(6):2177–2181
66. Sasaki M, Fang Z, Fukushima Y, Adschiri T, Arai K (2000) Dissolution and hydrolysis of cellulose in subcritical and supercritical water. Ind Eng Chem Res 39(8):2883–2890
67. Dolan R, Yin S, Tan Z (2010) Effects of headspace fraction and aqueous alkalinity on subcritical hydrothermal gasification of cellulose. Int J Hydrogen Energy 35(13):6600–6610
68. Muangrat R, Onwudili JA, Williams PT (2010) Alkali-promoted hydrothermal gasification of biomass food processing waste: a parametric study. Int J Hydrogen Energy 35(14): 7405–7415
69. Yanik J, Ebale S, Kruse A, Saglam M, Yueksel M (2007) Biomass gasification in supercritical water (Part I): effect of the nature of biomass. Fuel 86(15):2410–2415
70. Fang Z, Minowa T, Fang C, Smith RL Jr, Inomata H, Kozinski JA (2008) Catalytic hydrothermal gasification of cellulose and glucose. Int J Hydrogen Energy 33(3):981–990
71. Takahashi H, Kori T, Onoki T, Tohji K, Yamasaki N (2008) Hydrothermal processing of metal based compounds and carbon dioxide for the synthesis of organic compounds. J Mater Sci 43(7):2487–2491
72. Li Y, Xu GH, Liu CJ, Eliasson B, Xue BZ (2001) Co-generation of syngas and higher hydrocarbons from CO_2 and CH_4 using dielectric-barrier discharge: effect of electrode materials. Energy Fuels 15(2):299–302
73. Yuksel A, Sasaki M, Goto M (2011) Electrolysis reaction pathway for lactic acid in subcritical water. Ind Eng Chem Res 50(2):728–734
74. Jin FM, Zhang GY, Jin YJ, Watanabe Y, Kishita A, Enomoto H (2010) A new process for producing calcium acetate from vegetable wastes for use as an environmentally friendly deicer. Bioresour Technol 101(19):7299–7306

Chapter 3
Hydrothermal Conversion of Lignin and Its Model Compounds into Formic Acid and Acetic Acid

Xu Zeng, Guodong Yao, Yuanqing Wang and Fangming Jin

Abstract With the fast depletion of fossil fuels, the development of effective approach for the production of value-added chemicals from renewable resources is strongly desired. Lignin is a main constituent of lignocellulosic biomass, which accounts for 15–30 % by weight and 40 % by energy. Therefore, lignin conversion has significant potential as a source for the sustainable production of chemicals and fuels. However, lignin has received little attention because of its highly crosslinked macromolecule structure and chemical properties. Hydrothermal technology has received much attention in the treatment of organic wastes and biomass conversion because of the unique inherent properties of high temperature water. Here, some recent studies on hydrothermal conversion of lignin and its model compounds into value-added chemicals such as formic acid and acetic acid are presented. Phenol and syringol are mainly introduced as lignin model compounds. It will be useful in exploring the potential approaches for the lignin conversion into useful chemicals and fuels.

3.1 Introduction

The Earth's sustainable development is threatened, which is caused by the imbalance of the rapid consumption of fossil fuels by anthropogenic activities and the slow rate of fossil fuel formation. Therefore, chemicals and energy, supplied by renewable resources such as biomass, are expected to increase in the foreseeable future [1]. Extensive methods have been developed for the rapid conversion of biomass into chemicals and fuels. For example, Dumesic's group at the University of Wisconsin-Madison has proposed a series of catalytic conversion of biomass to

---

X. Zeng · G. Yao · Y. Wang · F. Jin (✉)
School of Environmental Science and Engineering, Shanghai Jiao Tong University, Shanghai 200240, P.R. China
e-mail: fmjin@sjtu.edu.cn

liquid fuels [2–6]. Huber developed new, rapid method for producing low-cost biofuels from biomass in one simple step in which cellulose is directly heated to 400–600 °C with the addition of zeolite catalysts [7]. On the other hand, biomass consists of three basic components: cellulose, hemicellulose, and lignin [8]. Cellulose, a linear polymer of glucose [9], and hemicelluloses, an amorphous, which are the main parts of biomass, have been studies widely. Compared with cellulosic biomass, lignin remains relatively underutilized to its potential. Thus, the conversion of lignin into value-added chemicals has become a promising topic.

Lignin, one of the most abundant biomacromolecules, accounts for up to 40 % of the dry biomass weight [10], which is a highly crosslinked macromolecule which consists of various oxygen and carbon bridges between alkylated methoxyl-phenol rings [11], mainly including guaiacyl propane, syringyl propane, and phenyl propane structures [12]. The most complete structural model of lignin was proposed by Adler in 1977 (Fig. 3.1) [13]. With its unique structure, a wide variety of chemicals and fuels are potentially obtainable. Among several conversion processes including pyrolysis, acid hydrolysis, and enzymatic hydrolysis, hydrothermal oxidation is receiving increased attention because water behaves as an outstanding reaction medium at high temperature and high pressure [14]. Lignin thermal-cracking studies using temperatures of 250–600 °C have demonstrated the potential of generating low molecular weight feedstocks for further processing [15]. Extensive research in this area was carried out [16–18]. Recently, biomass and phenolic compounds have been chosen as the subject of numerous studies both in supercritical and subcritical conditions [19, 20]. It has been shown that lignin can be easily broken down into phenol and phenol derivatives (phenolic compounds) in hydrothermal oxidation [21]. Oxygen concentration or temperature has a positive influence on the wet air oxidation of phenols [22]. Nevertheless, very few studies suggested reaction pathways.

Some researchers focused on the mechanism investigation of some phenolic model compounds, for example, catechol [23], guaiacol [24], vanillic acid [25], and diphenylether [26]. The three monolignols are shown in Fig. 3.2. Although there have been a lot of studies on the conversion of lignin model compounds, up to now, the knowledge of reaction pathway of lignin model compounds is still very limited.

Jin has performed a series of studies on hydrothermal oxidation of lignin and its model compounds, which showed that formic acid and acetic acid could be easily formed [27–31]. Formic acid is an important organic chemical, which is used as a leather-tanning agent, a concrete cure accelerator, and an animal feed additive. Formic acid in its Na or Ca salt form has been proposed as an environmentally friendly road deicer [32, 33]. Furthermore, formic acid has been paid much attention as a raw material for hydrogen production [34, 35]. More importantly, recent research has demonstrated that formic acid has the potential to power fuel cells for electricity generation for automobiles [36–38]. Acetic acid is also an important chemical. Acetic acid production by hydrothermal oxidation of various carbohydrate biomasses has been studied to produce calcium/magnesium acetate (CMA) which is known as an environmentally friendly deicer [39]. Extensive

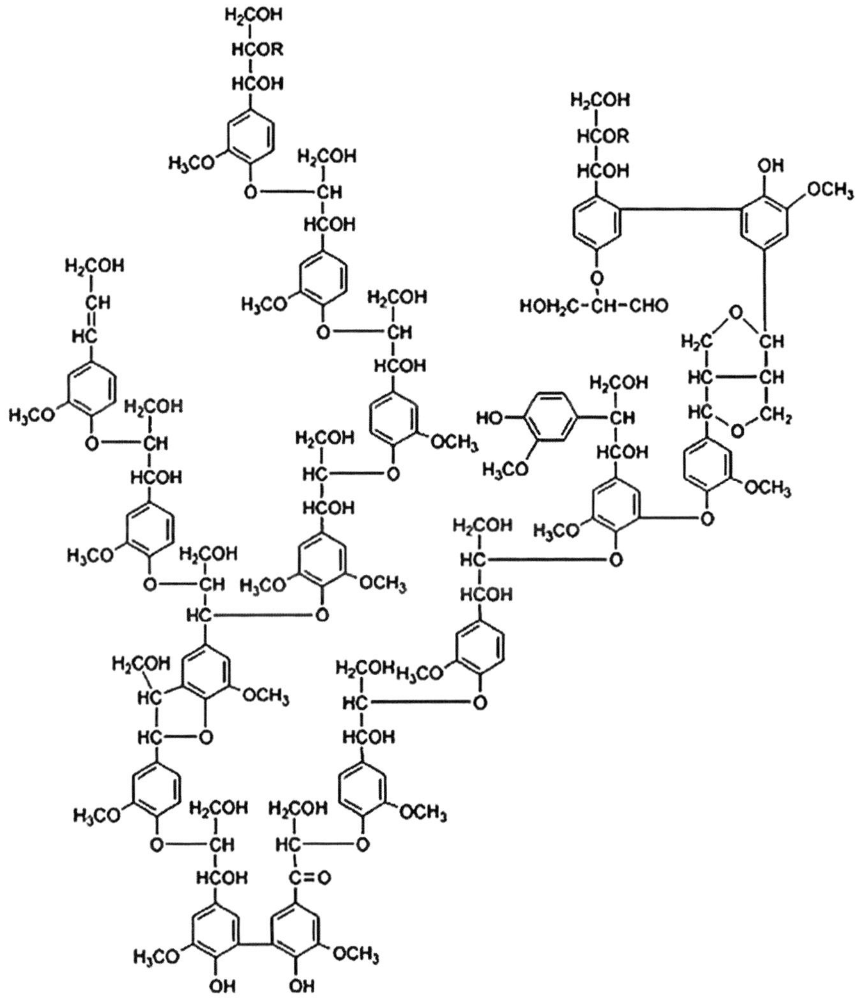

3 Hydrothermal Conversion of Lignin and Its Model Compounds

Fig. 3.1 Adler's structural model of lignin

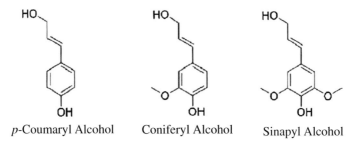

Fig. 3.2 Three monolignols, the building blocks of lignin

studies have been carried out to enhance the yields of acetic acid from carbohydrate biomass [40]. However, nowadays acetic acid for industrial production is obtained principally from natural gas. Although many attempts have been made to ferment biomass, their reaction rate is generally low.

Here, the approaches and strategies for the hydrothermal conversion of lignin and its model compounds into formic acid and acetic acid are introduced. We believe that it will provide a perspective that would prompt the effective utilization of lignin biomass.

3.2 Production of Formic Acid and Acetic Acid from Lignin Model Compounds

Because lignin is a highly crosslinked macromolecule, researches always chose some phenolic model compounds to reduce complexity. Jin has conducted a series of studies on the hydrothermal oxidation of lignin model compounds [41]. Here, the studies on phenol, and syringol (2,6-dimethoxyphenol), as lignin model compounds, for the production of formic acid and acetic acid are presented.

3.2.1 Effects of Reaction Conditions on the Yields of Formic Acid and Acetic Acid Without the Addition of Alkali

Oxygen supply was considered first because it influenced the reaction process directly. Figure 3.3a presents the effects of oxygen supply on the yields of formic acid and acetic acid. As shown in this figure, oxygen supply influenced the yields significantly, which means that it is very important to find the suitable additive ratio to acquire yields of formic acid and acetic acid as high as possible. The suitable value of oxygen supply existed in the hydrothermal oxidation of phenol and syringol. This might be because of oxidative decomposition of formic acid by excess oxygen supply. Because the yields of formic acid and acetic acid from syringol were higher than those from phenol, and higher oxygen supply influenced the yields significantly, it means that syringol is easier to be hydrothermal oxidized for the production formic acid and acetic acid than phenol. Probably the reason is due to the methoxy on the benzene. Therefore, for phenol, it needs oxygen supply to acquire optimum yields of formic acid and acetic acid. For syringol, it only needs oxygen supply of 40 %.

To examine the effect of reaction temperature, experiments were performed by varying reaction temperature from 260 to 320 °C for 90 s. As shown in Fig. 3.3b, the yields of formic acid and acetic acid increased with the reaction temperature changing from 260 to 300 °C, and the values decreased when the temperature changed to 320 °C. It suggested that higher temperature was a plus factor for the

Fig. 3.3 Effects of the oxygen supply (**a** 300 °C, 30 s), reaction temperature (**b** 30 s, oxygen supply 60 % (Phenol), oxygen supply 40 % (Sringol)), and reaction time (**c** 300 °C, oxygen supply 60 % (Phenol), oxygen supply 40 % (Sringol)) on the yields of formic acid and acetic acid

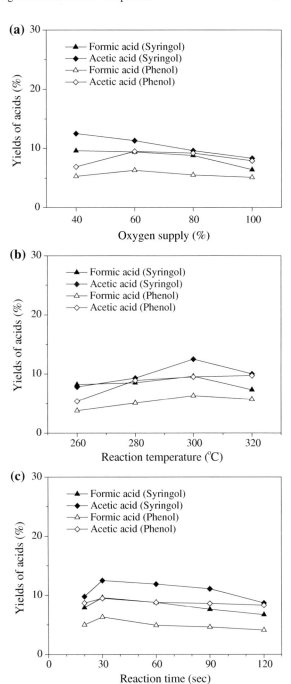

production of formic acid and acetic acid, but the yields decreased when the temperature was too high due to the decomposition of the acids. Compared with the yields from phenol, higher yields of formic acid and acetic acid could be obtained from syringol.

Figure 3.3c shows the effects of reaction time on the yields of formic acid and acetic acid at 300 °C. As seen in the figure, the yields of formic acid and acetic acid increased promptly with the reaction proceeded. The highest yields were acquired with the reaction time of 30 s while the yields decreased with the extension of reaction time. This may imply that longer reaction time would reduce the yields of formic acid and acetic acid due to the further reaction.

3.2.2 Improvement of the Yields of Formic Acid and Acetic Acid with the Addition of Alkali

The effect of alkaline is very important. Many reports claim that alkali can accelerate the decomposition of organic compounds [42, 43]. In our previous study involving the formation mechanism of formic acid and acetic acid in subcritical water without the addition of NaOH, we found that the conversion mechanisms were very different. Thus, the addition of alkali may also be an effective way to acquire higher yields of formic acid and acetic acid.

Figure 3.4 shows the effect of NaOH on the yields of formic acid and acetic acid. Compared with the addition, the increase of the yields of formic acid and acetic acid was not significant with the addition of NaOH 0.5 $mol \cdot l^{-1}$. But the yields increased significantly with the addition of 1.0 $mol \cdot l^{-1}$ NaOH, especially the yield of formic acid. When the concentration of NaOH increased to 1.5 $mol \cdot l^{-1}$, the yield of formic acid changed little. It should be notable that optimal yields were obtained at the concentration of 1.5 $mol \cdot l^{-1}$ NaOH at 300 °C for 30 s with the additive ratio of H_2O_2 60 %, because the yield of acetic acid was the highest. These results indicate that the addition of NaOH can improve significantly the yields of acetic acid, which is caused by the prevention of further oxidative decomposition of acetic acid. The case is different for formic acid, which is probably due to the different production pathway.

Figure 3.5 shows the effect of reaction temperature, time, and oxygen supply on the yields of acetic acid and formic acid. The highest total yield of formic and acetic acids can reach about 25 %, which occurred at temperature of 300 °C, time 60 s, and oxygen supply 60 % with 1.5 $mol \cdot l^{-1}$ NaOH. Comparing the yields of formic acid and acetic acid with or without alkali addition, it was found that the optimum condition of reaction temperature, time, and oxygen supply was different. The reason may be that the additional alkali changed the main production pathway and subsequently influenced the yields of formic acid and acetic acid with different trends.

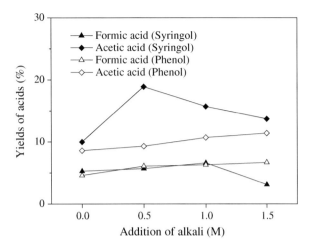

Fig. 3.4 Effects of alkali (300 °C, 30 s, oxygen supply 60 % (Phenol), oxygen supply 40 % (Sringol)) on the yields of formic acid and acetic acid

3.2.3 Possible Explanation for the Production of Formic Acid and Acetic Acid from Lignin Model Compounds

To get high yields of acids, the mechanism of hydrothermal oxidation of lignin model compounds should be known. We reported the mechanism study of hydrothermal oxidation of lignin model compounds. Based on the intermediate products identified for the samples, acetic acid was obtained not only from direct oxidation of lignin model compounds but also from further oxidation of intermediate products. Therefore, it is possible to increase the yield of acetic acid from lignin by controlling reaction pathways.

3.2.3.1 Identification of Intermediate Products

Although there are many studies on the mechanism of the oxidation of phenol and substituted phenols, the mechanism of the oxidation of phenols is extremely complex and is not yet fully understood. Identification of the intermediate products is an essential prerequisite in the investigation of reaction mechanism. In our study, the intermediate products after reaction for three lignin model compounds were identified in detail by GC/MS, HPLC [26].

Figure 3.6 displays the GC/MS chromatograms of liquid samples obtained at a temperature of 300 °C, a reaction time of 60 s, and an oxygen supply of 50 %. A lower oxygen supply is helpful to easily get the initial oxidation products. As shown in Fig. 3.6, in the case of syringol, 3-methoxy-1,2-benzenediol and 2,6-dimethoxy-1,4-benzenediol were detected as major intermediate products. With phenol, major products detected were 1,2-benzenediol and 1,4-benzenediol. Beside these substituted phenols, acetic acid and formic acid were identified in all cases. Identification of 1,2-benzenediol, 1,4-benzenediol, acetic acid, and formic

Fig. 3.5 Effects of reaction temperature on the yields of formic acid and acetic acid (**a** 90 s, oxygen supply 60 %), reaction time (**b** 300 °C, oxygen supply 60 %), and the oxygen supply (**c** 300 °C, 90 s) (1.5 mol·l^{-1} NaOH)

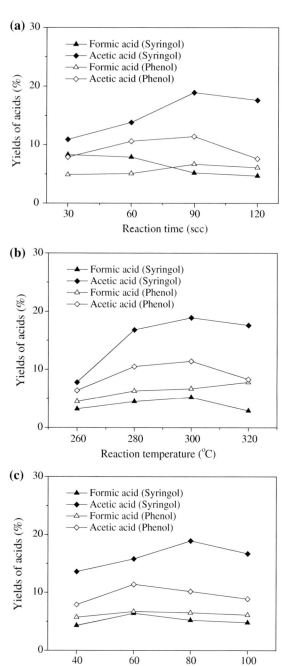

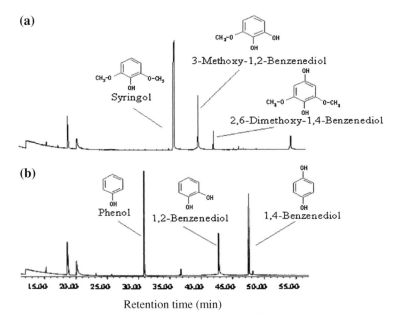

Fig. 3.6 HPLC chromatograms of the oxidation products for lignin model compounds at a temperature of 300 °C reaction time 60 s, and oxygen supply 70 %. **a** Syringol, **b** Phenol

acid was obtained by matching both the mass spectrum and the GC retention time for each compound with those of the authentic compounds. Identification of other intermediate products was performed only by matching the mass spectrum, because authentic standards were not commercially available. In these cases, a good match between the mass spectra of a product and reference spectra stored in the computer library was obtained. Because of the limitation of the GC analysis, not all the intermediates were detected.

Figure 3.7 shows HPLC chromatograms for liquid samples after the reaction of phenol and syringol at 300 °C and for 60 s, with a 70 % oxygen supply. Selection of an oxygen supply at 70 % was helpful to detect further oxidation products such as low molecular weight carboxylic acids. Many of the compounds formed are seen in Fig. 3.7. Among these compounds, peaks labeled with 1–12 represent compounds identified. These compounds were mainly a variety of lowmolecular weight carboxylic acids with 1–6 carbon atom(s), including unsaturated dicarboxylic acids (muconic acid, glutaconic acid, maleic acid, and fumaric acid), saturated dicarboxylic acids (succinic acid, malonic acid, and oxalic acid) and saturated monocarboxylic acids (acetic acid, formic acid). Although the peaks of glutaconic acid and oxalic acid are not clearly seen in Fig. 3.7, they were clearly identified with a shorter reaction time of 10 s for all three model compounds (data not shown). Besides the low molecular weight carboxylic acids, substituted phenols of 1,2- benzenediol, 1,4-benzenediol and 1,2,4-benzenetriol were detected in the case of phenol. Most of the intermediate products described above have also

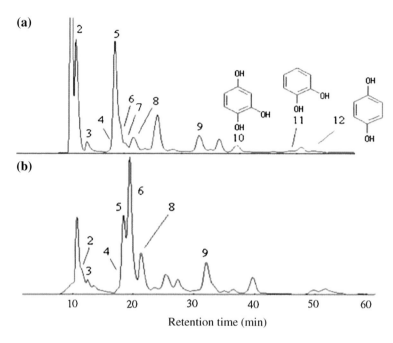

Fig. 3.7 HPLC chromatograms of the oxidation products for lignin model compounds at a temperature of 300 °C, reaction time 60 s, and oxygen supply 70 %. **a** Phenol, **b** Syringol. *1* Oxalic acid, *2* Maleic acid, *3* Malonic acid, *4* Succinic acid, *5* Fumaric acid, *6* Formic acid, *7* Glutaconic acid, *8* Acetic acid, *9* Muconic acid, *10* 1,2,4-Benzenetriol, *11* 1,2-Benzenediol, *12* 1,4-Benzenediol

been reported as major products in the wet air oxidation or supercritical wet air oxidation of phenol and/or substituted phenols [44–47]. However, the formation of glutaconic acid as an intermediate product has not yet been reported.

In order to study the mechanism of the oxidation of lignin and its model compounds, and thus make clear the reason for a lower yield of acetic acid from lignin or phenolic compounds, intermediate products from the oxidation of model compounds of lignin were identified in detail. To get more and precise information for an investigation of the oxidation mechanisms of lignin model compounds, identifying not only the intermediate products obtained directly from the oxidation of lignin model compounds but also further oxidation products of the intermediate products from the lignin model compounds are important. So, a series of oxidation experiments with 1,2-benzenediol, 1,4-benzenediol, and 1,2,4-benzenetriol as well as all low molecular weight carboxylic acids identified (see Fig. 3.7) were performed at a temperature of 300 °C, a reaction time of 60 s, and an oxygen supply of 70 %. The intermediate products from 1,2-benzenediol, 1,4-benzenediol, and

… 3 Hydrothermal Conversion of Lignin and Its Model Compounds

Table 3.1 Hydrothermal oxidation products of ring-opening products for lignin model compounds (d: detected)

Materials \ Products	1	2	3	4	5	6	7	8	9
1 Muconic acid		d	d	d	d	d	d	d	d
2 Glutaconic acid			d	d	d	d	d	d	d
3 Maleic acid				d		d	d	d	d
4 Fumaric acid			d			d	d	d	d
5 Succinic acid								d	d
6 Malonic acid								d	d
7 Oxalic acid									d
8 Acetic acid									d
9 Formic acid									

1,2,4-benzenetriol were almost the same as those from phenol. The intermediate products from ring-opening products or low molecular weight carboxylic acids are summarized clearly in Table 3.1.

From Table 3.1, we can see clearly the skeleton relationship of the oxidation process. On the basis of the products obtained in the oxidation reaction of model compounds and further oxidation products of the intermediate products from model compounds, hydrothermal oxidation pathways of lignin model compounds are discussed.

3.2.3.2 Hydrothermal Oxidation Pathways of Lignin Model Compounds

Subsequently, pathways after ring-opening reactions were proposed on the basis of the identification of intermediate products from lignin model compounds and further oxidation products from ring-opening products. Figures 3.8 and 3.9 show the proposed pathways before and after ring-opening reactions for three lignin model compounds.

For the pathways before ring-opening reactions, it was assumed that first phenol and syringol were oxidized into their ortho and para compounds. It should be noted that for syringol, oxidation occurred only at the para position, and 3-methoxy-1,2-benzenediol from syringol could be hydrolysis products rather than oxidation products, because 3-methoxy-1,2-benzenediol was not found as an intermediate product in the oxidation of guaiacol. That the oxidation of syringol hardly occurs at the ortho position would probably because they have a substituent group at the ortho position. The 1,2-benzenediol and 1,4-benzenediol produced may be further

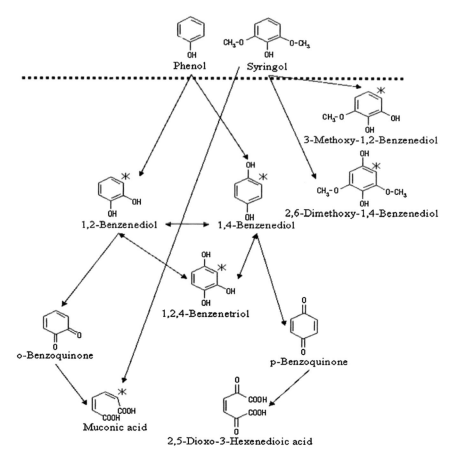

Fig. 3.8 Proposed pathways to ring-opening reactions for lignin model compounds (*: compounds identified in this study). Adapted with permission from ref. [48]. Copyright 2006 Springer

oxidized to o-benzoquinone and p-benzoquinone, which are subsequently oxidized to the corresponding unsaturated dicarboxylic acids, muconic acid and 2,5-dioxo-3-hexenedioic acid, by ring-opening reactions. Additionally, 1,2,4-benzenetriol was found in the oxidation experiments with 1,2-benzenediol and 1,4-benzenediol, which implies that 1,2-benzenediol and 1,4-benzenediol were also oxidized to 1,2,4-benzenetriol. However, a further oxidation mechanism of 1,2,4-benzenetriol and OCH_3-substituted benzenediols such as 2-methoxy- 1,4-benzenediol, 3-methoxy-1,2-benzenediol and 2,6-dimethoxy- 1,4-benzenediol is still unclear. To the best of our knowledge, there are no reports concerning the oxidation mechanism of OCH_3-substituted benzenediols and phenol attached at more than two sites.

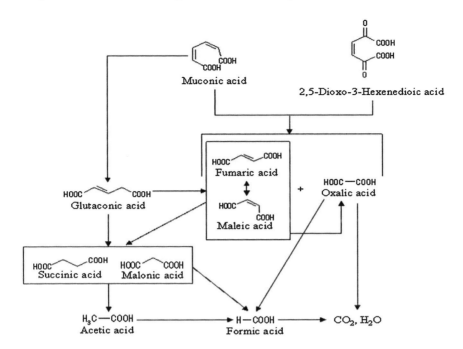

Fig. 3.9 Proposed pathways after ring-opening reactions for lignin model compounds. Adapted with permission from ref. [48]. Copyright 2006 Springer

As shown in Fig. 3.9, unsaturated dicarboxylic acids with six carbon atoms, muconic acid, and 2,5-dioxo-3-hxenedioic acid may be directly oxidized to unsaturated dicarboxylic acids with two carbon atoms less, maleic and fumaric acids, and with four carbon atoms less, oxalic acid. The reason that lignin model compounds cannot produce a large amount of acetic acid may be that the products after ring-opening reactions, unsaturated dicarboxylic acids, may not produce a large amount of acetic acid. In the oxidation of the saturated dicarboxylic acids except oxalic acid, the acetic acid yield was higher and, especially for malonic acid, the acetic acid yield reached about 50 %. Quantitative analyses for the oxidation samples of three model compounds showed that the amount of maleic and fumaric acids was much higher than that of saturated carboxylic acids, succinic and malonic acids. This may be the reason why lignin model compounds or lignin cannot produce a large amount of acetic acid. Even so, these results may give us some suggestions about how to improve the yield of acetic acid. That is, increasing the formation of saturated dicarboxylic acids and glutaconic acid would enhance the acetic acid yield.

Table 3.2 Element composition of alkali lignin

Element	C	H	N	O
Value (%)	61.26	5.90	0.87	31.25

3.3 Hydrothermal Conversion of Alkaline Lignin into Formic Acid and Acetic Acid

Hydrothermal conversion of alkali lignin showed that considerable high yields of formic acid and acetic acid were attained [49].

The element components of alkali lignin are shown in Table 3.2:

3.3.1 Production of Formic Acid and Acetic Acid by the Usual Hydrothermal Oxidation

Experiments were performed over a wide range of conditions with temperature varying from 260 to 320°C, oxygen supply varying from 60 to 120 %, and reaction time varying from 30 to 150 s to test whether formic or acetic acid can be produced in large quantity.

First, experiments were performed by varying reaction temperature from 260 to 320 °C with the additive ratio of oxygen supply 100 % for reaction time 120 s. As shown in Fig. 3.10a, the yield value of acetic acid increased with the reaction temperature from 260 to 300 °C, and the value decreased when the temperature changed to 320 °C. The biggest yield of acetic acid was 12.3 %. In contrast, the higher yield of formic acid was acquired at 280 °C, which was 4.9 %. It suggested that higher temperature was a plus factor for the production of formic acid and acetic acid, but the yields decreased when the temperature was too high because of the decomposition process.

Figure 3.10b presents the effects of oxygen supply on the yields of formic acid and acetic acid in the hydrothermal reaction. The additive ratio of oxygen supply varied from 60 to 120 % at 300 °C for 120 s. For the four H_2O_2 supplies, the highest yield of formic acid was 4.1 % with the additive ratio of H_2O_2 60 %. That of acetic acid was 12.3 % with the additive ratio of oxygen supply 100 %. Compared with formic acid, the acetic acid was the major product. These results further indicate that the oxidative decomposition of formic acid is easier than that of acetic acid. The change regularity of the yield of acetic acid was not in agreement with that of formic acid. Probably it is hard to acquire the highest yields of formic and acetic under the same reaction condition.

Figure 3.10c shows the effects of reaction time on the yields of formic acid and acetic acid at 300 °C with the additive ratio of oxygen supply 100 %. As seen in the figure, the yield of acetic acid increased gradually with prolonging the reacting time while the yield value of formic acid decreased at the same progress.

Fig. 3.10 Effects of reaction temperature on the yields of formic acid and acetic acid (**a** 120 s, oxygen supply 100 %), oxygen supply (**b** 300°C, 120 s), and the reaction time (**c** 300°C, oxygen supply 100 %)

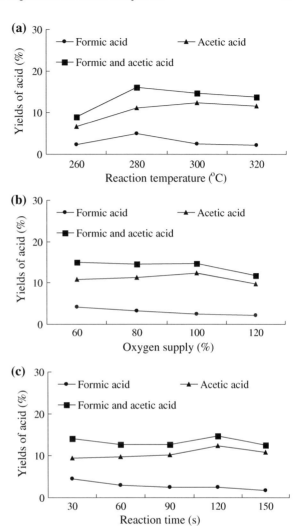

The highest yield of acetic acid was 12.3 % with the reaction time 120 s. For formic acid, the value is 4.5 % with 30 s. This may imply that higher yield of formic acid can be acquired with a short reaction time. For acetic acid, high yield needs longer reaction time.

3.3.2 Improvement of the Yields of Formic Acid and Acetic Acid with the Addition of Alkali

In this section, the alkaline hydrothermal conversion of real lignin into formic acid and acetic acid was introduced. The addition of alkali may be an effective way to prevent oxidative decomposition of formic acid. Though many reports claim that alkali can accelerate the decomposition of organic compounds, our previous studies have demonstrated that alkali can prevent organic compounds from being oxidized [50]. In our study, we have performed the experiments over a wide range of conditions with temperature varying from 260 to 340 °C, oxygen supply from 40 to 120 %, NaOH addition from 0 to 2.0 M, and reaction time varying from 30 to 150 s [51]. In this section, the additional amount of alkali was introduced, and then the effects of reaction conditions on the formic acid and acetic acid yield under alkaline hydrothermal conditions were discussed.

The effects of alkali amount on the yields of formic acid and acetic acid could be seen in Fig. 3.11. These results suggested that a higher yield of formic acid could be realized when the oxidative decomposition of formic acid was inhibited by additional alkali. However, the function of alkali was unclear now. Some researchers thought that alkaline may change the reaction mechanism, and the metal ions also have some effects on the oxidation reaction [52]. However, experimental results showed that the production of formic acid and acetic acid was accelerated significantly.

The effects of reaction conditions under alkaline hydrothermal conditions are introduced in the next section. First, experiments were performed by varying reaction temperature from 260 to 340 °C with the oxygen supply 100 % for 90 s. As shown in Fig. 3.12a, the change trends of yields of formic acid and acetic acid with temperature are similar. As the temperature increases from 260 to 340 °C, the yields first increase and then decrease. However, for the different small molecule acids, the highest yield is obtained at different temperatures. At 280 °C, the highest yield of 9.4 % for formic acid can be achieved, while the maximum yield of 23.8 % is achieved at 300 °C for acetic acid. The results suggested that there existed the different optimal temperatures in the alkaline hydrothermal conversion of lignin to different small molecule acids.

The effects o oxygen supply, varied from 40 to 120 %, at 300 °C for 90 s are shown in Fig. 3.12b. The highest yield of formic acid was 8.3 % with the oxygen supply 80 %. That of acetic acid was 23.8 % with the oxygen supply 100 %. The yields of formic and acetic acid were much lower with the oxygen supply 40 %. The reason for that was that the reaction was mainly the oxidation decomposition of lignin molecule in initial stage of the reaction and there were not much formic acid or acetic acid produced. With the increase of the oxygen supply, the intermediates, namely a lot of small molecule organic chemicals, were further oxidized, and so more formic and acetic acid were produced. Meanwhile, the excessive oxygen supply may also cause the oxidation of formic and acetic acid, so the yields decreased.

Figure 3.12c shows the effects of reaction time on the acid yields at 300 °C with the oxygen supply 100 %. As seen in Fig. 3.12c, the yield of acetic acid increased gradually with prolonging the reaction time, and then decreased slightly from 90 to 150 s while the yield of formic acid decreased at the same progress. The highest yield of acetic acid was 23.8 % with the reaction time 90 s. For formic acid, the biggest value is 10.3 % for 30 s. This may imply that higher yield of formic acid

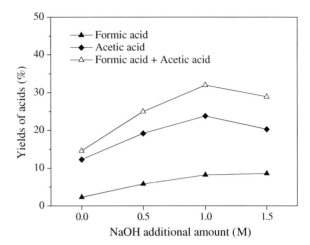

Fig. 3.11 Effects of NaOH on the yields of formic acid and acetic acid (with oxygen supply 100 % at 300°C for 90 s)

can be acquired with a short reaction time. For acetic acid, the situation is just the opposite. The reason for that may be the different pathways of the production of formic and acetic acid.

3.3.3 Possible Explanation for the Production of Formic Acid and Acetic Acid from the Oxidation of Alkaline Lignin

Oxidation pathways of lignin for formic acid and acetic acid formation were illustrated. As reported by Jin et al., formic acid was probably formed in the oxidation of lignin compound by the rupture of benzene ring, and the products were further oxidized. The key step in the process was the formation of small molecular organic acid. Some products, such as maleic acid, malonic aicd, propionic acid, oxalic acid, etc., were formed. Some experimental studies thought the oxidation of maleic acid largely prevailed over thermal decomposition [53]. There is a lack of information to doubtless affirm that this is a true pathway for phenolic compound oxidation under the current experimental conditions [54].

Proposed pathways of hydrothermal conversion of lignin into formic acid are showed in Fig. 3.13. Figure 3.13 shows the GC/MS chromatogram of a sample. Some intermediate products were labeled in the figure. The main process of production of formic acid and acetic acid formed unsaturated dicarboxylic acids, such as maleic acid, fumaric acid, levulinic acid and butanedioic acid, as showed in Fig. 3.8. There were some other saturated dicarboxylic acids produced as intermediate products.

Suzuki [55] has done some research of lignin model compounds. The results were similar to the hydrothermal oxidation of alkali lignin in this study. The main

Fig. 3.12 Effects of reaction time on the yields of formic acid and acetic acid (**a** NaOH 1.0 M, 300 °C, oxygen supply 100 %), reaction temperature (**b** NaOH 1.0 M, 90 s, oxygen supply 100 %) and oxygen supply (**c** NaOH 1.0 M, 300 °C, 90 s)

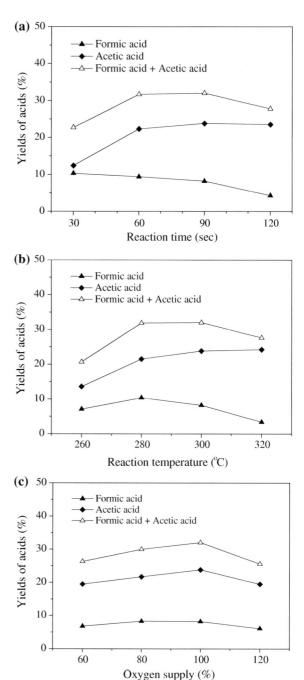

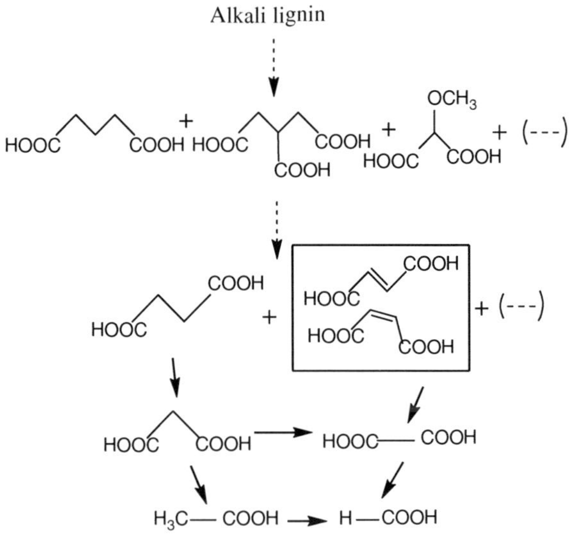

Fig. 3.13 Proposed reaction pathways of hydrothermal oxidation of alkali lignin

process of production of formic acid and acetic acid formed unsaturated dicarboxylic acids, such as maleic acid, fumaric acid, levulinic acid, and butanedioic acid, as shown in Fig. 3.13. During the hydrothermal oxidation reaction of alkali lignin, acetic acid had the highest yield among all carboxylic acid because acetic acid was the hardest to be oxidized. Knowing that saturated dicarboxylic acids and unsaturated dicarboxylic acid of glutaconic acid with 5 carbon atoms can produce a higher yield of acetic acid, so it is possible to increase the yield of acetic acid by controlling reaction pathways to enhance the formation of saturated dicarboxylic acids and unsaturated dicarboxylic acid of glutaconic acid with 5 carbon atoms. There were some other saturated dicarboxylic acids produced as intermediate products.

3.4 Conclusions

In summary, some recent studies on the hydrothermal conversion of lignin and its model compounds into value-added chemicals, i.e., formic acid and acetic acid were introduced. As nascent works, results showed that hydrothermal reactions can efficiently and rapidly convert lignin into useful value-added chemicals. Hydrothermal technology could be expected to be used to develop green chemical processes and new technologies, using which renewable products can be produced. Based on the intermediate products identified by GC/MS and HPLC, reaction pathways of lignin are presented. It is the authors' hope that some of the materials compiled in this chapter will be useful in building a fundamental understanding for exploring the potential approaches of lignin utilization.

Acknowledgments The authors gratefully acknowledge the financial support of the National Natural Science Foundation of China (No. 21277091 and No. 21077078), the National High Technology Research and Development Program of China (863 Program; No. 2009AA063903).

References

1. Zakzeski J, Bruijnincx PC, Jongerius AL, Weckhuysen BM (2010) The catalytic valorization of lignin for the production of renewable chemicals. Chem Rev 110:3552–3599
2. Kunkes EL, Simonetti DA, West RM, Serrano-Ruiz JC, Gartner CA, Dumesic JA (2008) Catalytic conversion of biomass to monofunctional hydrocarbons and targeted liquid-fuel classes. Science 322:417–421
3. Chheda JN, Huber GW, Dumesic JA (2007) Liquid-phase catalytic processing of biomass-derived oxygenated hydrocarbons to fuels and chemicals. Angew Chem Int Ed 46:7164–7183
4. Huber GW, Chheda J, Barrett CB, Dumesic JA (2005) Production of liquid alkanes by aqueous-phase processing of biomass-derived carbohydrates. Science 308:1446–1450
5. Chheda JN, Huber GW, Dumesic JA (2004) Renewable alkanes by aqueous-phase reforming of biomass derived oxygenates. Angew Chem Int Ed 43:1549–1551
6. Bond JQ, Alonso DM, Wang D, West RM, Dumesic JA (2010) Integrated catalytic conversion of g-valerolactone to liquid alkenes for transportation fuels. Science 327:1110–1114
7. Garlson TP, Vispute TP, Huber GW (2008) Green gasoline by catalytic fast pyrolysis of solid biomass derived compounds. ChemSusChem 1:397–400
8. Pandey MP, Kim CS (2011) Lignin depolymerization and donversion: a deview of thermochemical methods. Chem Eng Technol 34:29–41
9. Hendriks AT, Zeeman G (2007) Pretreatments to enhance the digestibility of lignocellulosic biomass. Bioresour Technol 100:10–18
10. Effendi A, Gerhauser H, Bridgwater AV (2008) Production of renewable phenolic resins by thermochemical conversion of biomass: a review. Renew Sustain Energ Rev 12(8):2092–2116
11. Lin SY, Dence CW (1992) Methods in lignin chemistry. Springer, Berlin
12. Barbie J, Charon N, Dupassieux N (2012) Supercritical water biomass gasification process as a successful solution. Biomass Bioenergy 46:479–491
13. Adler E (1977) Lignin—past, present and future. Wood Sci Technol 11:169–218
14. Shaw RW, Brill YB, Clifford AA (1991) Supercritical water, a medium for chemistry. Chem Eng News 23:26–39
15. Britt PF, Buchanan AC, Cooney MJ, Martineau DR (2000) Flash vacuum pyrolysis of methoxy-substituted lignin model compounds. J Org Chem 65(5):1376–1389
16. Jin FM, Kishita A, Enomoto H (1999) Oxidation of garbage in supercritical water. Haikibutsu Gakkaishi 10:257–266 Japanese
17. Girisuta A, Janssen LP, Heeres HJ (2006) A kinetic study on the conversion of glucose to levulinic acid. IChemE 84:339–349
18. Akiya N, Savage PE (2002) Roles of water for chemical reaction in high-temperature water. Chem Rev 102:2725–2750
19. Quintanilla A, Menéndez N, Tornero J, Casas JA, Rodríguez JJ (2008) Surface modification of carbon-supported iron catalyst during the wet air oxidation of phenol: influence on activity, selectivity and stability. Appl Catal B 81:105–114
20. Wu Q, Hu X, Yue P (2003) Kinetics study on catalytic wet air oxidation of phenol. Chem Eng Sci 58:923–928
21. Jin FM, Cao JX, Kishida H, Moriya T, Enomoto H (2007) Impact of phenolic compounds on hydrothermal oxidation of cellulose. Carbohydr Res 342:1129–1132

22. Kolaczkowski ST, Beltran FJ, Mclurgh DB, Rivas FJ (1997) Wet air oxidation of phenol. Trans IChemE 75:257–265
23. Wahyudiono A, Sasaki M, Goto M (2009) Conversion of biomass model compound under hydrothermal conditions using batch reactor. Fuel 88:1656–1664
24. Lawson JR, Klein MT (1985) Influence of water on guaiacol pyrolysis. Ind Eng Chem Res 24:203–208
25. Gonzalez G, Salvado J, Montane D (2004) Reactions of vanillic acid in sub and supercritical water. J Supercrit Fluids 31:57–66
26. Penninger JM, Kersten RJ, Baur HC (1999) Reactions of diphenylether in supercritical water e mechanism and kinetics. J Supercrit Fluids 16:119–132
27. Jin FM, Zhou Z, Kishita A, Enomoto H, Kishida H, Moriya T (2007) A new hydrothermal process for producing acetic acid from biomass waste. Chem Eng Res Des 85(2):201–206
28. Jin FM, Zhou Z, Moriya T, Kishida H, Higashijima H, Enomoto H (2005) Controlling hydrothermal reaction pathways to improve acetic acid production from carbohydrate biomass. Environ Sci Technol 39(6):1893–1902
29. Jin FM, Kishita A, Moriya T, Enomoto H (2000) Kinetics of oxidation of food wastes with H_2O_2 in supercritical water. J Supercrit Fluids 19:251–262
30. Jin FM, Yun J, Li GM, Kishita A, Tohji K, Enomoto H (2008) Hydrothermal conversion of carbohydrate biomass into formic acid at mild temperatures. Green Chem 10:612–615
31. Jin FM, Cao JX, Kishita A, Moriya T, Enomoto H (2004) Effect of lignin on acetic acid production in wet oxidation of lignocellulosic wastes. Chem Lett 33(7):910–911
32. Bang SS, Johnston D (1998) Environment effects of sodium acetate/formate deicer. Environ Contam Toxicol 35:580–587
33. Palmer DA (1987) Formiates as alternatives deicers. Transp Res Rec 1127:34–36
34. Yu J, Savage PE (1998) Decomposition of formic acid under hydrothermal conditions. Ind Eng Chem Res 37:2–10
35. Akiya N, Savage PE (1998) Role of water in formic acid decomposition. AIChE J 44:405–415
36. Weber M, Wang JT, Wasmus S, Savinell RF (1996) Formic acid oxidation in a polymer electrolyte fuel cell. J Electrochem Soc 143:158–160
37. Uhm S, Chung ST, Lee J (2008) Characterization of direct formic acid fuel cells by impedance studies. J Power Sources 178:34–43
38. Rice C, Ha S, Masel RI, Waszczuk P, Wieckowski A, Barnard T (2002) Direct formic acid fuel cells. J Power Sources 111:83–89
39. Jin FM, Kishita A, Moriya T, Enomoto H, Sato N (2002) A new process for producing Ca/Mg acetate deicer with Ca/Mg waste and acetic acid produced by wet oxidation of organic waste. Chem Lett 31(1):88–89
40. Jin FM, Zhou Z, Moriya T, Kishda H, Higashijima H, Enomoto H (2005) Controlling hydrothermal reaction pathways to improve acetic acid production from carbohydrate biomass. Environ Sci Technol 39:1893–1902
41. Lu M, Zeng X, Cao J, Huo Z, Jin FM (2011) Production of formic and acetic acids from phenol by hydrothermal oxidation. Res Chem Intermed 37:201–209
42. Kojima Y, Fukuta T, Yamada T, Onyango MS, Bernardo EC, Matsuda H, Yagishita K (2005) Photolytic hydrogen peroxide oxidation of 2-chlorophenol waste water. Water Res 39:29–36
43. Chang CJ, Li SS, Ko CM (1995) Catalytic wet oxidations of phenol- and p-chlorophenol-contaminated waters. J Chem Technol Biotechnol 64:245–252
44. Martino CJ, Savage PE (1997) Reactions at supercritical conditions: applications and fundamentals. Ind Eng Chem Res 36:1391–1400
45. Thornton TD, Savage PE (1990) Phenol oxidation in supercritical water. J Supercrit Fluids 3:240–248
46. Joglekar HS, Samant SD, Joshi JB (1991) Kinetics of wet air oxidation of phenol and substituted phenols. Wat Res 25:135–145
47. Portela JR, Nebot E, Delaossa EM (2001) Kinetic comparison between subcritical and supercritical water oxidation of phenol. Chem Eng J 81:287–299

48. Suzuki H, Cao J, Jin FM, Kishita A, Enomoto H (2006) Wet oxidation of lignin model compounds and acetic acid production. J Mater Sci 41:1591–1597
49. Zeng X, Jin FM, Cao JL, Yin GD, Zhang YL, Zhao JF (2010) Second international symposium on aqua science, water resource, and low carbon energy. AIP Conf Proc 1251:384–387
50. Jin FM, Kishita A, Moriya T, Enomoto H (2000) Kinetics of oxidaiton of food wastes with H_2O_2 in supercritical water. J Supercrit Fluids 19:251–262
51. Zeng X, Cao J, Jin FM, Huo ZB (2010) Production of formic acid and acetic acid from lignin by alkaline hydrothermal oxidation. In: The 2nd international solvothermal & hydrothermal association conference , 27 July 2010
52. Yan X, Jin FM, Enomoto H (2004) Use of biomass pyrolysis oils. In: 14th International Conference on the Properties of Water and Steam, pp 724–728
53. Shende RV, Levec J (2000) Subcritical aqueous-phase oxidation kinetics of acrylic, maleic, fumaric, and muconic acids. Ind Eng Chem Res 39:40–49
54. Eftaxias J, Fonta A, Fortuny J, Giralt A, Fabregat F (2001) Kinetic modelling of catalytic wet air oxidation of phenol by simulated annealing. Appl Catal B 33:175–190
55. Suzuki H, Cao J, Jin FM, Kishita A, Enomoto H (2004) Hydrothermal oxidation of lignin model compounds. In: Proceedings of 2nd international worship on water dynamics, Tohoku University, Sendai, Japan, pp 11–12

Chapter 4
Production of Lactic Acid from Sugars by Homogeneous and Heterogeneous Catalysts

Ayumu Onda

Abstract Lactic acid (2-hydroxypropionic acid, $CH_3CHOHCOOH$) is one of the platform chemicals derived from biomass. It is used in the food industry and in the manufacture of biodegradable plastics and useful chemicals. Recently, various examinations were carried out not only by fermentation but also by the chemical methods using heterogeneous and homogenous catalysts. This chapter focuses on the chemical processes with heterogeneous catalysts in lactic acid and lactate ester productions from sugars. Brønsted basic catalysts and Lewis acid catalysts gave lactic acid in high yields. In the lactic acid productions from triose, lactic acid ester is obtained with high yields of nearly 100 % in alcohols around 100 °C using Sn-β zeolite, Sn–carbon–silica, and H-USY catalysts. In the lactic acid production from hexose, lactic acid ester or a lactate salts was obtained from glucose, fructose, and sucrose with the comparatively high selectivity of about 50 % by several catalytic processes, that were in water around 50 °C using heterogeneous basic catalysts, such as activated hydrotalcite catalyst, in hydrothermal water around 300 °C using homogeneous basic catalysts, such as NaOH and $ZnSO_4$, and in alcohols around 160 °C using heterogeneous Lewis acid catalysts, such as Sn-β zeolite.

4.1 Introduction

Carbohydrates are the largest fraction of biomass resources, and various strategies for their efficient use as a commercial feedstock are being established in the interest of supplementing and replacing petroleum resources [1–3]. Ethanol is one representative biomass-derived compound, almost all ethanol produced globally is manufactured by fermentation routes from carbohydrates. The ethanol fermentation

---

A. Onda (✉)
Research Laboratory of Hydrothermal Chemistry, Faculty of Science, Kochi University, Kochi, Japan
e-mail: aonda@kochi-u.ac.jp

F. Jin (ed.), *Application of Hydrothermal Reactions to Biomass Conversion*, Green Chemistry and Sustainable Technology, DOI: 10.1007/978-3-642-54458-3_4,

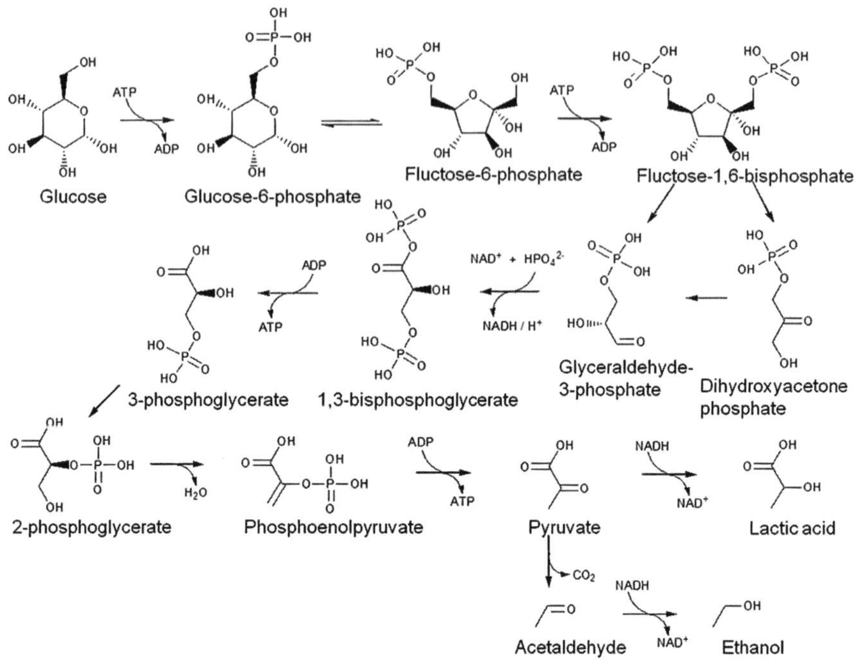

Fig. 4.1 Fermentation pathways from glucose into lactic acid and ethanol. Reproduced from Ref. [4] by permission of John Wiley and Sons Ltd

is a method using glycolytic metabolism. As shown in Fig. 4.1, various compounds can be produced in the glycolytic system, and lactic acid is one of them [4].

Lactic acid (2-hydroxypropionic acid, CH_3-CHOHCOOH) is a commodity chemical. It is a natural organic acid with a long history of use in the food and nonfood industries, including the cosmetic and pharmaceutical industries, and for the production of oxygenated chemicals, plant growth regulators, and special chemical industries [5–7]. Currently, lactic acid is a feedstock for the manufacture of biodegradable plastics such as poly (lactic acid). Poly lactic acid has many uses in surgical structure, drug delivery systems, and disposal consumer product [8, 9]. Lactic acid market growth is increasing. Furthermore, as shown in Fig. 4.2, lactic acid is one of the platform chemicals that can be used to synthesize a wide variety of useful products by chemical and biotechnological routes [3].

Over 90 % of the current commercial production of lactic acid is performed via fermentation. However, previously lactic acid was partially manufactured by the hydrolysis of lactonitrile, and presently chemical routes from carbohydrates for lactic acid productions are also investigated.

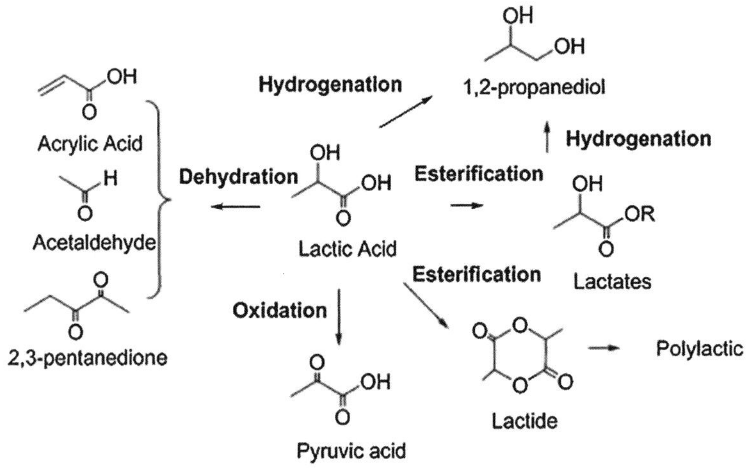

Fig. 4.2 Useful compounds derived from lactic acid. Reprinted with the permission from Ref. [3]. Copyright 2007 American Chemical Society

4.2 Fermentation Routes

Lactic acid can be produced either by chemical synthesis or by microbial fermentation. Chemical synthesis from petrochemical resources always results in racemic mixture of D, L-lactic acid [10]. Microbial lactic acid fermentation offers an advantage in terms of the utilization of renewable carbohydrate biomass, low production temperature, low energy consumption, and the production of optically high pure lactic acid by selecting an appropriate strain [11, 12]. Presently, almost all lactic acid produced globally is manufactured by fermentation routes using lactic acid bacteria.

The fermentation is carried out under anaerobic conditions to convert glucose into lactic acid instead of CO_2 and H_2O, and it can produce two lactic acid molecules from one glucose molecule. The conventional fermentation processes for producing lactic acid from natural carbohydrates, such as cellulosic biomass, include the following four main steps [7]:

(1) Collection and pretreatment of natural plants.
(2) Enzymatic hydrolysis: depolymerizing polysaccharides to fermentative sugars, such as glucose and xylose, by means of hydrolytic enzymes.
(3) Fermentation: metabolizing the sugars to lactic acid, generally by lactic acid bacteria.
(4) Separation and purification of lactic acid: purification of lactic acid to meet the standards of commercial applications.

There are some bottlenecks in each step. In the first step, the bottleneck is the substrate cost because of collections of biomass and the energies for breaking down natural biomass matrix.

With regard to the second step, generally enzymatic hydrolysis for depolymerization of natural carbohydrates needs another reactor from the fermentation of sugars into lactic acid because of different reaction conditions, such as pH and temperature. And, the other bottlenecks are in saccharization of lignocellulose. [3, 9] A great potential lies in the usage of nonedible cellulose, as it is considered to be a key substrate in the future chemical industry as soon as its hydrolysis becomes economically feasible [13–16].

In the third step, lactic acid bacteria have their optimum productivity in the pH range of 5–7. When lactic acid is produced, the pH of the fermentation media lowers causing inhibition of the microbial culture. Therefore, alkali sources, such as $Ca(OH)_2$, $CaCO_3$, NH_4OH or $NaOH$, are added continuously in the reactor during fermentation. The final concentration of formed lactate salts is less than about 10 wt%. After fermentation, the suspension is filtered for biomass residue removal, and sulfuric acid is added to the aqueous calcium lactate solution to gain free LA and to precipitate calcium ions as low-value insoluble salts as $CaSO_4$ or gypsum. About equivalent amounts of gypsum are formed to that of lactic acid. After removal of gypsum, the filtrate is further purified with ion exchange columns.

The fourth step is mainly composed of esterification, distillation, and hydration. Impure lactic acid is esterified with methanol and ethanol and purified via distillation. In this way, after hydrolysis of the lactate ester with water, highly pure aqueous solutions of LA are obtained.

One of the advantages of fermentation methods is to produce a desired stereoisomer, optically pure L-(+)- or D-(−)-lactic acid, whereas chemical methods produce racemic lactic acid [17]. For example, in order to prepare poly lactic acid with the excellent properties, the stereoisomer is very important. Compared to chemical synthesis, the biotechnological process for lactic acid production offers the other advantages: production temperature and energy consumption [21].

The fermentative production of lactic acid is not the scope of this chapter. The reader is referred to other works for extensive details on the fermentation [7, 9, 17, 18].

4.3 Conventional Chemical Reaction Routes via Lactonitrile

The output of lactic acid was around 5,000 kg per year [9] in 1950s, around 25,000 metric tons per year in 1980s, and 40,000 metric tons per year in 1990s. And at present, it is approximately 260,000 tons per year, which is almost produced entirely via fermentation routes. In contrast, until 1990s, lactic acid was partially manufactured by the hydrolysis of lactonitrile by strong acid [9]. Lactonitrile was produced as a by-product of acrylonitrile production. The advantage of the lactonitrile route is a simple procedure, rather than fermentation described above. In contrast, the drawbacks of the chemical process from lactonitrile are that only the racemic mixture of D-type and L-type of lactic acid is formed and the production of the raw material, i.e., lactonitrile, depends on the acrylonitrile industry.

4.4 Hydrothermal Methods: Catalytic Effects of Alkalis and Salts

Presently, the major production process of lactic acid is the fermentation of glucose from starch hydrolysis by using genetically modified enzymes [7, 9, 17–19]. The biotechnological process has some disadvantages such as limited space-time yield, reactor control for fermentations (temperature and pH), and population control regulations of the microorganisms. Some chemical processes from saccharides into lactate were reported in hot water or aqueous solutions with alkali and/or metal salts [20–22].

4.4.1 Hydrothermal Method with Alkalis

De Bruijn et al. reported the hydrothermal alkaline degradation behavior of sugars in 1980s [20, 23]. After 2000, Jin et al. reported the conversions of carbohydrates into lactic acid under harsh hydrothermal conditions at temperature higher than 250 °C with or without homogeneous or heterogeneous catalysts [24].

The formation of lactic acid from glucose was observed in the hydrothermal conditions without any additives at 300 °C [22] and was greatly facilitated by the addition of a base [24]. The lactic acid yield increases to 27 and 20 % at an initial OH^- concentration of 2.5 M NaOH and 0.32 M of $Ca(OH)_2$ at 300 °C [25, 26]. A similar dependence of the lactic acid yield on alkaline concentration was observed for fructose and xylose [24]. With fructose, a lactic acid yield of about 45 % was obtained. The order of lactic acid yield was fructose > glucose > xylose.

4.4.2 Hydrothermal Methods with Salts

Bicker et al. investigated the influence of Ni-, Co-, Cu-, and $ZnSO_4$ on the conversion of sugars in subcritical flowing water at 200–360 °C [21]. The concentration of salts as catalysts was 400 ppm. They found that transition metal ions have a great influence on the product selectivity and $ZnSO_4$ salts showed highest catalytic performance for lactic acid production. With $ZnSO_4$ as the catalyst, the lactic acid yield was increased to 42 % starting from glucose and 48 % starting

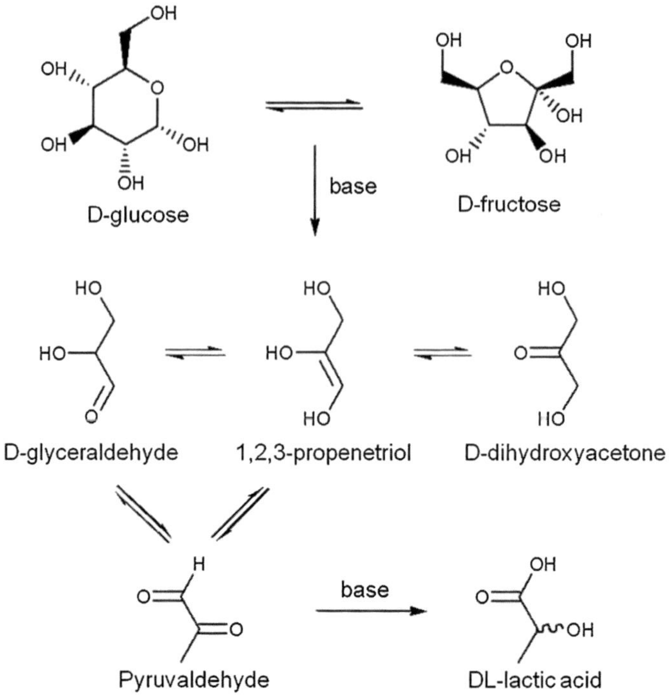

Fig. 4.3 Conversion of D-glucose into lactic acid over base catalysts. Reprinted from Ref. [21], Copyright 2005, with permission from Elsevier

from fructose on the carbon base at 300 °C. They also suggested that this conversion process occurs in a very complex reaction mechanism (Fig. 4.3), with the last step being the Zn(II)-catalysed Cannizzaro-type reaction of pyruvaldehyde to lactic acid.

In conclusion, these chemical processes are simpler than the fermentation processes and are flexible as far as the limitation of reaction conditions is concerned, which indicates the possibility to increase the lactate supply. However, disadvantages of these chemical processes are the high concentration of alkaline, the high reaction temperature, and the large amounts of by-products.

4.5 Heterogeneous Catalysts

The use of heterogeneous catalysts, i.e., solid catalysts, offers tremendous advantages as reaction media can easily be separated from the solid catalysts, which can be easily reused and regenerated.

4.5.1 Basic Catalysis for Lactic Acid Productions

Chemical processes from saccharides into lactate were also reported in alkaline aqueous solutions [27, 28] or in hot water with or without metal salts [21, 22, 29]. These chemical processes are simpler than the biotechnological processes, flexible as far as the limitation of reaction conditions, and can be carried out by using conventional chemical plants, is concerned, which indicates the possibility to increase the lactate supply. The reaction mechanism of alkaline degradation of glucose into lactate includes a reverse aldol condensation from C_6 compounds to C_3 compounds, as shown in Fig. 4.3 [21, 30]. Our group focused on activated hydrotalcite, which has been reported to have Brønsted-base sites and to show catalytic activity for aldol condensations [31–33], which are inverse reactions of the reverse aldol condensation in Fig. 4.3. Consequently, our group first demonstrated the transformation of D-glucose to produce lactic acid over activated hydrotalcite catalysts and the determination of accessible Brønsted-base sites on the catalyst.

4.5.1.1 Lactic Acid Production from Glucose Over Activated Hydrotalcite Solid Base Catalysts

Our group has undertaken the present work with the aim of studying the transformation of D-glucose to produce lactic acid over activated hydrotalcite catalysts and the determination of accessible Brønsted-base sites on the catalyst [34].

Solid Brønsted Base Hydrotalcite catalysts
Hydrotalcite-like compounds, also known as Layered double hydroxides of clay-like minerals, have recently received much attention [35] in view of their potential usefulness as adsorbents, anion exchangers, and basic catalysts [32]. These compounds are represented by the formula $[M^{2+}_{(1-x)}M^{3+}_x(OH)_2]^{x+}(A^{m-}_{x/m}) \cdot nH_2O$, typically $[Mg^{2+}_{(1-x)}Al^{3+}_x(OH)_2]^{x+}(CO_3^{2-}_{x/2}) \cdot nH_2O$ ($x = 0.2 - 0.4$). Hydrotalcite (HT) was prepared by the method described previously [33]. A mixed aqueous solution of $Mg(NO_3)_2 \cdot 6H_2O$ and $Al(NO_3)_3 \cdot 9H_2O$ was added slowly to a mixed aqueous solution of NaOH and Na_2CO_3. The obtained as-synthesized hydrotalcite sample (HT_{as}) had the Mg/Al atomic ratio equal to 2.1 (Table 4.1), which was close to the ratio in the mother solution, and had a HT-platelet structure according to X-ray diffraction (Fig. 4.4a). It was heated in an inert gas flow to 573 and 723 K to prepare the heated samples (HT573 and HT723). The calcination at 723 K resulted in marked increases of the surface area and the total pore volume, to 237 m^2 g^{-1} and 0.52 cc g^{-1} (Table 4.1), and the disappearance of the layered structure. The high surface area and large pore volume of HT723 might be attributed to the removal of interlayer water molecules and carbon dioxide from the carbonate anion present in the brucite layer [33, 36–38]. The heated samples were rehydrated by distilled deionized water at room temperature and dried at room temperature under vacuum, which resulted in the rehydrated samples, HT573$_{rehydr}$ and HT723$_{rehydr}$, generating the reconstruction

Table 4.1 The relationships between basic sites and lactic acid production on solid catalysts

Catalyst	Basic sites		Lactic acid yield[a]	Sugar conversion[a]
	Titration OH/Al	CO_2 adsorption OH/Al	C-%	%
Control	–	–	0.4	4
HTas	0.017	0.009	0.9	5
HT623	0.034	0.035	1.4	10
HT723 (0.3 g)	0.36	0.038	11	28
HT723 (0.6 g)	–	–	20	57
MgO	–	–	1.7	8
γ-Al_2O_3	–	–	0.4	5

[a] Reaction conditions: glucose 25 mmol L^{-1}, NaOH 50 mmol L^{-1}, flow rate 0.1 mL min^{-1}, 323 K, time on stream 8 h, catalyst 300 mg. Reprinted from Ref. [34], Copyright 2008, with permission from Elsevier

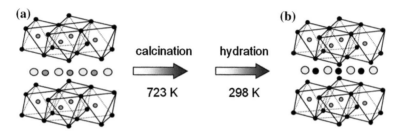

Fig. 4.4 a $Mg_6Al_2(OH)_{16}CO_3 \cdot 4H_2O$: Mg-Al hydrotalcite with carbonate anions, and **b** activated Mg-Al hydrotalcite with Brønsted-basic hydroxyl ions

of the layered HT structure (Fig. 4.4b). The layered structure is recovered to a large extent with hydroxyl ions (Brønsted-base sites) incorporated in the interlayer [31, 32, 39]. The images of as-synthesized hydrotalcite, heated hydrotalcite, and rehydrated hydrotalcite are shown in Fig. 4.5.

The accessible Brønsted-base sites on these hydrotalcite catalysts were determined by the ion-exchange method and the conventional CO_2 adsorption method. An aqueous solution of sodium gluconate or sodium chloride (0.20 mol L^{-1}) was added to slurry of the HT sample with a small amount of water. The OH^- ions at Brønsted-base sites of the HT are ion-exchanged with gluconate or chloride ions and the corresponding number of sodium hydroxide ions is formed in the solution. After filtration, the produced sodium hydroxide was determined by titration with an aqueous HCl solution. These basic sites determined by the ion-exchange method with sodium gluconate might be accessible for glucose. As shown in Table 4.1, the molar ratio of exchangeable OH^- ions to aluminum atoms in the rehydrated HT_{as} ($HT_{as\text{-}rehydr}$) was estimated to be about 2 %. The number of OH^- sites of $HT573_{rehydr}$, whose OH/Al molar ratio was 3 %, is almost double those of HT_{as}. In contrast, the number of OH^- sites of $HT723_{rehydr}$, whose OH/Al molar

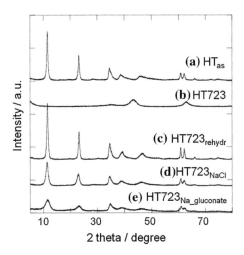

Fig. 4.5 XRD patterns of hydrotalcite samples. Reprinted from Ref. [34], Copyright 2008, with permission from Elsevier

ratio was 36 %, increased more than ten times. The determined OH/Al molar ratios were less than 40 %, which would indicate that gluconate ions might not completely intercalate into the interlayers of the HT platelets. The X-ray diffraction peaks of HT723$_{NaCl}$ (Fig. 4.4d) and HT723$_{Na\_gluconate}$ (Fig. 4.4e) were broad. During the ion-exchange, both HT samples were white powders but not or little dissolve in the aqueous solutions. The ion-exchange with sodium salts was reversibly at least six times using NaOH aqueous solution.

The rehydrated HT samples (HT$_{as}$, HT573$_{rehydr}$, and HT723$_{rehydr}$) were dried at 393 K under vacuum, and then the base sites were determined by the conventional CO_2 adsorption method in a glass vacuum system [33]. The both dried HT573$_{rehydr}$ and dried HT723$_{rehydr}$ samples have almost the same number of accessible basic sites for CO_2 as shown in Table 4.1. The number of accessible basic sites of dried HT573$_{rehydr}$ for CO_2 was almost the same as that of HT573 for anions in water phase. It is considered that a certain number of interlayer Brønsted-base sites of the HT platelets are accessible in water, although only Brønsted-base sites near extra-surface of the HT platelets are accessible in gas phase.

Catalytic conversions of glucose with or without NaOH
The reactions of D-glucose were done in the aqueous solution with D-glucose (25 mmol L^{-1}) over pretreated HT catalysts at 323 K. The HT$_{as}$ samples (0.30 g) were in situ pretreated at 323 (HT$_{as}$), 573 (HT573), or 723 K (HT723) under flowing nitrogen gas in a fixed-bed flow reactor. An aqueous solution of D-glucose (25 mmol L^{-1}) with or without NaOH (50 mmol L^{-1}) was introduced to the reactor.

In the case of reactions without NaOH, the HT723 catalyst showed significantly catalytic activity for the lactate production. However, magnesium ions were simultaneously eluted and the catalytic activity decreased immediately, which might be due to the acidic property of produced lactic acid and the coking of products, respectively. In the case of fermentations, a stoichiometric amount of

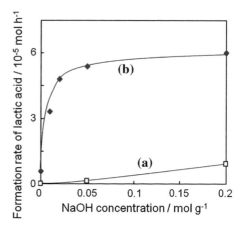

Fig. 4.6 Lactic acid production from glucose without (**a**) or with activated hydrotalcite HT723 catalyst (**b**)

alkali is added for pH-regulation [10]. Therefore, low concentration of alkali was introduced into the reactant solution for the inhibition of leaching with produced organic acids and for the removal of products from the HT surface. The concentration of alkali depended on the concentration of sugar as substrate, because one D-glucose molecule can be converted into two lactic acid molecules to be neutralized by alkali.

Effects of NaOH concentration on the glucose conversion over solid basic catalysts

Figure 4.6 shows effects of NaOH concentration on the lactic acid formation with or without HT723 catalyst. In the case of reactions without solid catalyst, the lactic acid formation rate was quite low even in 0.2 mol L^{-1} NaOH aqueous solution. In the case of reactions with HT723 catalyst, the lactic acid formed even in neutral water, and its formation rate was markedly higher than those of reactions without solid catalyst in aqueous solutions with more than 0.02 mol L^{-1} NaOH concentration. In addition, no leaching of magnesium and aluminum ions in the filtrate after reactions with NaOH was detected by ICP measurement. The NaOH would play the role to neutralize produced lactic acids and to remove products from the surface, and also it might be like a catalytic promoter as supplying hydroxyl group and its concentrated adsorption at the active sites of the HT723 catalysts during the reaction.

Reaction conditions: glucose 25 mmol L^{-1}, flow rate 0.1 mL min^{-1}, 323 K, time on stream 8 h. catalyst 300 mg.

In the case of reactions with NaOH (0.05 mol L^{-1}) aqueous solution as shown in Table 4.1, the blank reaction show that the conversions of D-glucose were less than 5 % and the yield of lactic acid was less than 0.5 C-% for the reaction of D-glucose. The HT_{as} and HT573 catalysts showed the small catalytic activity as well as MgO and Al_2O_3 catalysts. In contrast, the HT723 catalyst showed significant activity for the lactic acid production from glucose. The lactate yield was 11 C-%. A linear dependence of lactic acid yield on the catalyst content was also obtained. The other products were unknown but water soluble organic compounds,

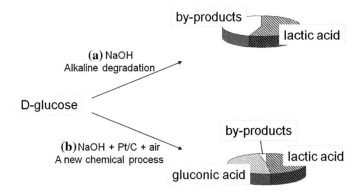

Fig. 4.7 Reaction of D-glucose in the conventional alkaline degradation (**a**) and in the proposed new chemical process (**b**)

because after reactions the TOC values of all solutions were more than 90 C-% of the reactant D-glucose solution.

At 30 h of time-on-stream over the HT723 catalyst (300 mg), the lactate yield was 7.5 C-% and the catalyst changed the color from white to brown. Although the catalyst was deactivated probably by the coking which might be due to by-products via the aldol condensation reactions, the TON was more than three. Consequently, it was clarified that the activated hydrotalcite has catalytic activity for the lactic acid production from D-glucose as a solid base catalyst.

4.5.1.2 Glucose to Lactic Acid and Other Useful Compounds Under Alkaline Solutions with Solid Catalyst

The significant disadvantage of the chemical processes described above is the large amount of by-products whose total yields is about 40 C-% or more. Many kinds of compounds were reported as the by-products, such as carboxylic acids, C_6 isomers, acidic aldol condensation products, dehydration products, and unidentified compounds. Although it is difficult to increase the yield of lactic acid to almost 100 C-% in the alkaline degradation of monosaccharides, it is concerned to be able to produce lactic acid and a useful chemical compound simultaneously, instead of complex by-products.

So, our group demonstrated a one-pot reaction for production of lactic acid and gluconic acid from D-glucose using metal supported catalysts with an alkaline aqueous solution, which is expected to be a new highly atomic-effective chemical process from renewable saccharides into useful chemicals, that is, the aim of this study is to reduce complex by-products in the alkaline degradations of D-glucose (Fig. 4.7) [40]. Gluconic acid is used in the food and pharmaceutical industries and a biodegradable chelating agent in admixing concrete and in removing calcareous and rust deposits from metal surfaces. Various works dealt with the study of

Table 4.2 The effects of support materials on the reaction of D-glucose over 5 wt% platinum catalysts[a]

Catalysts	Surface area m^2/g	Conversion /%	Yield /C-%					
			GluA	LA	GlyA	AceA	ForA	Others
Control	–	100	–	55	1.7	2.7	2.5	38
Pt/Al$_2$O$_3$	136	100	13	57	1.3	2.4	1.3	25
Pt/SiO$_2$	404	100	17	50	1.9	2.6	2.3	25
Pt/MgO	4.2	100	23	57	1.6	2.5	1.1	15
Pt/C	1200	100	45	43	2.9	2.9	0.8	6

Reprinted from Ref. [40], Copyright 2008, with permission from Elsevier
GluA (gluconic acid), LA (lactic acid), GlyA (glycolic acid), AceA (acetic acid), ForA(formic acid)
[a] Glucose (0.25 mmol) were added to 5.0 mL of NaOH (1.0 mol L^{-1}) aqueous solution with 50 mg of catalyst in a batch reactor bubbled by air flow (20 mLmin^{-1}). Reaction time and temperature were 2 h and 353 K, respectively

glucose oxidation into gluconic acid over platinum, palladium, gold, and these bimetallic catalysts [41–46].

Reaction procedures

The metal-supported catalysts were prepared by an impregnation method. The metal contents were 5 wt%. Activated carbon, silica-gel, magnesia, and γ silica-alumina were used as supports. After drying overnight at 333 K, the supported catalysts were treated under flowing hydrogen at 573 K. Catalytic reactions were performed in a polypropylene copolymer (PPCO) batch reactor (40 ml) equipped with two tubes for gas inlet and gas outlet of reflux condenser. The typical reaction procedure was as follows: 5.0 mL of NaOH (1.0 mol L^{-1}) aqueous solution was introduced in the reactor. The solution was stirred, and it was bubbled by flowing air at 353 K. And then, 45 mg of D-glucose and 50 mg of a solid catalyst were introduced simultaneously, and the reaction was started. After reaction, the aqueous solution was separated from the solid catalyst by the filtration and cooled in an ice bath.

Effects of catalytic supports

Table 4.2 shows the product selectivity for the reactions of D-glucose over platinum-supported catalysts in the 1.0 M NaOH aqueous solution at 353 K for 2 h. The reaction without metal catalysts (control) resulted in 55 C-% yield of lactic acid, which was the alkaline degradation catalyzed by sodium hydroxide. The total yields of lower organic acids, as acetic, glycolic and formic acids, was about 7 C-%, and the yields of others was about 38 C-% which were unidentified by HPLC and GC-MS analysis. The TOC value of the resultant aqueous solution was almost completely equivalent to the introduced glucose, which indicated that most of by-products were water soluble organic compounds.

To increase the atomic-efficiency for the reaction of glucose into useful chemicals, with remaining the high yield of lactic acid, platinum-supported catalysts were added as solid oxidation catalysts in the alkaline solution under air

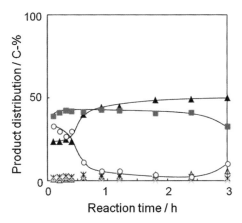

Fig. 4.8 Changes in product distribution during the reaction of glucose using Pt/C catalyst with 1.0 M NaOH at 353 K. Reaction conditions: D-glucose 0.25 mmol, Catalyst 50 mg, NaOH aqueous solution 5.0 mL, Air flow 20 mL min^{-1}. (*Filled square*) lactic acid; (*Filled triangle*) gluconic acid; (×) C$_2$ carboxylic acids; (+) formic acid; (*Open circle*) others. Reprinted from Ref. [40], Copyright 2008, with permission from Elsevier

bubbling. The simultaneous formations of D-gluconic acid and D, L-lactic acid were expected. The platinum contents on various supports were about 5 wt%. Table 4.2 shows the yield of lactic acid was between 43 and 57 C-%, which little depended on the kinds of supports. In contrast, although the yield of gluconic acid increased with the addition of any platinum catalysts, it significantly depended on the kinds of supports and increased as follows:

$$Pt/Al_2O_3 < Pt/SiO_2 < Pt/MgO < Pt/C.$$

The yield of unknown by-products (others) decreased in the above order. The selectivity might be due to the oxidation activity of platinum catalysts. According to the XRD data in Fig. 4.8, the dispersion of platinum on supports probably increased with the following order;

$$Pt/MgO < Pt/Al_2O_3, \; Pt/SiO_2 < Pt/C.$$

It was considered that the high activity of Pt/C might be due to relatively high dispersion of platinum on activated carbon with high surface area. However, the dispersion of platinum could not account for that Pt/MgO catalyst showed the relatively high yield of gluconic acid among the metal-oxide support we tested. The basicity of magnesium oxide and/or the interactions between platinum and supports might affect the catalytic activity. Consequently, the addition of Pt/C catalyst into the alkaline degradation under air resulted in the highest atomic-efficiency for the conversion of D-glucose into 43 C-% yield of lactic acid and 45 C-% yield of gluconic acid.

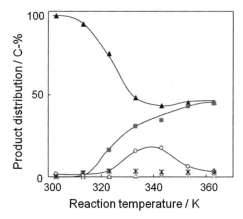

Fig. 4.9 Effect of reaction temperature on product distribution for the reaction of glucose using the Pt/C catalyst with 1.0 M NaOH for 2 h. Reaction conditions: D-glucose 0.25 mmol, Catalyst 50 mg, NaOH aqueous solution 5 ml, Air flow 20 mL min^{-1}. (■) lactic acid; (▲) gluconic acid; (×) C_2 carboxylic acids; (+) formic acid; (○) others. Reprinted from Ref. [40], Copyright 2008, with permission from Elsevier

Changes in the product distribution as a function of reaction time

Figure 4.8 shows the product distribution as a function of reaction time in the reaction of D-glucose over Pt/C catalyst in a 1.0 mol L^{-1} of NaOH aqueous solution at 353 K. At 10 min of reaction time, D-glucose disappeared completely and the yield of lactic acid, gluconic acid, and others were 39, 24, and 33 C-%, respectively. The TOC value of the resultant aqueous solution was equivalent to the introduced glucose, which indicated that most of others were water soluble. Between the reaction time of 20 and 40 min, the product distribution was almost constant. After the reaction time of 40 min, the yield of gluconic acid increased from 24 C-% (40 min) to 40 C-% (60 min) and to 49 C-% (180 min). As shown in Fig. 4.9, the yield of others decreased with increase of the yield of gluconic acid, which suggested the conversion of the others into gluconic acid. The yield of lactic acid was almost constant to be 43 C-% after the reaction time of 20 min. At 180 min of reaction time, the yield of lactic acid and gluconic acid was 42 and 49 C-%, respectively, and the yields of lower carboxylic acids (glycolic, acetic, and formic acids) and others were about 6 C-% and about 3 C-%, respectively.

Most of the others were water soluble organic compounds (WSOCs), and they seemed to form with the formation of lactic acid and were further converted into gluconic acid after an induction period between 10 and 40 min. Probably, during the induction period, WSOCs (A) formed with lactic acid from D-glucose might be converted into WSOCs (B) easily oxidized into gluconic acid. Both WSOCs were not identified by HPLC and GC-MS, which were not oligosaccharides and sugar alcohols. In the previous reports, the alkaline degradation of monosaccharides results in the formation of unidentified products, besides of the well-characterized carboxylic acids equal and lower than C_6, which can be assigned as carboxylic

Table 4.3 The reactions of various monosaccharides and gluconic acid using Pt/C with a NaOH aqueous solution[a]

Reactants	Conversion /%	Yield /C-%					
		C_5-C_6 acids	LA	GlyA	AceA	For A	Others
Xylose	100	20	27	6.2	1.8	1.5	43
Glucose	100	45	43	2.9	2.9	0.8	6
Fructose	100	29	48	3.7	2.2	1.1	16
Mannose	100	30	48	3.3	1.7	1.2	15
Galactose	100	43	16	5.1	1.9	1.9	32

Reprinted from Ref. [40], Copyright 2008, with permission from Elsevier
LA (lactic acid), GlyA (glycolic acid), AceA (acetic acid), ForA (formic acid)
[a] Monosaccharide (0.25 mmol) were added to 5.0 mL of NaOH (1.0 mol L^{-1}) aqueous solution with 50 mg of catalyst in a batch reactor bubbled by air flow (20 mLmin^{-1}). Reaction time and temperature were 2 h and 353 K, respectively

acids higher than C_6 [27]. In this study, the others were highly selectively oxidized into gluconic acid, which suggested that the others might have similar composition to glucose and gluconic acid, such as these dimmers and/or isomers.

Effects of reaction temperature on the product distribution

Figure 4.9 shows the product distribution as a function of the reaction temperature in the reaction of D-glucose using Pt/C catalyst in 1.0 mol L^{-1} of NaOH aqueous solution for 2 h. The TOC values of the resultant aqueous solutions after the reaction at 303–363 K were equivalent to the introduced glucose. The reaction below 323 K favored the production of gluconic acid, which corresponded with previous reports about the oxidation of glucose over noble metal catalysts [42]. The yield of gluconic acid decreased with increasing the yields of lactic acid and others over 323 K, however, the yield of others decreased over 353 K. The yields of lactic acid, gluconic acid, and others were 44 C-%, 46 C-%, and about 3 C-%, respectively, at 363 K. Consequently, the reaction over 353 K resulted in the highly atomic-effective conversion of D-glucose into lactic acid and gluconic acid, whose yields were about 44 and 46 C-%, respectively.

Catalytic conversion of several sugars

Table 4.3 shows the product yields for the reactions of monosaccharides and gluconic acid using Pt/C catalysts in the NaOH aqueous solution at 353 K for 2 h. The reactions of fructose and mannose resulted in about 48 C-% yield of lactic acid as high as that in the reaction of glucose, but lower yield of C_6 aldonic acids and higher yield of others than those of glucose. In contrast, the reaction of galactose resulted in 43 C-% yield of C_6 aldonic acid as high as that of glucose, but lower yield of lactic acid and higher yields of glycolic acid (5 C-%) and others than those of glucose. The reaction of xylose, i.e., an aldopentose, resulted in about 20 and 27 C-% yields of C_5 aldonic acid and lactic acid, respectively. The both yields were lower and the yield of others was higher than those of glucose. Although it is considered that glycolic acid is produced as a pair product of lactic acid in the xylose alkaline degradation, the yield of glycolic acid was 6 C-% which was much less than 27 C-% of the lactic acid yield but was significantly higher than that in the reaction of C_6 sugars.

D-glucose, D-fructose, and D-mannose were favorable for the alkaline degradations into lactic acid compared to D-galactose and D-xylose, which was corresponded to the previous papers [47, 48]. D-glucose ((2R, 3S, 4R, 5R)-2,3,4,5,6-pentahydroxyhexanal) and D-mannose ((2S, 3S, 4R, 5R)-2,3,4,5,6-pentahydroxyhexanal) were favorable for the alkaline degradations into lactic acid compared to D-galactose ((2R, 3S, 4S, 5R)-2,3,4,5,6-pentahydroxyhexanal), which might be due to alternate actual geometry of the third and the forth asymmetric carbons in monosaccharides. And, fructose is reversibly converted into glucose in an alkaline aqueous solution.

Although de Bruijn et al. reported that the oxidation rates of glucose and galactose were not much higher than those of mannose and xylose over 5 % Pt/Al_2O_3 under pH 9 at 313 K [49], the yields of aldonic acids depended on the kinds of monosaccharides as shown in Table 4.3, for which the following two causes could account; the first was the competitive reaction of the air oxidation with the alkaline degradation and the second was the different reactivity of others into gluconic acid. The high selectivity for C_6 aldonic acid in the reaction of D-galactose might be due to the first cause because of low yield of lactic acid. The low yield of C_6 aldonic acid in the reaction of D-mannose and D-fructose might be due to the second cause and might be improved by the extended reaction time.

In conclusion, the conventional alkaline degradation of D-glucose into lactic acid involved a large amount of by-products including unidentified compounds. In contrast, a new one-pot reaction of D-glucose resulted in yielding about 45 C-% of lactic acid and about 45 C-% of gluconic acid using Pt/C catalysts with a 1.0 mol L^{-1} of sodium hydroxide aqueous solution at 353 K under flowing air.

4.5.2 Lewis Acid Catalysis for Lactate Ester Productions

In both routes of fermentations and chemical methods described above, alkali sources are added to neutralize lactic acid, which results in the production of lactate salts. Lactate salts were transformed into lactic acid by using large amounts of mineral acids. In order to overcome this drawback, lactate esters were produced directly in sugar conversions by using Lewis acid catalysts.

4.5.2.1 Lactate Esters from Trioses

FAU zeolite catalysts
Seles et al. reported the conversion of trioses into alkyl lactates (entry 3 in Table 4.4) [50]. For various Y type zeolite catalysts with FAU type structure, the catalysts eventually were grouped based on their catalytic selectivity to form either dialkyl acetal or alkyl lactate (Fig. 4.10). The former product would be formed by

4 Production of Lactic Acid from Sugars

Table 4.4 The catalytic conversion of trioses: dihydroxyacetone (DHA) in alcohols

Entry	Catalyst[c]	Substrate conc. [M][b]	Alcohol	Catalyst conc. [mg mL^{-1}]	T [°C]	T [h]	Yield [%] di-acetal	Yield [%] alkyl lactate	Ref.
1	SnCl$_4$·5H$_2$O[a]	0.625	MeOH	10 mol%[a]	90	2	9	82	51
2	SnCl$_4$·5H$_2$O[a]	0.625	EtOH	10 mol%[a]	90	1	–	84	51
3	Y type zeolite ZF210	0.4	EtOH	40	90	6	28	65	50
4	H-USY-zeolite (6)	0.25	MeOH	16	115	24	–	96	52
5	H-USY-zeolite (6)	0.25	MeOH	16	115	1	5	75	52
6	Sn-β zeolite (125)	0.25	MeOH	16	80	24	0	99	54
7	Amorphous SiAl (10)	0.4	EtOH	40	90	6	1	22	56
8	Sn-MCM-41 (49)	0.4	EtOH	40	90	6	2	98	57
9	Sn-montmorillonite	0.25	EtOH	10	150	15	0	97	58
10	Sn-montmorillonite	0.25	EtOH	10	120	1	22	53	58
11	Sn-graft-MCM-41(460)	0.2	EtOH	20	90	6	3	23	59
12	Sn-CSM-(15 %C)[d, e]	0.2	EtOH	20	90	6	0	100	59
19	Sn-CSM-(16 %C)[d]	0.2	C$_8$OH	20	90	6	–	83	59
20	Sn-CSM-(16 %C)[d]	0.2	C$_{14}$OH	20	90	6	–	54	59

Reproduced from Ref. [55] by permission of The Royal Society of Chemistry

[a] Homogeneous catalyst: in that case the catalyst concentration is in mol% to the substrate. [b] Monomeric form. [c] Value in brackets = bulk Si/Metal ratio. [d] Si to Sn ratio of the Sn-MCM part (= 85 wt% of the composite) is 460. [e] Post-oxidised at 300

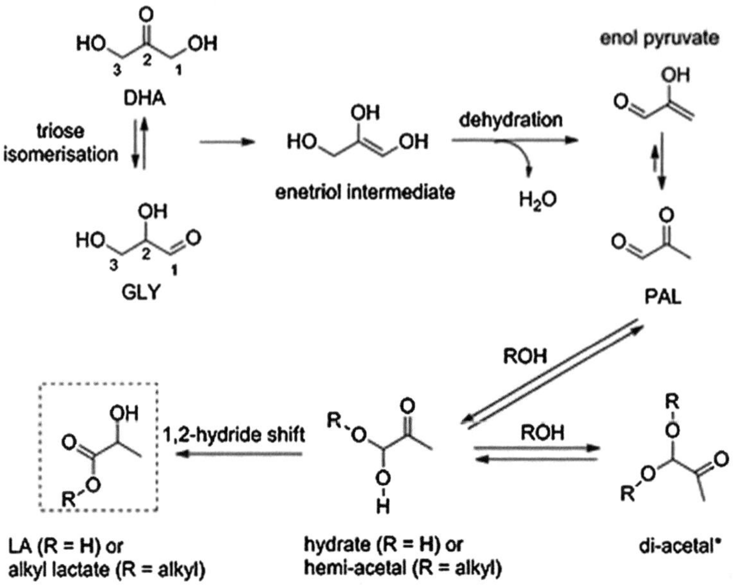

Fig. 4.10 Mechanism for converting trioses into alkyl lactates in alcohols (R = alkyl) or into lactic acid in water (R = H). Reproduced from Ref. [55] by permission of The Royal Society of Chemistry

strong Brønsted acidity, while the latter would result from Lewis acidity. As consequently, ethyl lactate was thus produced with a yield of 65 % in ethanol with a Y type zeolite catalyst in DHA conversion [50], which was comparable with those over $SnCl_4$ homogeneous catalyst (entry 2 in Table 4.4) [51].

Taarning et al. also performed a screening with several zeolite catalysts, such as USY zeolites, H-ZSM-5, H-BEA, H-MOR, a sulfated zirconia catalyst, and H-montmorillonite catalyst, both in water and methanol [52]. The H-USY with a bulk Si/Al ratio equal to 6 (H-USY6) outperformed all other catalysts and it showed markedly better catalytic performance than the other H-USY with a Si/Al ratio equal to 30 (H-USY30). A methyl lactate yield of 96 % at 99 % conversion was obtained at 115 °C after 24 h of reaction over H-USY6 catalyst (entry 4, Table 4.4). The yield of 75 % was already obtained after 1 h (entry 5). The ratios of Brønsted to Lewis acid sites were measured to be 1.8 for H-USY6 and 5.6 for H-USY30 catalysts, respectively, by an FTIR using pyridine and an NH_3-TPD. On the other hand, addition of water generally lowered reaction rates, yield and lactate selectivity in the reaction media. In addition, the stability of the zeolite and its deactivation process were monitored in a promising continuous flow mode setup. It was found that lactic acid in water destroyed the catalyst structure even at a concentration as low as 0.3 M. In contrast, this was not observed for methyl lactate in methanol. Another cause for zeolite deactivation in water is catalyst coking, mainly attributed to an intermediate of pyruvic aldehyde. A kinetic study with H-USY6 catalysts in water showed an apparent energy barrier of 53 and

61 kJ mol^{-1} for dihydroxy acetone dehydration and pyruvic aldehyde to lactic acid reaction, respectively. These values are clearly lower than those obtained with soluble Al^{3+} salts [53].

Sn-β zeolite catalysts

Taarning et al. prepared Al-free Sn-β zeolite catalysts according to a previous report. They proved that in particular Lewis acidic β zeolites were very active for the DHA conversion [54]. A comparison of a series of zeolites with Ti, Zr, and Sn incorporated in the framework revealed a correlation of the catalytic activity with the Lewis acid strength, β zeolites with tin showed the strongest in the series (entry 6, Table 4.4). The Sn-β zeolites showed the catalytic selectivity to alkyl lactate, whereas, conventional Al-β zeolite and Brønsted acidic ion-exchange resin catalysts yielded only dialkylacetals, which suggested that strong Brønsted acidic zeolites are selective toward acetals and Lewis acidic zeolites are selective toward alkyl lactates. Interestingly, steaming a parent Al-β zeolite to produce extra-framework Al enhanced the lactate yield in regard to the pristine H–Al-β [54]. These results led to the same conclusion as with USY6 versus USY30 study [52]. In addition, the Sn-β zeolite catalyst performed more active per Sn site than the homogeneous halides: the initial turnover rate of methyl lactate formation was 45 mol mol^{-1}Sn h^{-1}, whereas a value of 4.2 mol mol^{-1}Sn h^{-1} was calculated for homogeneous Sn^{IV}Cl$_4$·5H$_2$O catalyst.

Other heterogeneous catalysts

Several researchers investigated conversion of torioses to lactic acid under mild conditions with heterogeneous catalysts [55]. Pescarmona et al. screened widely heterogeneous catalysis with a high-throughput equipment [56]. They focused on different aluminosilicates with various types and strengths of both Brønsted and Lewis acidity, since that appeared critical in the performance of H-USY zeolites. An amorphous silica-alumina (entry 7, Table 4.4) presented very selective results at low temperature, but insufficient activity.

Li et al. reported the triose reaction over substituted MCM-41 materials and Sn-MCM-41 appeared to be very selective to ethyl lactate and high catalytic activity (entries 8, Table 4.4) [57]. For Sn-MCM-41 the initial turnover was around 8 mol mol^{-1}Sn h^{-1}. The catalytic advantage of Sn-MCM-41 was ascribed to a combination of Lewis acid and mild Brønsted acid sites. The authors demonstrated the rate-accelerating effect of Brønsted acidity on the formation of lactate esters, suggesting dehydration to be rate limiting with Sn-MCM-41, because the presence of Brønsted acidity in Sn-MCM-41 is somehow correlated to the presence of Sn and Sn-free MCM-41 did not exhibit such high acidity. However, such bifunctional working hypothesis of Sn-MCM-41 was still not clear because of co-existence of both acid sites.

Onaka et al. reported that Sn-exchanged montmorillonite catalyst showed very high ethyl lactate yields for DHA conversion in ethanol at 150 °C, despite the strong Brønsted acidity of the catalyst (entry 9, Table 4.4). This example nicely illustrates the aforementioned advantageous effect of a high reaction temperature, since the yield is much lower at 120 °C (entry 10, Table 4.4) [58].

Table 4.5 Overview of total released CO_x content and the initial dihydroxyacetone (DHA) conversion rate over catalysts

	Catalyst	Total content CO_x^a	Initial DHA conversion rate[b]
	mmol g^{-1}	mmol g^{-1} h^{-1}	
1	Sn-Si-MCM-41	–	22
2	Sn-Si-CSM-773-11.9	0.25	36
3	Sn-Si-CSM-773-16.4	0.38	54
4	Sn-Si-CSM-773-24.3	0.55	65
5	Sn-Si-CSM-1073-15.7	0.24	39
6	Sn-Si-CSM-773-15.3/O$_2$ 473	0.76	68
7	Sn-Si-CSM-773-14.9/O$_2$ 573	1.02	108

Reprinted with the permission from Ref. [59]. Copyright 2012 American Chemical Society
[a] The values for "Total content CO_x" were obtained from the TPD-MS experiment in heating from 303 to 1073 K at 10 K min^{-1}, [b] the initial conversion rates were typically calculated from the kinetic profiles, i.e. relationships between conversion and reaction time

The bifunctional catalytic mechanism for alkyl lactate synthesis from trioses has recently been revealed by de Clippel and Dusselier et al. [59]. They used Sn grafted carbon–silica composite catalysts to alter the number of Brønsted acid sites independently from that of the Lewis acid sites. Lewis acidity was provided by grafting a mesoporous silica like MCM-41 with isolated Sn^{IV}. The oxygen containing functional groups like carboxylic acids and phenols functioned as weak Brønsted acid sites. The density of the acid sites was controlled by the carbon loading, the degree of oxidation (depending on the pyrolysis temperature) and an optional post-synthesis oxidation. For DHA conversion to ethyl lactate at 90 °C in ethanol, the carbon-free pure siliceous Sn-grafted MCM-41 catalyst (entry 11 in Table 4.4) showed high ethyl lactate selectivity but low catalytic activity on Sn basis (TOF = 41 mol mol^{-1}Sn h^{-1}). In contrast, introduction of the weak Brønsted acidic carbon in the mesopores seriously boosted the catalyst activity without compromising the excellent ethyl lactate selectivity (TOF = 289 mol mol^{-1}Sn h^{-1}). The best result was obtained with a composite containing 15 % of carbon, which was pyrolysed at 500 °C and post-treated at 300 °C under oxygen flow (entry 12 in Table 4.4). Table 4.5 shows the initial DHA conversion rate against the total amount of COx released by heating the composite catalyst. The amount of released COx was estimated to be the total weak Brønsted acidity. Both the amount of released COx and the initial conversion rate correlated linearly, pointing for the accelerating effect of the estimated weak Brønsted acidity on the Lewis acid catalyzed conversion of DHA. From these results, the authors suggested that the bifunctional desige was focused as to maximize alkyl lactate and lactic acid formation rate by balancing the number of weak Brønsted acid sites for a given content of Sn, whereas the strong Brønsted acidity triggered unwanted competitive reactions.

4.5.2.2 Lactate Esters from Hexoses

The direct conversion of glucose, fructose, and sucrose to methyl lactate in methanol at 160 °C in an autoclave reactor was reported by Holm et al. [60]. Brønsted acidic zeolite H-Al-β catalyst gave almost no yield of methyl lactate and catalyzes the dehydration of the sugars, leading to HMF derivatives and methyl levulinate from fructose and predominantly methyl- d-pyranoside from glucose and sucrose, which were in agreement with early reports by Rivalier et al. [61]. On the other hand, the Lewis acidic Sn-β zeolite catalysts were found to induce high selectivity toward methyl lactate of 43, 44, and 64 % yields from glucose, fructose, and sucrose, respectively. Nanocrystalline SnO_2 is inactive, whereas $SnCl_4$ shows moderate selectivity toward methyl lactate of 31 % yields from sucrose. The nonacidic Si-β zeolite did not improve the yield of methyl lactate (5 and 6 % yield from glucose and sucrose, respectively), indicating that the catalytic ability is related to the Lewis acidity of Sn ions incorporated into the zeolite structure. Roman-Leshkov and Davis reported that the Sn-β catalyst is unique among its fellow Zr, Ti, Si, and H–Al-β zeolites for this sugar conversion [62]. The capability of Sn ions to activate carbonyls and to shift a hydride from one carbon to the adjacent one is not only useful in the internal Cannizzaro reaction of the hemiacetal of pyruvic aldehyde to lactic acid, but also in the ketose-to-aldose isomerisation of sugars in water.

de Clippel and Dusselier et al., for instance, also tested their bifunctional Sn-based carbon-silica composite materials for the conversion of hexoses [59]. Low fructose and glucose working concentrations are indeed important as to avoid competitive side reactions such as methylation to stable methyl sugars and dehydration ultimately to insoluble humins. An excess amount of Brønsted acids would be not desired. With regard to the reaction mechanism, the one-pot conversion of hexoses into lactate esters may be regarded in part as a synthetic glycolysis including an isomerisation of glucose to fructose and the splitting into trioses, and in part as a synthetic glyoxalase system including formation of lactic acid from pyruvic aldehyde, followed by esterification of lactic acid with methanol or ethanol.

4.6 Lactic Acid from Cellulose

Lactic acid productions for oligosaccharide or a polysaccharide, except for sucrose, require the saccharization process in many cases of both of fermentation and chemical methods.

Currently, some examinations which obtained lactic acid from cellulose directly were also investigated. Although cellulose saccharization methods in previously chemical routes were the acidic hydrolysis using mineral acids and solid acid catalysts and the hydrothermal depolymerization under subcritical water higher than 250 °C and supercritical water, there is a study example of the lactic acid productions by the combination of the latter depolymerization and basic-catalyzed glucose conversions and the direct lactic acid generation by solid acid catalysts.

Jin et al. reported that the hydrothermal reaction without catalyst at 300 °C gave lactic acid from not only glucose but also cellulose [22]. They also reported that the hydrothermal reactions with NaOH or Ca(OH)$_2$ gave lactic acid of 20 % yields and further additions of Zn, Ni, and activated carbon powders in the hydrothermal reactions with NaOH increased the lactic acid yield to be 42 % from cellulose at 300 °C [24].

Dos Santos et al. reported that microcrystalline cellulose was treated under hydrothermal conditions with dibutyl tin dilaurate, which resulted in yielding 11 % lactic acid and converting 24 % cellulose at 190 °C for 8 h [63].

Chambon et al. showed that solid Lewis acid catalysts, such as tungstated zirconia and tungstated alumina, exhibited a remarkable promoting effect on the cellulose depolymerisation and gave a 27 % LA yield at 190 °C for 24 h [64]. Liu et al. used a solid base catalyst MgO to achieve similar yields in methanol at 200 °C for 20 h [65].

4.7 Summary and Future Plan

Although lactic acid has a great possibility as a future platform compound, the production process has many subjects. Recently, various examinations were carried out not only by fermentation but by the chemical method using a solid catalyst. In this chapter, lactic acid and lactate ester productions from sugars are explained focusing on recent chemical processes using the solid catalysts.

In the lactic acid productions from triose, lactic acid ester is obtained with high yields of nearly 100 % in alcohol solvents. In the lactic acid production from hexose, lactic acid ester or a lactate salts was obtained from glucose and fructose with the comparatively high selectivity of about 50 % by heterogeneous or homogeneous basic catalysts in warm water and by Lewis acid catalysts in alcohol solvents.

The further improvement in the lactic acid yield from glucose is desired for future sustainable societies. Moreover, as for the selective formations of intermediate compounds from sugars, the developments of effective solid catalysts are also desired. Probably, for these realizations, it will be required to clear in detail of each elementary step and reaction conditions, like the glycolytic system by bacterial methods. Moreover, exploitation of a catalyst process which obtains lactic acid from polysaccharides, such as cellulose, directly is also desired.

References

1. Huber GW, Iborra S, Corma A (2006) Synthesis of transportation fuels from biomass: chemistry, catalysts, and engineering. Chem Rev 106:4044–4098
2. Danner H, Braun R (1999) Biotechnology for the production of commodity chemicals from biomass. Chem Soc Rev 28:395–405
3. Corma A, Iborra S, Velty A (2007) Chemical routes for the transformation of biomass into chemicals. Chem Rev 107:2411–2502
4. Conn EE, Stumpf PK, Bruening G (1987) Doi RH Outlines of biochemistry, 5th edn. Wiley, New York
5. Oshiro M, Hanada K, Tashiro Y, Sonomoto K (2010) Efficient conversion of lactic acid to butanol with pH-stat continuous lactic acid and glucose feeding method by clostridium saccharoperbutylacetonicum. Appl Microbiol Biotechnol 87:1177–1185
6. Tashiro Y, Kaneko W, Sun Y, Shibata K, Inokuma K, Zendo T, Sonomoto K (2011) Continuous d-lactic acid production by a novel thermotolerant Lactobacillus delbrueckii subsp lactis QU 4. Appl Microbiol Biotechnol 89:1741–1750
7. Abadel-Rahman MA, Tashiro Y, Sonomoto K (2011) Lactic acid production from lignocellulose-derived sugars using lactic acid bacteria: overview and limits. J Biotechnol 156:286–301
8. Adnan AFM, Tan IKP (2007) Isolation of lactic acid bacteria from Malaysian foods and assessment of the isolates for industrial potential. Bioresour Technol 98:1380–1385
9. John RP, Nampoothiri KM, Pandey A (2007) Fermentative production of lactic acid from biomass: an overview on process developments and future perspectives. Appl Microbiol Biotechnol 74:524–534
10. Hofvendahl K, Hahn-Hägerdal B (2000) Factors affecting the fermentative lactic acid production from renewable resources. Enzyme Microb Technol 26:87–107
11. Ilmen M, Koivuranta K, Ruohonen L, Suominen P, Penttila M (2007) Efficient production of l-lactic acid from xylose by Pichia stipites. Appl Environ Microbiol 73:117–123
12. Pandey A, Soccol CR, Rodriguez-Leon JA, Nigam P (2001) Production of organic acids by solid state fermentation. In: solid state fermentation in biotechnology: fundamentals and applications, Asiatech Publishers, New Delhi
13. Van de Vyver S, Geboers J, Jacobs PA, Sels BF (2011) Recent advances in the catalytic conversion of cellulose. ChemCatChem 3:82–94
14. Rinaldi R, Schüth F (2009) Acid hydrolysis of cellulose as the entry point into biorefinery schemes. ChemSusChem 2:1096–1107
15. Singhvi M, Joshi D, Adsul M, Varma A, Gokhale D (2010) d-(−)-Lactic acid production from cellobiose and cellulose by Lactobacillus lactis mutant RM2-24. Green Chem 12:1106–1109
16. Onda A, Ochi T, Yanagisawa K (2008) Selective Hydrolysis of cellulose into glucose over solid acid catalysts. Green Chem 10:1033–1037
17. Datta R, Henry M (2006) Lactic acid: recent advances in products, processes and technologies—a review. J Chem Technol Biotechnol 81:1119–1129
18. Okano K, Zhang Q, Yoshida S, Tanaka T, Ogino C, Fukuda H, Kondo A (2010) d-Lactic acid production from cellooligosaccharides and β-glucan using genetically modified l-lactate dehydrogenase gene-deficient and endoglucanase-secreting Lactobacillus plantarum. Appl Microbiol Biotechnol 85:643–650
19. Lunt J (1998) Large-scale production, properties and commercial applications of polylactic acid polymers. Polym Degrad Stab 59:145–152
20. de Bruijn JM, Kieboom APG, van Bekkum H (1986) Alkaline degradation of monosaccharides III. Recl Trav Chim Pays-Bas 105:176–183
21. Bicker M, Endres S, Vogel LO (2005) Catalytic conversion of carbohydrates in subcritical water: a new chemical process for lactic acid production. J Mol Catal A: Chem 239:151–157

22. Jin F, Zhou Z, Enomoto H, Moriya T, Higashijima H (2004) Conversion mechanism of cellulosic biomass to lactic acid in subcritical water and acid–base catalytic effect of subcritical water. Chem Lett 33:126–127
23. de Bruijn JM, Kieboom APG, van Bekkum H (1987) Alkaline degradation of monosaccharides Part VII. A mechanistic picture. Starch/Staerke 39(1):23–28
24. Jin F, Enomoto H (2011) Rapid and highly selective conversion of biomass into value-added products in hydrothermal conditions: chemistry of acid/base-catalysed and oxidation reactions. Energy Environ Sci 4:382–397
25. Yan X, Jin F, Tohji K, Moriya T, Enomoto H (2007) Production of lactic acid from glucose by alkaline hydrothermal reaction. J Mater Sci 42:9995–9999
26. Yan X, Jin F, Tohji K, Kishita A, Enomoto H (2010) Hydrothermal conversion of carbohydrate biomass to lactic acid. AIChE J 56:2727–2733
27. de Bruijn JM, Kieboom APG, van Bekkum H, van der Poel PW (1986) Reactions of monosaccharides in aqueous alkaline solutions. Sugar Technol Rev 13:21–52
28. Yang BY, Montgomery (1996) Alkaline degradation of glucose: effect of initial concentration of reactants. Carbohydr Res 280: 27–45
29. Theander O (1988) Aqueous, high-temperature transformation of carbohydrates relative to utilization of biomass. Adv Carbohydr Chem Biochem 46:273–326
30. Ellis AV, Wilson MA (2002) Carbon exchange in hot alkaline degradation of glucose. J Org Chem 67:8469–8474
31. Rao KK, Gravelle M, Valente JS, Figueras F (1998) Activation of Mg–Al hydrotalcite catalysts for aldol condensation reactions. J Catal 173:115–121
32. Climent MJ, Corma A, Iborra S, Velty A (2004) Activated hydrotalcites as catalysts for the synthesis of chalcones of pharmaceutical interest. J Catal 221:474–482
33. Roelofs JCAA, Lensveld DJ, van Dillen AJ, de Jong KP (2001) On the structure of activated hydrotalcites as solid base catalysts for liquid-phase aldol condensation. J Catal 203:184–191
34. Onda A, Ochi T, Kajiyoshi K, Yanagisawa K (2008) Lactic acid production from glucose over activated hydrotalcites as solid base catalysts in water. Catal Comm 9:1050–1053
35. Miyata S (1980) Physico-chemical properties of synthetic hydrotalcites in relation to composition. Clays Clay Miner 28:50–56
36. Suzuki E, Ono Y (1988) Aldol condensation reaction between formaldehyde and acetone over heat-treated synthetic hydrotalcite and hydrotalcite-like compounds. Bull Chem Soc Jpn 61:1008–1010
37. Pesic L, Salipurovic S, Markovic V, Vucelic D, Kagunya W, Jones W (1992) Thermal characteristics of a synthetic hydrotalcite-like material. J Mater Chem 2:1069–1073
38. Prescott HA, Li ZJ, Kemnitz E, Trunschke A, Deutsch J, Lieske H, Auroux A (2005) Application of calcined Mg–Al hydrotalcites for Michael additions: an investigation of catalytic activity and acid–base properties. J Catal 234:119–130
39. Winter F, Xia X, Hereijgers BPC, Bitter JH, van Dillen AJ, Muhler M, de Jong KP (2006) On the nature and accessibility of the Brønsted-base sites in activated hydrotalcite catalysts. J Phys Chem B 110:9211–9218
40. Onda A, Ochi T, Kajiyoshi K, Yanagisawa K (2008) A new chemical process for catalytic conversion of d-glucose into lactic acid and gluconic acid. Appl Catal A 343:49–54
41. de Wit G, de Vlieger JJ, Kock-van Dalen AC, Heus R, Laroy R, van Hengstum AJ, Kieboom APG, van Bekkum H (1981) Catalytic dehydrogenation of reducing sugars in alkaline solution. Carbohydr Res 91:125–138
42. Abbadi A, van Bekkum H (1995) Effect of pH in the Pt-catalyzed oxidation of d-glucose to d-gluconic acid. J Mol Catal A 97:111–118
43. Besson M, Lahmer F, Gallezot P, Fuertes P, Flèche G (1995) Catalytic oxidation of glucose on Bismuth-promoted palladium catalysts. J Catal 152:116–121
44. Wenkin M, Touillaux R, Ruiz P, Delmon B, Devillers M (1996) Influence of metallic precursors on the properties of carbon-supported bismuth-promoted palladium catalysts for the selective oxidation of glucose to gluconic acid. Appl Catal A 148:181–199

45. Biella S, Prati L, Rossi M (2002) Selective oxidation of D-glucose on gold catalyst. J Catal 206:242–247
46. Mirescu A, Prüße U (2006) Selective glucose oxidation on gold colloids. Catal Comm 7:11–17
47. Raharja S, Rigal L, Gaset A, Barre L, Chornet E, Videl, PF (1994) Biomass for energy, environment, agriculture and industry. Proceeding of the eighth European biomass conference vol 2. pp 1420–1427
48. Raharja S, Rigal L, Videl PF (1997) Alkaline oxidation of sugar: thermochemical conversion of xylose from hemicellulose into lactic acid. Dev thermochem biomass convers 1:773–782
49. Mirescu A, Prüße U (2007) A new environmental friendly method for the preparation of sugar acids via catalytic oxidation on gold catalysts. Appl Catal B 70:644–652
50. Janssen KPF, Paul JS, Sels BF, Jacobs PA (2007) Glyoxylase biomimics: zeolite catalyzed conversion of trioses. Stud Surf Sci Catal 170:1222–1227
51. Hayashi Y, Sasaki Y (2005) Tin-catalyzed conversion of trioses to alkyl lactates in alcohol solution. Chem Comm 21: 2716–2718
52. West RM, Holm MS, Saravanamurugan S, Xiong J, Beversdorf Z, Taarning E, Christensen CH (2010) Zeolite H-USY for the production of lactic acid and methyl lactate from C_3-sugars. J Catal 269:122–130
53. Rasrendra CB, Fachri BA, Makertihartha IBGN, Adisasmito S, Heeres HJ (2011) Catalytic conversion of dihydroxyacetone to lactic acid using metal salts in water. ChemSusChem 4:768–777
54. Taarning E, Saravanamurugan S, Spangsberg HM, Xiong J, West RM, Christensen CH (2009) Zeolite-catalyzed isomerization of triose sugars. ChemSusChem 2:625–627
55. Dusselier M, Van Wouwe P, Dewaele A, Makshina E, Sels BF (2013) Lactic acid as a platform chemical in the biobased economy: the role of chemocatalysis. Energy Environ Sci 6:1415–1442
56. Pescarmona PP, Janssen KPF, Stroobants C, Molle B, Paul JS, Jacobs PA, Sels BF (2010) A high-throughput experimentation study of the synthesis of lactates with solid acid catalysts. Top Catal 53:77–85
57. Li L, Stroobants C, Lin K, Jacobs KA, Sels BF, Pescarmona PP (2011) Selective conversion of trioses to lactates over Lewis acid heterogeneous catalysts. Green Chem 13:1175–1181
58. Wang J, Masui Y, Onaka M (2011) Conversion of triose sugars with alcohols to alkyl lactates catalyzed by Brønsted acid tin ion-exchanged montmorillonite. Appl Catal B 107:135–139
59. de Clippel F, Dusselier M, Van Rompaey R, Vanelderen P, Dijkmans J, Makshina E, Giebeler L, Oswald S, Baron GV, Denayer JFM, Pescarmona PP, Jacobs PA, Sels BF (2012) Fast and selective sugar conversion to alkyl lactate and lactic acid with bifunctional carbon-silica catalysts. J Am Chem Soc 134:10089–10101
60. Holm MS, Saravanamurugan S, Taarning E (2010) Conversion of sugars to lactic acid derivatives using heterogeneous zeotype catalysts. Science 328:602–605
61. Rivalier P, Duhamet J, Moreau C, Durand R (1995) Development of a continuous catalytic heterogeneous column reactor with simultaneous extraction of an intermediate product by an organic solvent circulating in countercurrent manner with the aqueous phase. Catal Today 24:165–171
62. Roman-Leshkov Y, Davis ME (2011) Activation of carbonyl-containing molecules with solid Lewis acids in aqueous media. ACS Catal 1:1566–1580
63. dos Santos JB, da Silva FL, Altino FMRS, da Silva Moreira T, Meneghetti MR, Meneghetti SMP (2013) Cellulose conversion in the presence of catalysts based on Sn(IV). Catal Sci Technol 3:673–678
64. Chambon F, Rataboul F, Pinel C, Cabiac A, Guillon E, Essayem N (2011) Cellulose hydrothermal conversion promoted by heterogeneous Brønsted and Lewis acids: remarkable efficiency of solid Lewis acids to produce lactic acid. Appl Catal B 105:171–181
65. Liu Z, Li W, Pan C, Chen P, Lou H, Zheng X (2011) Conversion of biomass-derived carbohydrates to methyl lactate using solid base catalysts. Catal Comm 15:82–87

Chapter 5
Catalytic Conversion of Lignocellulosic Biomass to Value-Added Organic Acids in Aqueous Media

Hongfei Lin, Ji Su, Ying Liu and Lisha Yang

Abstract The transition from today's fossil-based economy to a sustainable economy based on renewable biomass is driven by the concern of climate change and anticipation of dwindling fossil resources. Although biofuels are the central theme of the transition, biomass resources cannot completely replace petroleum. It is projected that biofuels can only supply up to 30 % of today's transportation fuel market even if all available domestic biomass resources are used for the production of liquid fuels. Therefore, transformation of biomass into high-value-added chemicals is advantageous to secure optimal use of the abundant, but limited, biomass resources from the economical and ecological perspective. Industry is increasingly considering bio-based chemical production as an attractive area for investment. The potential for chemical and polymer production from biomass is substantial. The US Department of Energy recently issued a report which listed 12 chemical building blocks considered as potential building blocks for the future. Organic acids (e.g., succinic, lactic, levulinic acid, etc.) are among the widely spread "platform-molecules," which may be further converted into possibly derivable high-value-added chemicals. The transition from a fossil chemical industry to a renewable chemical industry will likewise depend on our ability to focus research and development efforts on the most promising alternatives. In this chapter, we review the emerging technologies on catalytic conversion of biomass to value-added organic acids in aqueous media.

H. Lin (✉) · J. Su · Y. Liu · L. Yang
Department of Chemical and Materials Engineering, University of Nevada, Reno, NV 89557, USA
e-mail: HongfeiL@unr.edu

5.1 Introduction

Today, the world is facing environmental and economic issues caused by our heavy dependence on fossil fuels for energy and chemical production. Biomass has been considered to be a renewable feedstock that has the potential to replace fossil fuels and related petrochemicals [40, 45]. Lignocellulosic materials consist of cross-linked biopolymers including cellulose (35–50 %), hemicelluloses (25–30 %), and lignin (25–30 %), which compose the bulk of all higher plants and have evolved complex structural and chemical properties providing microbial and chemical resistance. In the context of the biorefinery, these chemical and structural barriers to deconstruction must be overcome to reduce the conversion costs. One of the desirable routes for the biorefining is to selectively convert lignocellulosic biomass into value-added chemicals or into platform compounds, which can be easily converted to versatile chemicals or fuels in the subsequent step [39].

In general, the reaction solvents used for biomass conversion can be summarized into three groups: organic solvents [57, 78, 116, 122], ionic liquids [21, 114, 163], and water [7, 22, 95, 109, 135, 140]. Among these solvents, organic solvents require significant waste management. Ionic liquids are quite expensive at present and need costly separation. In contrast, water is a convenient solvent for biomass deconstruction because the properties of water, such as dielectric constant, density, and ionic product, can be adjusted with the temperature and pressure. There are numerous studies [11, 151, 25, 51, 60, 64, 68, 70, 71, 97, 98, 104, 124, 149, 152, 151] on the biomass conversion in hot temperature water (HTW) and supercritical water (SCW). HTW is broadly defined to be subcritical liquid water (150 °C < T < 374 °C), and SCW is water at temperatures and pressures above the critical point (T > 374 °C and P > 221 bar) [3, 120]. Especially, HTW becomes less dense, and the dielectric constant and the level of hydrogen bonding decrease in the temperature range of 180–300 °C compared to water at room temperature. As a result, HTW behaves like many organic solvents with a high solubility for biomass-derived molecules. In addition, HTW becomes a stronger electrolyte by dissociation, resulting in higher concentrations of H^+ and OH^- that can catalyze chemical reactions [3].

Biocatalytic as well as chemocatalytic methods have been applied for the conversion of biomass molecules [101]. The thermal instability of carbohydrates is a major problem, and thus biocatalysis operated at lower temperatures has proven to be applicable. However, the biological processes have disadvantages such as sluggish kinetics. On the other hand, chemical catalysis often presents improved process design options, resulting in higher productivity and lower costs. Indeed, chemocatalysis has proven to be scalable and can supply numerous low-cost products from petroleum [55]. Acid/base catalysis is the most widely used way to convert biomass into useful organic acids. For example, mineral acids can remarkably promote hydrolysis and dehydration reactions and yield levulinic acid from cellulose. On the other hand, isomerization and retro-aldol condensation reactions are favorable in basic environment, which can generate low molecular weight organic acid products like lactic acid, formic acid, acetic acid, etc. [13, 42, 101, 155]. Besides acid/base catalysis, catalytic oxidation can also convert biomass molecules to organic acids [6, 15, 27, 34, 58, 59, 61, 72–74, 80, 85, 103, 156, 159]. Herein we summarize the current emerging technologies of catalytic conversion of lignocellulosic biomass to organic acids.

5.2 Carbohydrates to Carboxylic Acids

There are two types of carbohydrates, C5 and C6 sugars, which account for 65–85 % of lignocellulosic biomass in two forms of biopolymers, cellulose and hemicellulose. Cellulose is a linear homopolysaccharide of D-glucose (C6 sugar) units linked through β-(1, 4) glycosidic linkages. Hemicellulose is any of several heteropolymers (matrix polysaccharides), such as arabinoxylans, present along with cellulose in almost all plant cell walls with a random, amorphous structure [121]. Xylose, the main building block of xylan, comprises about 25 % of the dry weight biomass of some plant species such as hardwoods and agricultural residues [75]. Hydrolysis of carbohydrate biomass followed by sugar degradation can produce organic acids by an acid-catalyzed hydrolysis reaction [96]. Recently, our group has demonstrated that aqueous phase partial oxidation is an alternative pathway that directly converts cellulose to carboxylic acids [85].

5.2.1 Levulinic Acid

Levulinic acid, also known as 4-oxopentanoic acid or γ-ketovaleric acid, is a short-chain fatty acid having a ketone carbonyl group and an acidic carboxylic group. Although levulinic acid has been known for almost 150 years [79], its current status is still an expensive specialty chemical in limited volume. However, levulinic acid is a versatile platform chemical with numerous potential uses, which can be used as textile dye, antifreeze, animal feed, coating material, solvent, food flavoring agent, pharmaceutical compounds, and hydrocarbon fuels [16, 19, 20, 28, 77, 79, 105, 123]. Biomass feed stocks have been used to synthesize levulinic acid and the most widely used approach is the controlled dehydrative treatment of carbohydrates with acid catalyst. Many sugar materials such as glucose, sucrose, fructose, and cellulosic biomass materials including wood, starch, cane sugar, grain sorghum, and agricultural wastes have been used to produce levulinic acid [8, 17, 19, 23, 29, 46–48, 50, 53, 56, 65, 76, 77, 107, 160, 108, 112, 127, 134, 136, 139, 153, 155, 160].

5.2.1.1 Homogeneous Acid Catalysis

Beginning with Saeman in 1945 [117], cellulose hydrolysis has been wildly used as the following equation.

$$\text{Cellulose} \xrightarrow{K_{hyd}} \text{Glucose} \xrightarrow{K_{deh}} \text{HMF} \xrightarrow{K_{reh}} \text{LevulinicAcid} + \text{FormicAcid}$$

Basic chemical transformations that enable this conversion are hydrolysis of cellulose to carbohydrates and subsequent selective dehydration and rehydration to lead to equal moles of levulinic acid and formic acid [96]. The most widely used reaction pathway for the acid-catalyzed hydrolysis of cellulose to levulinic acid by mineral acid is given in Fig. 5.1. In the first step, the polymer chains of cellulose are broken down into lower molecular weight fragments and ultimately to glucose by the acid catalyst. The glucose is then dehydrated to HMF, which is further rehydrated to form equal moles of levulinic and formic acid. The highest levulinic acid yield about ~80 mol% can be obtained using mineral acid as catalyst in 150–240 °C. Besides the anticipated products, small amounts of glucose reversion products such as isomaltose, levoglucosan, gentiobiose, and furfural were detected in the aqueous phase, but the maximum amount of glucose-reversion products was very low [50].

The production of levulinic acid is dependent on both the feedstock and processing conditions. Feed stocks range from sugars (glucose, fructose, sucrose), starch, cellulose, sawdust, woods, paper and pulp, and various agricultural residues (bagasse, rice hulls, corn stalks, sorghum grains, cotton stems, wheat and rice straw) [49]. Acid concentration, temperature, solvent concentration, and residence time are four main processing conditions that affect levulinic acid yields [112]. There exists a critical acid concentration limit found to vary between 3.5 and 10 wt% depending on other processing conditions and the feedstock [31, 50, 115, 129]. The levulinic yield increases with acid concentration until reaching the critical value. On the other hand, too high acid concentration leads to side reactions and repolymerization of products and intermediates. Temperature has a strong effect on the yield of levulinic acid. The optimum temperature was found to be 200–220 °C. [31, 46, 47, 160]. The amount of water solvent can affect the hydrolysis process and the optimal concentration was found to be ~90 wt%. Lastly, the effect of residence time was found to be dependent on the cellulose content of the biomass feedstock [146]. Recently, Dumesic and co-workers use gamma-valerolactone (GVL) as cosolvent and 0.2 M H_2SO_4 as catalyst, the cellulosic fraction of lignocellulosic biomass can be converted into levulinic acid (LA) up to mole yield of 68 % at 170 °C [5].

The drawbacks of homogenous acid-catalyzed processes are the high corrosivity requiring special materials for reactor construction and costly mineral acid recovery. Also significant amounts of waste are produced raising the environmental concern.

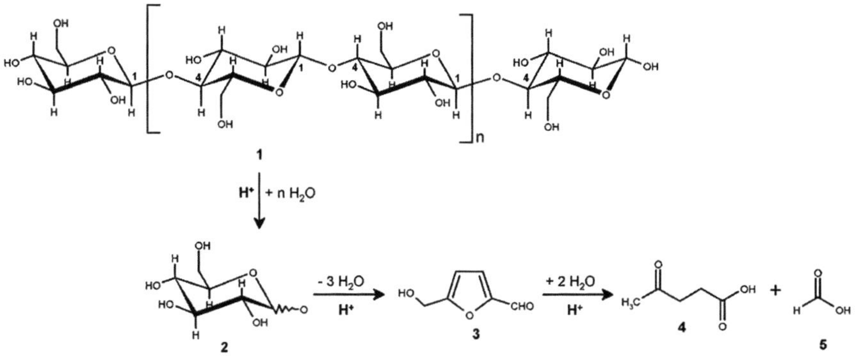

Fig. 5.1 Acid-catalyzed hydrolysis reaction of cellulose to levulinic acid. (Reprinted with the permission from Ref. [50]. Copyright 2007 American Chemical Society)

5.2.1.2 Heterogeneous Catalysts

Heterogeneous catalysts have the inherent advantage of easy recovery but are not commonly used for direct conversion of cellulose to levulinic acid due to sluggish solid–solid mass transfer. In contrast, reactions of water-soluble sugars with solid acid catalysts have been extensively studied [8, 14, 65, 78, 87, 88, 115]. In general, the heterogeneous catalytic system could convert water-soluble sugars to organic acids by shape-selective, solid acid-catalyzed processes at low temperatures. Lourvanij and Rorrer [87] reacted aqueous solutions of glucose (12 wt%) with HY-zeolite powder in a well-mixed batch reactor at temperatures ranging from 110 to 160 °C and catalyst concentrations ranging from 2 to 20 g/150 mL. Under these conditions, 15–20 % molar yields of levulinic acid and ∼30 % formic acid were obtained. Since the reaction stoichiometry of HMF rehydration requires equimolar production of formic acid and levulinic acid, they ascribed the difference to be the reactions of levulinic acid within the Y-zeolite and suggested that molecular-sieving acid catalysts which have pore sizes slightly larger than the glucose molecule (i.e., >10 Å) might promote glucose conversion. However, their successive studies using large pore-size pillared montmorillonite catalysts again only promoted the formation of formic acid (up to 56 mol% at 150 °C) but not the levulinic acid yield [88]. When using fructose as the feedstock, a molar yield of 67 % of levulinic acid was obtained after 15 h over a LZY zeolite (Y-type Faujasite) catalyst at 140 °C [65], which is significantly higher than that with glucose as the feedstock.

Wang et al. studied the production of levulinic acid directly from cellulose with sulfated TiO_2 as a solid acid catalyst and obtained a levulinic acid molar carbon yield of up to 38 % [83]. Chen and co-workers described the use of a solid super acid ($S_2O_8^{2-}/ZrO_2$–SiO_2–Sm_2O_3) to produce levulinic acid from rice straw [32]. At their optimal conditions they obtained a molar yield of 70 % from the cellulose portion of the steam-exploded rice straw. The catalyst stability was found to be a concern: a 50 and 67 % loss of activity following the second and third recycling use of the catalyst, respectively. Van de Vyver et al. showed that levulinic acid can be produced directly from cellulose at a ~30 % molar yield at 190 °C using sulfonated high surface area polymer as the solid acid catalyst [139]. Similarly, Lai et al. used sulfonated mesoporous silica modified with magnetic iron oxide particles (Fe_3O_4–SBA–SO_3H) to produce levulinic acid from cellulose and a molar yield of 45 % was obtained after 12 h at 150 °C. The catalyst was able to be separated and recycled after reaction when a magnetic field was applied. More sulfonated solid acid catalysts include acidic ion-exchange polymer resins such as Nafion and Amberlyst. Lucht and co-workers found relatively low levulinic acid yield of 5 % from cellulose with Nafion SAC-13 [53]. They further studied the effect of NaCl on these reactions with Nafion SAC-13 as the catalyst and found the yield of levulinic acid increased to 72 % upon incorporation of 25 % NaCl solution. They ascribed the dramatic increase in yield to sodium chloride, which is effective in interrupting the hydrogen bonding network at high temperatures and pressures [110]. Weingarten et al. have investigated the ZrP catalyst with a high Brønsted to Lewis acid ratio and achieved 16.8 % levulinic acid yield at 220 °C. They found that ZrP showed a predominantly higher HMF selectivity and a lower levulinic acid selectivity compared to Amberlyst 70, which could be due to the presence of Lewis acid sites on ZrP [142].

Besides solid acid catalysts, TiO_2 and ZrO_2 were commonly used in carbohydrate conversion due to their excellent hydrothermal stability. Qi et al. studied the production of 5-HMF from glucose and fructose catalyzed by TiO_2 and ZrO_2 under microwave irradiation [111]. They found that the addition of TiO_2 and ZrO_2 increased the conversion of fructose and glucose. In the reactions, ZrO_2 acted as base, and could promote the isomerization of glucose to fructose, whereas TiO_2 could not only enhance the 5-HMF yield from glucose, but also promote the isomerization of glucose to fructose. Both TiO_2 and ZrO_2 promoted the formation of 5-HMF from fructose and could effectively suppress the rehydration of 5-HMF into levulinic acid and formic acid. Figure 5.2 shows the acid/base bifunctionality of TiO_2 and ZrO_2.

Our group recently presents an alternative pathway, aqueous phase partial oxidation (APPO), to produce levulinic acid directly from cellulose with lean air and water over inexpensive solid metal oxide catalysts [85]. The maximum obtainable yield of levulinic acid in the APPO process is ~60 mol%, comparable to the yields of levulinic acid which have been achieved via acid hydrolysis. Our findings reveal that the redox properties of the APPO catalyst play a key role and that super oxide radical anions are the possible active oxidant species in aqueous media, which distinguish APPO from acid hydrolysis. As shown in Fig. 5.3, hydrogen at the C1 position is abstracted by a superoxide radical, and then molecular oxygen is inserted into the C1 position and formed a new organic radical. The organic radical is combined with a proton to form hydroperoxy radical and leaves the carbonyl bond on the C1 carbon. The carbonyl bond is hydrated and the glycosidic bond connecting the two sugar units is cleaved. And the 3,4,5-trihydroxy-6-hydroxymethyl-tetrahydro-pyran-2-one is further hydrated to form

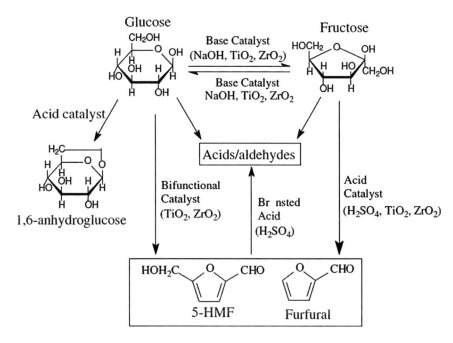

Fig. 5.2 Reaction pathways of glucose and fructose in hot-pressed water with TiO_2 and ZrO_2 catalysts. (Reprinted from Ref. [111], Copyright 2008, with permission from Elsevier)

gluconic acid, which is deoxygenated through an initial Hofer–Moest-type decarboxylation followed by a series of consecutive dehydration/rehydration reactions to form levulinic acid. The APPO process has the inherent merits of environmental friendliness, low humin formation, and ease of separation of catalyst, and thus it has the potential to replace the current acid hydrolysis process for commercial production of levulinic acid.

5.2.2 Lactic Acid

Lactic acid (2-hydroxy-propanoic acid) has both a hydroxyl group and a carboxylic acid group and can be used as a precursor to produce a wide range of chemicals. Thus it is one of the important biomass-derived organic acids. Lactic acid has recently been receiving much more attention as a commodity chemical with a growing market. It is applied in the food and pharmaceutical industries as mild acid flavor, pH regulator, or preservative. An emerging product is poly (lactic acid), PLA, used in manufacturing biodegradable plastics [13]. However, lactic acid supply is insufficient and the role of lactic acid in the future will rely on its manufacture costs [101].

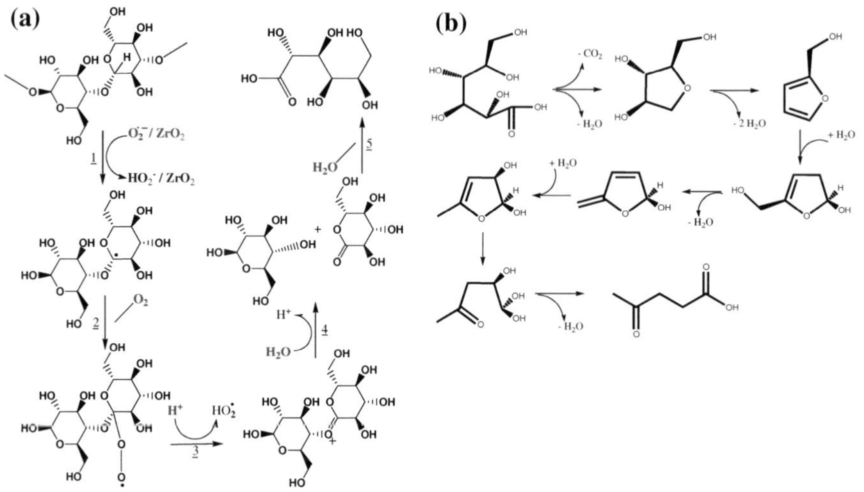

Fig. 5.3 Reaction pathway of **a** converting a cellobiose unit in cellulose to glucose and gluconic acid by superoxide radical anions. **b** Converting gluconic acid to levulinic acid by a Hofer–Moest-type decarboxylation reaction followed by consecutive dehydration/rehydration reactions. (Reproduced from Ref. [85] by permission of The Royal Society of Chemistry)

It has been on steady increase for production of lactic acid since the early 1990s all over the world. The mainstream production process of lactic acid is the fermentation of glucose from starch hydrolysis by using genetically modified enzymes [44, 63, 89, 91]. However, the fermentation process cannot directly utilize nonfood cellulosic biomass. Chemical processes that converting saccharides into lactic acids or lactates were investigated in hot water with alkali or metal salts [13]. These chemical processes are simpler and more flexible than the fermentation processes, which help to increase the lactic acid supply. However, these chemical processes use a high concentration of alkaline, operate at high reaction temperatures, and generate a large amount of by-products, all of which are undesirable.

Alkaline hydrothermal conversion of lignocellulosic biomass into lactic acid was discussed by many researchers. Yan et al. showed that both NaOH and $Ca(OH)_2$ can promote the formation of lactic acid in a hydrothermal reaction of glucose [147]. Lactic acid was obtained at a yield of 27 % at 300 °C for 1 min with the addition of NaOH. They also investigated the hydrothermal conversion of other carbohydrates including cellulose and starch to lactic acid using NaOH and $Ca(OH)_2$ as catalysts. Lactic acid obtained from cellulose and starch was almost in the same yield as that obtained from glucose under the same reaction conditions of 300 °C and 0.32 M $Ca(OH)_2$, but in the case of cellulose, a slightly longer reaction time of 1.5 min was required [147]. Ma et al. investigated alkaline hydrothermal experiments of glucose, fructose, and xylose in a batch reactor with the temperature range of 250–350 °C to examine the production of lactic acid [90]. Result showed that the formation of lactic acid could be strongly affected by the concentration of alkaline catalysts, such as NaOH, KOH, and Ca(OH)$_2$. The yield of lactic acid from glucose, fructose, and xylose followed the sequence of fructose > glucose > xylose with NaOH as alkaline catalyst. A high lactic acid yield of 42.7 % was obtained from fructose. Figure 5.4 shows the reaction mechanism of alkaline degradation of glucose into lactic acid including a reverse aldol condensation from C6 to C3 chemicals.

The use of inorganic salts as catalysts for the conversion of various sugars to lactic acid in aqueous solutions has been reported. Rasrendra et al. investigated the use of inorganic salts as catalysts for the reactions of D-glucose in aqueous solutions and found that Al^{3+} and Cr^{2+} salts gave the highest conversion of D-glucose [113]. The type of inorganic salt greatly affected the chemoselectivity. The major liquid product for the Al-salts is lactic acid with the hypothesis that the intermediate trioses like glyceraldehyde and dihydroxyacetone may be converted in high yields to lactic acid when using Al-salts. Bicker et al. found that by adding small quantities of metal ions such as Co^{2+}, Ni^{2+}, Cu^{2+}, and Zn^{2+} in subcritical water, the lactic acid yield was increased up to 42 % starting from sucrose and 86 % starting from dihydroxyacetone at 300 °C and 25 MPa. Zn^{2+} gave the best results with regard to the lactic acid yield [13]. Kong et al. [69] reported the influence of metal ions (Zn^{2+}, Ni^{2+}, Co^{2+}, and Cr^{3+}) on biomass decomposition in subcritical water (T = 300 °C). In comparison with a noncatalytic process, the lactic acid yield from glucose was 9.5 wt% for 400 ppm Co(II) catalyst at 300 °C and 2 mins. Besides metal salts, Zhang et al. [162] investigated the effect of Zn, Ni, and activated carbon on the yield of lactic acid from cellulose under alkaline hydrothermal conditions. The results showed that the yield of lactic acid can be improved greatly by adding Zn, Ni, and activated carbon. And the highest lactic acid yield was 42 % under this condition, which was much higher than that only using NaOH (15 %). Li et al. [82] also reported that the CuO can be reduced completely to Cu with cellulose as a reductant under alkaline hydrothermal condition.

The catalytic production of lactic acid from biomass-derived sugars over solid base catalysts was an effective process with low impact to the environment. Onda et al. [101] studied the activated hydrotalcites for the reaction of converting glucose to lactic acid in a flow reactor at 323 K in water media. In this study, the number of accessible Brønsted-base sites of hydrotalcite (Mg/Al = 2) is correlated to the calcination temperature, which had profound effects on the amount of Brønsted-base sites. The lactate yield was 20.3 C-% at 8 h of time-on-stream over the hydrotalcite catalyst calcined at 723 K. They also found under the same conditions, MgO and γ-Al$_2$O$_3$ catalysts showed little and almost no catalytic activity for the reaction, respectively. In contrast, Liu et al. demonstrated that MgO was an active solid base catalyst for the production of methyl lactate from glucose in methanol. The yield of methyl lactate reached 29 % under optimal reaction conditions [86]. In another interesting study, a series of Al–Zr mixed oxides with different molar ratios were examined with glucose reaction in hot-compressed water at 180 °C [155]. Al$_2$O$_3$ and ZrO$_2$ are well-known for their acid–base bifunctional property and stability. In their work, mixed oxides showed better

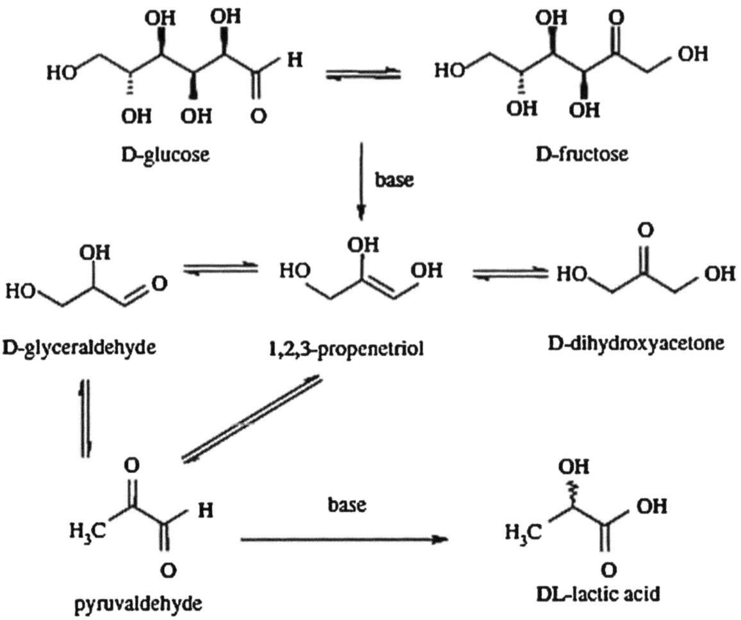

Fig. 5.4 Conversion of D-glucose into lactic acid over base catalysts. (Reprinted from Ref. [101], Copyright 2008, with permission from Elsevier)

catalytic activity to lactic acid compared to single oxide, Al–3Zr and 3Al–Zr showed high selectivity to lactic acid, ~33 %. Both moderate and strong base sites on the metal oxides are helpful for the formation of lactic acid.

Onda et al. also investigated the catalytic conversion of monosaccharides, such as D-glucose, D-fructose, D-mannose, D-galactose, and D-xylose, into lactic acid and aldonic acid in an alkaline media using supported noble metal catalysts under flowing air [102]. This one-pot reaction improved the carbon efficiency for the D-glucose conversion and decreased the by-products. It succeeded in yielding about 45 C-% of lactic acid and about 45 C-% of gluconic acid from D-glucose in via a one-pot reaction using Pt/C catalysts in NaOH aqueous solution at 353 K under flowing air. The reaction of xylose, an aldopentose, resulted in about 20 C-% and 27 C-% yields of C5 aldonic acid and lactic acid, respectively. Thus, D-Glucose, D-fructose, and D-mannose were favorable for the alkaline degradations into lactic acid as compared to D-galactose and D-xylose, which was speculated that the favorability might be due to alternate actual geometry of the third and fourth asymmetric carbons in monosaccharides.

Solid Lewis acids such as tungstated zirconia (ZrW) and tungstated alumina (AlW) have exhibited a remarkable promoting effect on the cellulose depolymerization and 27 and 18.5 mol% lactic acid yield were achieved on AlW and ZrW, respectively [30]. Moreover, tungsten-based Lewis acids exhibited a good stability and recyclability. The efficiency of the solid Lewis acids ZrW and AlW to

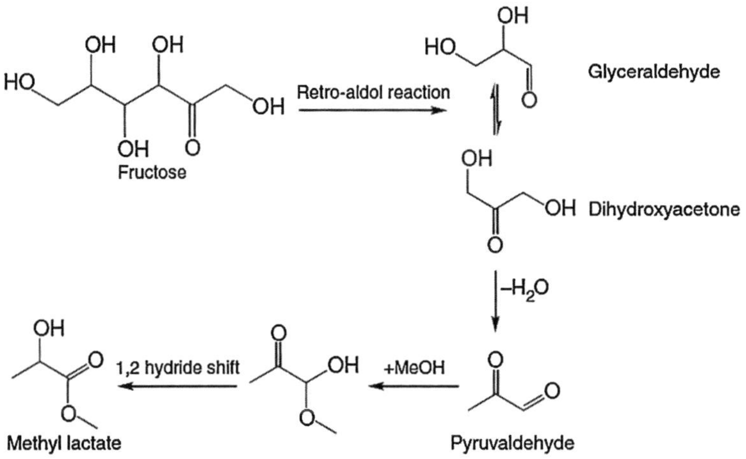

Fig. 5.5 Proposed reaction pathway for the conversion of fructose to methyl lactate. The reaction formally comprises a retro-aldol fragmentation of fructose and isomerization-esterification of the trioses. From [55]. Reprinted with permission from AAAS

produce lactic acid directly from crystalline cellulose was explained by a positive synergy between water autoprotolysis responsible of the cellulose depolymerization into soluble intermediates which are further converted on the solid Lewis catalyst surface. The catalytic conversion of mono- and disaccharides for the direct formation of methyl lactate using Lewis acidic zeotypes, such as Sn-Beta, was investigated by Taarning and co-workers [55]. In this process, resembling the alkaline degradation of sugars, sucrose, glucose, and fructose dissolved in methanol were converted into methyl lactate in yields up to 68 % at 160 °C. However, in contrast to the traditional alkaline degradation, a stoichiometric amount of base is not consumed in acid-catalyzed conversion of mono- and disaccharides. Nonetheless, the reaction pathway in the acid-catalyzed conversion is greatly sensitive to the kind of acid used: Brønsted acids catalyzed monosaccharide dehydration reactions primarily result in HMF and its decomposition products, while Lewis acidic zeotype catalysts lead to retro-aldol reaction of the monosaccharides and subsequent transformation to hydroxyl carboxylic acid derivatives. It is critical to diminish the catalytic effect of Brønsted acids to achieve a high selectivity in the Lewis acid-catalyzed process, for example, esters rather than carboxylic acids were formed by using methanol as solvent (Fig. 5.5).

5.2.3 Gluconic Acid

Gluconic acid is found naturally in fruit, honey, and wine, which can be used as a food additive like an acidity regulator, and it is also used in cleaning products where it dissolves mineral deposits especially in alkaline solution. The biodegradable gluconic acid is also used in paper and concrete production because of its good complex properties and stability against hydrolysis at high temperatures and pH values. Gluconic acid and its derivatives are produced by biochemical oxidation of glucose as the only industrial route [84]. The annual worldwide production is estimated to be 100,000 tons [133].

Except the biotechnological processes, chemical catalysis can also be used for the oxidation of carbohydrates [132]. Selective catalytic oxidation of D-glucose with heterogeneous catalyst is part of the principles of green chemistry to produce D-gluconic acid [148]. Heterogeneous catalysis shows excellent activity and selectivity for the oxidation of aldoses [10, 15, 93, 94]. The heterogeneous catalytic oxidations are mainly performed with air or oxygen in aqueous medium and in the presence of a supported noble metal catalyst. Many catalysts have been investigated in the catalytic oxidation conversion from glucose to gluconic acid, which can be summarized into three groups: Pt/Pd-based metal catalyst, modified Pt/Pd-based catalyst, and Au-based catalyst.

5.2.3.1 Pt/Pd-Based Metal Catalyst

Supported Pd and Pt catalysts showed high selectivity but they were easily deactivated with increasing conversion [2, 99]. The deactivation could be caused by the adsorption of reaction products and the oxygen poisoning on the active metal surface. The pH effect on the Pt-catalyzed selective heterogeneous oxidation of D-glucose to D-gluconic acid was studied by Abbdi and van Bekkum [2]. The oxidation reactions were performed in the pH range 2–9 in a batch reactor using 5 % Pt/C as the catalyst. They found that pH had a profound effect in the platinum-catalyzed oxidation of glucose to gluconic acid, because the degree of the inhibition of the catalytic activity was pH dependent. And in neutral and acidic mediums, it was obviously found the poisoning of the catalyst by reaction products and D-Gluconic acid was the main inhibiting species of the platinum catalyst during the glucose oxidation in acidic medium.

Recently, a green, simple, and energy-efficient route, an argon glow discharge plasma approach, was used to synthesize $Pd/\gamma-Al_2O_3$ catalysts, which were tested for selective oxidation of glucose to gluconic acid Liang et al. [84]. The plasma-reduced catalyst exhibited a higher activity than the conventional hydrogen thermally reduced catalyst. All the catalysts exhibit selectivities above 95 %. Another advantage was the high stability without leaching of active metal into the reaction solvent, which is attributed to the stronger metal–support interaction in plasma-reduced $Pd/\gamma-Al_2O_3$. Onda et al. prepared a new bifunctional $Pt/AC-SO_3H$ catalyst by the impregnation and sulfonation method [103]. The catalyst was highly water-tolerant and showed the catalytic properties for the hydrolysis of polysaccharides and sequentially the air oxidation into gluconic acid in the one-pot process under hot water. Under optimum conditions, 40 C-% gluconic acid yield was obtained using starch as the feedstock.

5.2.3.2 Modified Pt/Pd Catalyst: Bimetallic Catalysts

Bimetallic catalysts play an important role in many catalytic processes. For example, the Pt–Re, Pt–Ir, and Pt–Sn catalysts were very important in reforming naphtha; the Pt–Rh catalysts was used in the detoxication of car exhaust gases; Pd–Pt alloys showed high selectivity in oxidation process of hydrogen to hydrogen peroxide. Selectivities of supported Pd- or Pt-based catalysts can be increased by the addition of secondary metals such as Bi, Tl, Co, Sn, Pb, etc. for the oxidation of alcohols into aldehydes and carboxylic acids with oxygen in liquid solutions. This effect has been wildly investigated in the oxidation of glucose or other monosaccharides [1, 52, 67].

Bimetallic catalysts with other different metals were also investigated by Karski et al. [66]. The catalytic properties of Pd–Bi, Pd–Tl, Pd–Sn, and Pd–Co catalysts supported on carbon and silica were studied in the reaction of glucose oxidation to gluconic acid at 60 °C and pH = 9. Catalysts modified with Bi showed the best selectivity and activity. The increase in selectivity observed for bimetallic catalysts in the reaction was probably due to the compounds formed as a result of Pd and Bi interaction. The leaching problem of Bi, Tl, Sn, and Co during the oxidation process was connected with the chelating properties of gluconic acid.

Bismuth was also confirmed to drastically improve the catalytic activity of a Pd/C catalyst in the oxidation of glucose to gluconic acid by Wenkin et al. [143]. In their study, the catalytic activity of bimetallic Pd–Bi systems was improved by choosing appropriate catalyst precursors. The catalysts made from the acetate precursors display better performances. For catalyst stability, monometallic Bi/C catalysts were found to lose significantly larger amounts of bismuth than bimetallic Pd–Bi/C catalysts. Both glucose and gluconate appear as responsible for catalyst leaching. When Pd/C and Bi/C catalysts were used simultaneously, it was shown that the promoting role of bismuth was not obviously, which suggested that alloying Bi and Pd atoms in a given stoichiometric ratio might enhance the catalytic activity. One could suspect that this association involves an adsorbed organic species, probably glucose, gluconic acid, or some other oxygenated species derived from them, possibly in the form of a bismuth complex on the surface of palladium.

Although most bi- and trimetallic catalysts only improved the activity and selectivity with bismuth acting as a co-catalyst to prevent the oxygen poisoning of palladium [84]; however, no sufficient long-term stability and leaching of the second or third metal had been described, which may directly reduce the gluconate productivity.

5.2.3.3 Au-Based Catalyst

Metal oxide-supported gold catalysts showed excellent properties in the liquid-phase carbohydrate oxidation by lots of investigations [125, 126, 128, 131, 137, 141, 145, 156, 158, 164]. The gold catalysts can result in very high activity and

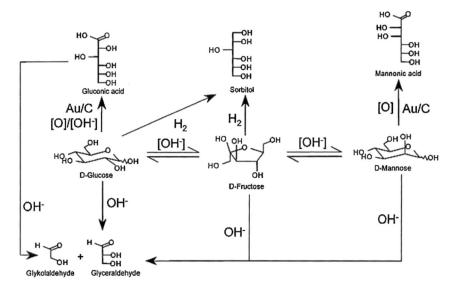

Fig. 5.6 Reaction network of the heterogeneously catalyzed D-glucose oxidation reaction in alkaline solution and at elevated temperatures. (Reprinted from Ref. [100], Copyright 2004, with permission from Elsevier)

selectivity to form the corresponding aldonic acids on the oxidation of various hexoses such as glucose, mannose, rhamnose, and galactose; pentoses such as xylose, arabinose, lyxose, and ribose; and di-/oligosaccharides such as cellobiose, lactose, melibiose, and maltotriose. Moreover, these gold catalysts showed excellent stability. An important feature of gold is to be active either in the form of supported metal or in the form of nanosized colloidal particles. Both forms of Au catalysts work in a wide range of pH, from alkaline to acidic conditions [35]. Previous studies on the supported Au catalysts have proven their superior catalytic performance for D-glucose oxidation by having better selectivity and higher activity to form D-gluconic acid compared to traditional platinum- or palladium-based catalysts. They are also less sensitive to chemical poisoning [148].

The heterogeneously catalyzed oxidation of D-glucose to D-gluconic acid over Au/C catalysts has been studied. A series of Au/C catalysts were prepared by the gold–sol method with different reducing agents [100]. All experiments were carried out in a semibatch reactor under pH control at atmospheric pressure, 30–90 °C, and pH 7.0–9.5. Nanosized gold particles in the range of 3–6 nm were deposited on Black Pearls and Vulcan-type carbons. The reaction conditions have great impact on the selectivity of the products in liquid-phase oxidation of D-glucose to D-gluconic acid. When the reaction temperature and pH were too high like T > 70 °C, pH > 9, a number of side reactions of D-glucose occurred, as illustrated in Fig. 5.6. In experiments with higher pH values (>7.0), fructose can be formed at any temperature. These reaction pathways are considered as isomerization reaction which leads to the formation of sorbitol. This phenomenon is significant evidence to prove the mechanism of oxidative dehydrogenation to skip the production of D-gluconic acid.

In another study, a gold catalyst supported on alumina, 0.3 % Au/Al_2O_3 catalyst prepared by the incipient wetness method and the long-term stability of gold catalyst was demonstrated in the continuous-flow liquid-phase glucose oxidation [133]. The continuous-flow glucose oxidation was carried out at 40 °C, pH 9, and 1 bar oxygen partial pressure. The gold catalyst showed very high activity and selectivity to gluconic acid higher than 98.5 % within its 110 days of operation. In catalyst regeneration, its activity could be completely restored by in situ regeneration with 2-propanol (70 vol%).

The bimetallic Au–M (M = Pt, Pd, Rh) catalysts and their properties in the oxidation of glucose under mild conditions were studied [35]. The activity of single metals was so weak that Au and Pt showed TOF = 51–60 h^{-1}, and Rh and Pd showed very low TOF < 2 h^{-1}, in contrast, the activity of the bimetallic particles resulted higher TOFs (max 924 h^{-1}) by combining Au with Pd and Pt. With the above catalysts, oxidation of glucose to free gluconic acid was evaluated working at the alkali conditions pH = 9.5 and 323 K, which resulted in remarkable improvement either in the case of monometallic or bimetallic gold catalysts. However, bimetallic particles appeared more stable toward agglomeration than monometallic gold particles, thus allowing higher conversions.

Nanoporous gold (NPG) made by dealloying Ag/Au alloys, were found to be novel unsupported Au nanocatalysts that exhibited effective catalytic activity and high selectivity (99 %) for the aerobic oxidation of D-glucose to D-gluconic acid at pH 7–9 and no byproduct such as fructose [148]. Different Sizes of NPG can affect catalytic activity, in their test, 6 nm NPG sample showed higher activity than those at 323 K in pH 9 alkaliñe solution. Additionally, the 30 nm sample was found to have better structure stability during the whole catalytic process. For the Catalyst stability, it is surprised that the long-term stored NPG exhibited almost the same activity for D-glucose oxidation as the freshly prepared one.

5.2.4 Formic Acid

The conversion of biomass to drop-in fuels must undergo a costly hydrogenation process. So it becomes important to make the hydrogenation from easier hydrogen provider. Formic acid can form hydrogen reversibly in the presence of metal catalysts. In order to satisfy the need, the conversion of cellulose to formic acid turns to be significant. There are two processes for cellulose to formic acid conversion. First, cellulose will be hydrolyzed to glucose. Then glucose will be oxidized to formic acid. Li et al. reported a catalytic air oxidation method with the $H_5PV_2Mo_{10}O_{40}$ catalyst. This heteropolyacid can be used as a bifunctional catalyst in the conversion of cellulose to formic acid with a yield of 35 %. And the yield of formic acid from glucose is up to 52 %. Jin et al. reported a higher yield of formic acid up to 75 % when using H_2O_2 as oxidant at a mild temperature of 25 °C in the presence of base [61, 147].

5.2.5 Glycolic Acid

Glycolic acid is a significant bulk fine chemical, which is consumed greatly in the United States annually. It has wide use in the leather, the oil and gas, and the textile industry. It can be used for adhesives and is a component in many personal care products. Glycolic acid is also the monomer to synthesize a variety of polyglycolic acid (PGA)-based polymers, including thermoplastic resins and biodegradable polymers for use in sutures.

Glycolic acid is commercially manufactured by reacting chloroacetic acid with sodium hydroxide. Reacting formaldehyde with carbon monoxide in the presence of an acid catalyst, such as HF, also produces glycolic acid. It can be made using glycolonitrile in an enzymatic process [12, 144]. Due to the extremely hazardous nature of reactants involved in these routes, other ways to produce glycolic acid have been investigated.

Recently, Zhang et al. [159] directly converted various cellulosic biomass to glycolic acid with heteromolybdic acids acting as multifunctional catalysts in a water medium and oxygen atmosphere. Four types of Keggin-type HPAs, including $H_3PW_{12}O_{40}$ (HPW), $H_3PMo_{12}O_{40}$ (HPM), $H_4SiW_{12}O_{40}$ (HSW), and $H_4SiMo_{12}O_{40}$ (HSM), were used as catalysts for the conversion of α-cellulose powder. The best result was achieved by $H_3PMo_{12}O_{40}$ with the production of glycolic acid of 49 %. Also, $H_3PMo_{12}O_{40}$ is even capable of converting raw cellulosic materials, such as bagasse and hay, to glycolic acid with yields of ~30 %. The strong Brönsted acidity of $H_3PMo_{12}O_{40}$ facilitates the hydrolysis of cellulose while the moderate oxidative activity allows selective oxidation of the aldehyde groups in the fragmentation products. Successive retro-aldol reactions dominate the fragmentation of monosaccharides generated from cellulose hydrolysis, resulting in high selectivity of glycolic acid. The author proposed a reaction pathway for the conversion of cellulose to glycolic acid and formic acid, as shown in Fig. 5.7.

5.3 Lignin to Phenolic Acids

Lignin is generally defined as polymeric natural products arising from three primary monomers: p-coumaryl alcohol (B1), coniferyl alcohol (B2), and sinapyl alcohol (B3), as shown in Fig. 5.8 [24, 41]. The enzyme-initiated polymerisation results in bonds of exceptional stability: biphenyl carbon–carbon linkages between aromatic carbons (5–5'), alkyl–aryl carbon–carbon linkages between an aliphatic

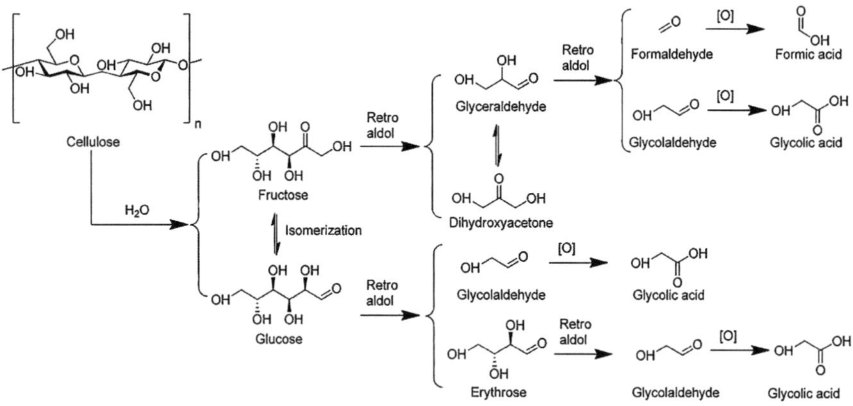

Fig. 5.7 Product yields from the conversion of different sacchariferous reactants using HPM as the catalyst. (Reprinted with the permission from Ref. [159]. Copyright 2007 American Chemical Society)

and aromatic carbon (β-1), and hydrolysis-resistant ether linkages. The only linkage which is relatively weak and hydrolysable is the α-aryl ether bond (β-O-4).

The utilization of lignin in the chemical industry has largely been limited because of the complicated chemical structure and persistent property of lignin [33]. Generally, there are two kinds of products for lignin conversion, chemical, and fuel. Catalytic oxidation has the potential to selectively convert lignin into various useful chemicals, and this methodology has rapidly progressed in the past several years. Figure 5.9 shows the oxidation reaction of different model compounds of lignin. Based on the different structure of the lignin precursors, the oxidation reaction pathways can be divided into three categories. First, p-coumaryl alcohol resembling lignin model compounds to 4-hydrobenzaldehyde [18, 26, 41, 43] and then 4-hydrobenzoic acid [106]; Second, coniferyl alcohol resembling lignin model compounds to vanillin [9, 81] and vanillic acid [36]; Third, sinapyl alcohol resembling lignin model compounds to syringaldehyde and syringic acid.

Partenheimer used a cobalt/manganese/zirconium/bromide (Co/Mn/Zr/Br) catalyst to oxidize five different lignin model compounds (4-hydroxy-benzaldehyde, 3-methoxy-4-acetoxy-benzaldehyde, 4-methoxy-toluene, 3-methoxy-4-hydroxy-toluene, 3-methoxy-4-acetoxy-toluene) in acetic acid solution by air. 18 products were identified via gas chromatography /mass spectrometry (GC/MS). The most valuable products from lignin were 4-hydroxybenzaldehyde, 4-hydroxybenzoic acid, 4-hydroxy-3-methoxybenzaldehyde (vanillin), 4-hydroxy-3-methoxybenzoic acid (vanillic acid), 4-hydroxy-3,5-dimethoxybenzaldehyde (syringaldehyde), and 4-hydroxy-3,5-dimethoxybenzoic acid (syringic acid). By the use of model compounds they demonstrated that (i) the presence of the phenolic functionality on an aromatic ring does inhibit the rate of reaction but that the alkyl group on the ring is still oxidized to the carboxylic acid; (ii) the masking of phenol by acetylation occurs at a reasonable rate in acetic acid; (iii) the alkyl group of the masked phenol

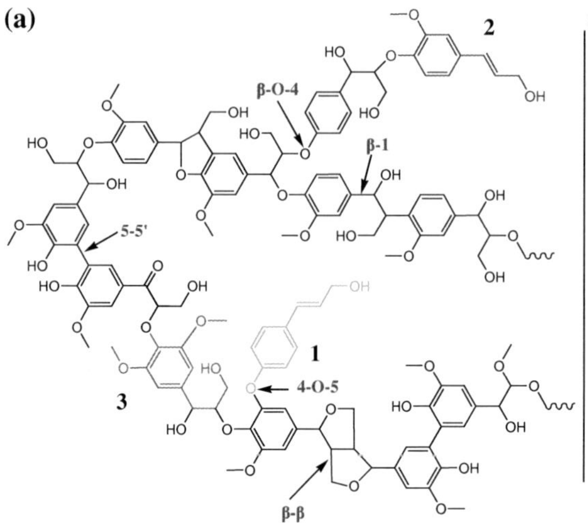

Fig. 5.8 The possible structure of lignin (**a**) and three common lignin monomers (**b**). **b1**, H-lignin (p-coumaryl alcohol); **b2**, G-lignin (coniferyl alcohol); **b3**, S-lignin (sinapyl alcohol) (Adapted from [41], Copyright 2000, with permission from Elsevier)

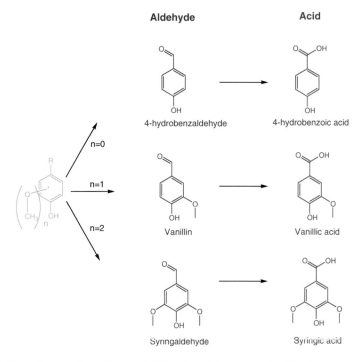

Fig. 5.9 Oxidation reaction pathway of different model compounds of lignin

is very readily oxidized; (iv) an acetic anhydride/acetic acid mixture is a good oxidation solvent; and (v) that a two-step acetylation/oxidation to the carboxylic acid is feasible [106].

Crestini et al. had reported a convenient and efficient application of heterogeneous methyl-rhenium trioxide (MTO) systems for the selective oxidation of lignin model compounds and lignin [36]. H_2O_2 was used as the oxidant. They found that the immobilized MTO catalysts are able to extensively oxidize both phenolic and nonphenolic, monomeric, and dimeric, lignin model compounds with H_2O_2. The diphenylmethane model reactant was found to be extensively oxidized. Immobilized MTO catalytic systems are potential candidates for the development of alternative chlorine-free and thus environmentally sustainable delignification processes.

Yokoyama et al. investigated the effect of lignin–biopolymer structure on the mechanism of its oxidative depolymerization by polyoxometalates (POMs) by reacting an equilibrated POM ensemble with a series of ring-substituted benzyl alcohols. Their results indicated that the reaction proceeds via successive oxidations of the benzylic carbon atom and aromatic-ring cleavage was not observed [150].

Besides, various catalysts such as base metals, noble metals, perovskite-type oxide catalyst and metal organic frameworks, etc., either heterogeneous or homogeneous, can be used to improve the efficiency of oxidation lignin and lignin-derived compounds.

5.3.1 Base Metals

Base metal catalysts mainly refer to one or mixture of Cu, Mn, Co, Ni, and other metals. The advantage of base transition metal catalysts is inexpensiveness; however, the catalytic activity of transition metal is relatively low. A mixture of copper and iron salts has been used as catalysts to improve the production of aldehydes (vanillin and syringaldehyde) from steam-explosion hardwood lignin by Wu et al. [54]. The experiment was carried out in a stirred autoclave reactor using pure oxygen as the oxidant and Cu sulfate and ferric chloride as catalysts. It was found that a mixture of copper and iron salts was more effective as a catalyst than copper or iron alone. The conversion and the yield of aldehydes increased sharply in a very short time with a higher rate of oxygen transfer. However, in order to obtain high yields of syringaldehyde, it was preferable to keep a high initial pressure of oxygen to rapidly break down the lignin and then lower the oxygen pressure to continue the oxidation of the lignin fragments but prevent the degradation of the syringaldehyde. In addition, higher pH (10.5–12) facilitated the ionization of hydroxyl groups associated with lignin structure. Based on the results, it was proposed that Fe^{3+}-lignin complexes were formed in situ and acted as oxygen carriers which increased the oxygen concentration in the reaction medium thus helping to speed up the ionic reaction, as well as the redox catalytic cycle of copper when present.

Catalytic wet oxidation of ferulic acid (a model lignin compound) was investigated by Bhargava and co-workers [4] in a continuously stirred Teflon-coated stainless steel autoclave with low temperature and pressure (100 °C, 172 kPa O_2). Several different homogeneous and heterogeneous catalysts have been studied. Homogeneous Cu^{2+} ions were observed to be the most active catalysts and followed by $Fe^{2+} > Mn^{2+} > Ce^{2+} > Bi^{2+} > Co^{2+} > Zn^{2+} > Mg^{2+} > Ni^{2+}$. Various copper-based heterogeneous catalysts were also prepared and tested. The most active catalyst was Cu-Ni-Ce-Al_2O_3 among nine heterogeneous catalysts studied; however, this catalyst also exhibited the highest extent of copper leaching. Cu-Mn-Al_2O_3, the second active catalyst only to Cu-Ni-Ce-Al_2O_3, was the most stable catalyst tested and leaching of Cu^{2+} was reduced considerably in the presence of Mn^{2+}.

Catalytic wet oxidation processes have been applied to treat paper and pulp effluent with Cu, Mn, Pd, Cu/Mn, and Cu/Pd as catalysts by Akolekar et al. [4]. The order of the catalytic materials in the TOC removal is as follows: Mn < Cu < Cu/Mn < Mn/Pd < Cu/Pd. A hardwood kraft lignin was oxidized in alkaline medium to obtain phenolic compounds (syringaldehyde, vanillin and its acids) by Villar et al. [138]. Oxygen was the oxidant employed, and copper(II), cobalt(II) salts and platinum–alumina were used as catalysts. However, catalysts did not increase the aldehyde yield. In contrast, cobalt(II) salts and the commercial platinum-alumina catalysts caused a clear reduction in aldehyde yield, which suggests that lignin oxidation is led mainly into low molecular weight products.

5.3.2 Noble Metals

Noble metal catalyst typically refer to one or more of Ru, Rh, Pt, Pd, Ir, Au, Ag, and other precious metals loaded on the catalyst supports. Although noble metal is expensive, the catalytic activity is better. Components of noble metals are more stable; therefore, the stability of noble metal catalyst mainly depends on the stability of the support [62].

The effectiveness of alumina-supported noble metal catalysts for the destructive oxidation of organic pollutants in effluent from a softwood kraft pulp mill has been studied by Zhang and Chuang [161]. Alumina-supported palladium catalysts were found to be more effective than alumina-supported platinum catalysts. However, addition of Ce improves the activity of alumina-supported Pt catalysts but inhibits the activity of alumina-supported Pd catalyst. It also indicated that wet catalytic oxidation may involve a free-radical chain mechanism, a redox mechanism, or an oxygen-transfer mechanism depending on the catalyst, nature of reactants, and reaction conditions. Sales et al. [118] studied the palladium-supported on γ-alumina as catalyst in the wet air oxidation of lignin obtained from sugarcane bagasse. A three-phase fluidized reactor, using atmospheric air as the oxidizer, was built for the purpose of producing aromatic aldehydes in continuous regime. The reactions in the lignin degradation and aldehyde production allowed validation of the kinetic model of series and parallel reactions with pseudo-first-order steps. The best yield in aromatic aldehydes was of 12 % with an air-flow rate of 1,000 L/h, a liquid-phase flow rate of 5 L/h, a catalytic loading of 4 wt% and a pressure of 4 bar. The data from the experimental results were compatible with those values obtained by the pseudo-heterogeneous axial dispersion model (PHADM) applied to the liquid phase [157]. They also studied a palladium as catalyst at the operating conditions 373–413 K under 2–10 bar oxygen partial pressure [119]. Vanillin and syringaldehyde were obtained in the continuous operation, with a feed concentration of lignin of 30 g/L and a liquid-phase flow rate of 5 L/h. Besson and co-worker [130] proved that Pt and Ru catalysts supported on titanium or zirconium dioxide showed strong potential for practical use in the catalytic wet air oxidation of industrial wastewaters.

5.3.3 Perovskite-Type Oxide Catalyst

Noble metals are expensive, greatly restricting their use in industry. Transition metal catalysts are usually homogeneous ions of transition metals with low catalytic activity, resulting in secondary pollution and high recycling costs, thus tremendously affecting the economics of their potential commercial applications. Therefore, developing perovskite catalysts is a possible way to overcome the high costs of noble metal catalysts and low efficiency of transition metal catalysts for the catalytic wet air oxidation of lignin.

Recently, Deng and co-workers prepared the perovskite-type oxides $LaMnO_3$ [37] and $LaCoO_3$ by the sol–gel method, which turned out to be efficient heterogeneous and recyclable catalysts for the wet aerobic oxidation of lignin to aromatic aldehydes. The rate of lignin conversion and the yield of each aromatic aldehyde were enhanced significantly by the catalytic process as compared to noncatalyzed processes. The perovskite-type oxide catalysts $LaMnO_3$ and $LaCoO_3$ also displayed distinctive stability of activity and structure in the oxidation of lignin after a series of successive recycles of catalytic reactions. They also studied the mechanism involving the reaction of lignin molecules with adsorbed oxygen surface sites $(Co(surf)^{3+}O^{2-})$, which was proposed on the basis of the experimental observations and the O 1 s and Co 2p XPS spectra of the catalyst, yielding the cycle $Co(surf)^{3+} \rightarrow Co(surf)^{2+} \rightarrow Co(surf)^{3+}O^{2-} \rightarrow Co(surf)^{3+}$ with the formation rate of intermediates quininemethide and hydroperoxide as the rate-determining step.

The research was continued by Zhang and Deng et al. The perovskite-type oxide catalysts $LaFe_{1-x}Cu_xO_3$ (x = 0, 0.1, 0.2) [157] and $LaCo_{1-x}Cu_xO_3$ (x = 0, 0.1, and 0.2) [38] were prepared by the sol–gel method. It was found that the Cu-substituted catalysts had higher lignin conversion and yield of each aromatic aldehyde as compared to the $LaCoO_3$ catalyst, and the catalyst activity of the Cu-substituted catalyst was improved with an increase in the Cu content.

A mechanism was stated: anion vacancies were generated after Cu incorporation into the $LaCoO_3$ catalyst, resulting in an increase in the oxygen surface adsorption ability of the catalyst, which promoted the adsorbed oxygen surface active site [Co(surf)$^{3+}$O^{2-}] species and also produced activated species [Cu(surf)$^{2+}$O^{2-}].

5.3.4 Metal Organic Frameworks

Recently, metal organic frameworks have attracted the scholars' concerns as catalysts for oxidation of lignin due to their structure diversity, high thermal and chemical stability. Zakzeski et al. [154] invented a novel heterogeneous catalyst Co-ZIF-9 for oxidation of several lignin-derived aromatics. The catalyst was produced by using Co nitrate and benzimidazole to obtain Co-zeoliticimidazolate framework. The effect of solvent, temperature, and NaOH addition on the Co-ZIF-9 catalytic system has been studied in a high pressure slurry reactor with oxygen (0.5 MPa) as an oxidant. Phthalan was converted to three different products: phthalide, phthalaldehyde, and small amounts of phthalic acid. And the catalyst exhibited excellent selectivity towards oxidation of the alcohol functionality of vanillyl alcohol, veratryl alcohol, and cinnamyl alcohol into corresponding aldehydes. It was observed that the presence of NaOH in the solution greatly enhanced the oxidation conversion, indicating that hydroxyl group was able to activate the surface oxygen. In addition, the catalytic activity was found to be stable under the reaction conditions employed, as supported by TGA analysis.

Four metal organic frameworks (MOFs) have been evaluated as possible catalysts for alkaline lignin oxidation by Masingale et al. [92]. The MOFs, containing a high mole ratio of Cu, Fe, or Cu+Fe, was prepared in a microwave-assisted synthesis. The experiments were conducted in a small stainless steel autoclave using 40 ml of 2 M NaOH and a known amount of MOF, CuO or Fe_2O_3 with an O_2 partial pressure of 0.3 MPa. The ratios of syringyl (S) to guiacyl (G) monomers for the three poplars with MOF as oxidant were similar to the S:G ratios obtained from nitrobenzene oxidation.

References

1. Abbadi A, Van Bekkum H (1995) Highly selective oxidation of aldonic acids to 2-ketoaldonic acids over Pt—Bi and Pt—Pb catalysts. Appl Catal A 124:409–417
2. Abbadi A, Van Bekkum H (1995) Effect of pH in the Pt-catalyzed oxidation of d-glucose to d-gluconic acid. J Mol Catal A Chem 97:111–118. doi:10.1016/1381-1169(94)00078-6
3. Akiya N, Savage PE (2002) Roles of water for chemical reactions in high-temperature water. Chem Rev 102:2725–2750
4. Akolekar DB, Bhargava SK, Shirgoankar I, Prasad J (2002) Catalytic wet oxidation: an environmental solution for organic pollutant removal from paper and pulp industrial waste liquor. Appl Catal A 236:255–262

5. Alonso DM, Wettstein SG, Mellmer MA et al (2013) Integrated conversion of hemicellulose and cellulose from lignocellulosic biomass. Energy Environ Sci 6:76–80. doi: 10.1039/c2ee23617f
6. An D, Ye A, Deng W et al (2012) Selective conversion of cellobiose and cellulose into gluconic acid in water in the presence of oxygen, catalyzed by polyoxometalate-supported gold nanoparticles. Chemistry (Weinheim an der Bergstrasse, Germany) 18:2938–2947. doi: 10.1002/chem.201103262
7. Asghari FS, Yoshida H (2006) Acid-catalyzed production of 5-hydroxymethyl furfural from D -fructose in subcritical water. Ind Eng Chem Res 45:2163–2173
8. Assary RS, Redfern PC, Hammond JR et al (2010) Computational studies of the thermochemistry for conversion of glucose to levulinic acid. J Phys Chem B 114:9002–9009. doi:10.1021/jp101418f
9. Avellar BK, Glasser WG (1998) Steam-assisted biomass fractionation. I. Process considerations and economic evaluation. Biomass Bioenergy 14:205–218
10. Baatz C, Thielecke N, Prüße U (2007) Influence of the preparation conditions on the properties of gold catalysts for the oxidation of glucose. Appl Catal B 70:653–660. doi:10.1016/j.apcatb.2006.01.020
11. Becker J, Toft LL, Aarup DF et al (2010) A high temperature, high pressure facility for controlled studies of catalytic activity under hydrothermal conditions. Energy Fuels 24:2737–2746. doi:10.1021/ef901584t
12. Ben-Bassat A, Walls AM, Plummer MA et al (2008) Optimization of biocatalyst specific activity for glycolic acid production. Adv Synth Catal 350:1761–1769. doi:10.1002/adsc.200800228
13. Bicker M, Endres S, Ott L, Vogel H (2005) Catalytical conversion of carbohydrates in subcritical water: A new chemical process for lactic acid production. J Mol Catal A Chem 239:151–157. doi:10.1016/j.molcata.2005.06.017
14. Bicker M, Hirth J, Vogel H (2003) Dehydration of fructose to 5-hydroxymethylfurfural in sub- and supercritical acetone. Green Chem 5:280–284. doi:10.1039/b211468b
15. Biella S, Prati L, Rossi M (2002) Selective oxidation of D-glucose on gold catalyst. J Catal 206:242–247. doi:10.1006/jcat.2001.3497
16. Bond JQ, Alonso DM, Wang D et al (2010) Integrated catalytic conversion of γ-Valerolactone to liquid alkenes for transportation fuels. Science (New York, NY) 327:1110–1114
17. Bozell JJ (2010) Connecting biomass and petroleum processing with a chemical bridge. Science 329:522–523. doi:10.1126/science.1191662
18. Bozell JJ, Hames BR (1995) Cobalt-Schiff base complex catalyzed oxidation of para-substituted phenolics. In: Preparation of benzoquinones. National Renewable Energy Laboratory, Colorado, pp 2398–2404
19. Bozell JJ, Moens L, Elliott DC et al (2000) Production of levulinic acid and use as a platform chemical for derived products. Resour Conserv Recycl 28:227–239
20. Braden DJ, Henao CA, Heltzel J et al (2011) Production of liquid hydrocarbon fuels by catalytic conversion of biomass-derived levulinic acid. Green Chemistry 13:1755–1765. doi: 10.1039/c1gc15047b
21. Brandt A, Gräsvik J, Hallett JP, Welton T (2013) Deconstruction of lignocellulosic biomass with ionic liquids. Green Chem 15:550–583. doi:10.1039/c2gc36364j
22. Brunner G (2009) Near critical and supercritical water. Part I. Hydrolytic and hydrothermal processes. J Supercrit Fluids 47:373–381. doi:10.1016/j.supflu.2008.09.002
23. Cabiac A, Guillon E, Chambon F et al (2011) Cellulose reactivity and glycosidic bond cleavage in aqueous phase by catalytic and non catalytic transformations. Appl Catal A 402:1–10. doi:10.1016/j.apcata.2011.05.029
24. Calvo-Flores FG, Dobado JA (2010) Lignin as renewable raw material. ChemSusChem 3:1227–1235. doi: 10.1002/cssc.201000157
25. Canakci M (2007) The potential of restaurant waste lipids as biodiesel feedstocks. Bioresour Technol 98:183–190. doi:10.1016/j.biortech.2005.11.022

26. Canevali C, Orlandi M, Pardi L et al (2002) Oxidative degradation of monomeric and dimeric phenylpropanoids: reactivity and mechanistic investigation. J Chem Soc, Dalton Trans 3007–3014. doi: 10.1039/b203386k
27. Carrettin S, McMorn P, Johnston P et al (2003) Oxidation of glycerol using supported Pt, Pd and Au catalysts. Phys Chem Chem Phys 5:1329–1336. doi:10.1039/b212047j
28. Case PA, Van Heiningen ARP, Wheeler MC (2012) Liquid hydrocarbon fuels from cellulosic feedstocks via thermal deoxygenation of levulinic acid and formic acid salt mixtures. Green Chem 14:85–89. doi: 10.1039/c1gc15914c
29. Cha J, Hanna M (2002) Levulinic acid production based on extrusion and pressurized batch reaction. Ind Crops Prod 16:109–118. doi:10.1016/S0926-6690(02)00033-X
30. Chambon F, Rataboul F, Pinel C et al (2011) Cellulose hydrothermal conversion promoted by heterogeneous Brønsted and Lewis acids: remarkable efficiency of solid Lewis acids to produce lactic acid. Appl Catal B 105:171–181. doi:10.1016/j.apcatb.2011.04.009
31. Chang C, Cen P, Ma X (2007) Levulinic acid production from wheat straw. Bioresour Technol 98:1448–1453. doi:10.1016/j.biortech.2006.03.031
32. Chen H, Yu B, Jin S (2011) Production of levulinic acid from steam exploded rice straw via solid superacid, S2O8(2-)/ZrO2-SiO2-Sm2O3. Bioresour Technol 102:3568–3570. doi:10.1016/j.biortech.2010.10.018
33. Collinson SR, Thielemans W (2010) The catalytic oxidation of biomass to new materials focusing on starch, cellulose and lignin. Coord Chem Rev 254:1854–1870. doi:10.1016/j.ccr.2010.04.007
34. Comotti M, Dellapina C, Falletta E, Rossi M (2006) Is the biochemical route always advantageous? The case of glucose oxidation. J Catal 244:122–125. doi:10.1016/j.jcat.2006.07.036
35. Comotti M, Della PC, Rossi M (2006) Mono- and bimetallic catalysts for glucose oxidation. J Mol Catal A: Chem 251:89–92. doi: 10.1016/j.molcata.2006.02.014
36. Crestini C, Caponi MC, Argyropoulos DS, Saladino R (2006) Immobilized methyltrioxo rhenium (MTO)/H2O2 systems for the oxidation of lignin and lignin model compounds. Bioorg Med Chem 14:5292–5302. doi:10.1016/j.bmc.2006.03.046
37. Deng H, Lin ÆL, Sun ÆY et al (2008) Perovskite-type Oxide LaMnO 3: An efficient and recyclable heterogeneous catalyst for the wet aerobic oxidation of lignin to aromatic aldehydes. Catal Lett 126:106–111. doi: 10.1007/s10562-008-9588-0
38. Deng H, Lin L, Liu S (2010) Catalysis of Cu-doped Co-based Perovskite-type oxide in wet oxidation of lignin to produce aromatic aldehydes. Energy Fuels 24:4797–4802. doi: 10.1021/ef100768e
39. Deng W, Wang Y, Zhang Q, Wang Y (2012) Development of bifunctional catalysts for the conversions of cellulose or cellobiose into polyols and organic acids in water. Catal Surv Asia 16:91–105. doi:10.1007/s10563-012-9136-1
40. Dodds DR, Gross RA (2007) Chemicals from biomass. Science 318:1250–1251
41. Dorrestijn E, Laarhoven LJJ, Arends IWCE, Mulder P (2000) The occurrence and reactivity of phenoxyl linkages in lignin and low rank coal. J Anal Appl Pyrol 54:153–192. doi:10.1016/S0165-2370(99)00082-0
42. Ellis AV, Wilson MA (2002) Carbon exchange in hot alkaline degradation of glucose. J Org Chem 67:8469–8474
43. Fabbri C, Aurisicchio C, Lanzalunga O (2008) Iron porphyrins-catalysed oxidation of α-alkyl substituted mono and dimethoxylated benzyl alcohols. Cent Eur J Chem 6:145–153. doi:10.2478/s11532-008-0005-8
44. Fan Y, Zhou C, Zhu X (2009) Selective catalysis of lactic acid to produce commodity chemicals. Catal Rev 51:293–324. doi:10.1080/01614940903048513
45. Farrell AE, Plevin RJ, Turner BT et al (2006) Ethanol can contribute to energy and environmental goals. Science 311:506–508. doi:10.1126/science.1121416
46. Fitzpatrick SW (1997a) Production of levulinic acid from carbohydrate-containing materials. US Patent 5,608,105

47. Fitzpatrick SW (1997b) Production of levulinic acid from carbohydrate-containing materials. US Patent 5608105
48. Geboers JA, Van de Vyver S, Ooms R et al (2011) Chemocatalytic conversion of cellulose: opportunities, advances and pitfalls. Catal Sci Technol 1:714. doi: 10.1039/c1cy00093d
49. Girisuta B (2007) Levulinic acid from lignocellulosic biomass. University of Goningen, Netherlands
50. Girisuta B, Janssen LPBM, Heeres HJ (2007) Kinetic study on the acid-catalyzed hydrolysis of cellulose to levulinic acid. Ind Eng Chem Res 46:1696–1708. doi:10.1021/ie061186z
51. Hammerschmidt A, Boukis N, Hauer E et al (2011) Catalytic conversion of waste biomass by hydrothermal treatment. Fuel 90:555–562. doi:10.1016/j.fuel.2010.10.007
52. Hayashi H, Sugiyama S, Katayama Y et al (1994) An alloy phase of Pd3Pb and the activity of Pb/Pd/C catalysts in the liquid-phase oxidation of sodium lactate to pyruvate. J Mol Catal 91:129–137. doi:10.1016/0304-5102(94)00026-3
53. Hegner J, Pereira KC, DeBoef B, Lucht BL (2010) Conversion of cellulose to glucose and levulinic acid via solid-supported acid catalysis. Tetrahedron Lett 51:2356–2358. doi:10.1016/j.tetlet.2010.02.148
54. Heitz M, Chornet E (1994) Improved alkaline oxidation process for the production of aldehydes (Vanillin and Syringaldehyde) from Steam-Explosion Hardwood Lignin. Ind Eng Chem Res 33:718–723
55. Holm MS, Saravanamurugan S, Taarning E (2010) Conversion of sugars to lactic acid derivatives using heterogeneous zeotype catalysts. Science 328:602–605. doi:10.1126/science.1183990
56. Horvat J, Klaic B, Metelko B, Sunjic V (1985) Mechanism of levulinic acid formation. Tetrahedron Lett 26:2111–2114
57. Huber GW, Chheda JN, Barrett CJ, Dumesic JA (2005) Production of liquid alkanes by aqueous-phase processing of biomass-derived carbohydrates. Science 308:1446–1450. doi: 10.1126/science.1111166
58. Ishida T, Kinoshita N, Okatsu H et al (2008) Influence of the support and the size of gold clusters on catalytic activity for glucose oxidation. Angew Chem Int Ed Engl 47:9265–9268. doi:10.1002/anie.200802845
59. Ishida T, Watanabe H, Bebeko T et al (2010) Aerobic oxidation of glucose over gold nanoparticles deposited on cellulose. Appl Catal A 377:42–46. doi:10.1016/j.apcata.2010.01.017
60. Jin F, Cao J, Kishida H et al (2007) Impact of phenolic compounds on hydrothermal oxidation of cellulose. Carbohydr Res 342:1129–1132. doi:10.1016/j.carres.2007.02.013
61. Jin F, Yun J, Li G et al (2008) Hydrothermal conversion of carbohydrate biomass into formic acid at mild temperatures. Green Chem 10:612. doi:10.1039/b802076k
62. Jing G, Luan M, Chen T (2012) Progress of catalytic wet air oxidation technology. Arab J Chem. doi:10.1016/j.arabjc.2012.01.001
63. John RP, Nampoothiri KM, Pandey A (2007) Fermentative production of lactic acid from biomass: an overview on process developments and future perspectives. Appl Microbiol Biotechnol 74:524–534. doi:10.1007/s00253-006-0779-6
64. Jomaa S, Shanableh A, Khalil W, Trebilco B (2003) Hydrothermal decomposition and oxidation of the organic component of municipal and industrial waste products. Adv Environ Res 7:647–653. doi: 10.1016/S1093-0191(02)00042-4
65. Jow J, Rorrer GL, Hawley MC, Lamport DTA (1987) Dehydration of d-fructose to levulinic acid over LZY zeolite catalyst. Biomass 14:185–194. doi: 10.1016/0144-4565(87)90046-1
66. Karski S, Paryjczak T, Witonska I (2003) Selective oxidation of glucose to gluconic acid over bimetallic Pd–Me Catalysts (Me = Bi, Tl, Sn, Co). Kinet Catal 44:678–682
67. Kimura H, Tsuto K, Wakisaka T et al (1993) Selective oxidation of glycerol on a platinum-bismuth catalyst. Appl Catal A 96:217–228. doi:10.1016/0926-860X(90)80011-3
68. Kobayashi N, Okada N, Hirakawa A et al (2009) Characteristics of solid residues obtained from hot-compressed-water treatment of woody biomass. Ind Eng Chem Res 48:373–379. doi:10.1021/ie800870k

69. Kong L, Li G, Wang H et al (2008) Hydrothermal catalytic conversion of biomass for lactic acid production. J Chem Technol Biotechnol 83:383–388. doi:10.1002/jctb
70. Kruse A, Krupka A, Schwarzkopf V et al (2005) Influence of proteins on the hydrothermal gasification and liquefaction of biomass 1. comparison of different feedstocks. Ind Eng Chem Res 44:3013–3020. doi:10.1021/ie049129y
71. Kumar S, Gupta RB (2009) Biocrude production from switchgrass using subcritical water. Energy Fuels 23:5151–5159. doi:10.1021/ef900379p
72. Kusema BT, Campo BC, Mäki-Arvela P et al (2010) Selective catalytic oxidation of arabinose—A comparison of gold and palladium catalysts. Appl Catal A 386:101–108. doi:10.1016/j.apcata.2010.07.037
73. Kusema BT, Mikkola J-P, Murzin DY (2012) Kinetics of l-arabinose oxidation over supported gold catalysts with in situ catalyst electrical potential measurements. Catal Sci Technol 2:423–431. doi:10.1039/c1cy00365h
74. Kusema BT, Murzin DY (2013) Catalytic oxidation of rare sugars over gold catalysts. Catal Sci Technol 3:297–307. doi:10.1039/c2cy20379k
75. Ladisch MR, Lin KW, Voloch M, Tsao GT (1983) Process considerations in the enzymatic hydrolysis of biomass. Enzyme Microb Technol 5:82–102. doi:10.1016/0141-0229(83)90042-X
76. Lai D, Deng L, Guo Q, Fu Y (2011) Hydrolysis of biomass by magnetic solid acid. Energy Environ Sci 4:3552–3557. doi:10.1039/c1ee01526e
77. Lange J-P, Price R, Ayoub PM et al (2010) Valeric biofuels: a platform of cellulosic transportation fuels. Angew Chem Int Ed Engl 49:4479–4483. doi:10.1002/anie.201000655
78. Lansalot-Matras C, Moreau C (2003) Dehydration of fructose into 5-hydroxymethylfurfural in the presence of ionic liquids. Catal Commun 4:517–520. doi:10.1016/S1566-7367(03)00133-X
79. Leonard RH (1956) Levulinic Acid as a Basic Chemical Raw Material. Ind Eng Chem 48:1330–1341
80. Li J, Ding D-J, Deng L et al (2012) Catalytic Air Oxidation of biomass-derived carbohydrates to formic acid. ChemSusChem 5:1313–1318
81. Li J, Gellerstedt G, Toven K (2009) Steam explosion lignins; their extraction, structure and potential as feedstock for biodiesel and chemicals. Bioresour Technol 100:2556–2561. doi:10.1016/j.biortech.2008.12.004
82. Li Q, Yao G, Zeng X et al (2012) Facile and green production of Cu from CuO using cellulose under hydrothermal conditions. Ind Eng Chem Res 51:3129–3136. doi:10.1021/ie202151s
83. Liang J, Wang L (2010) Production of levulinic acid from cellulose catalyzed by environmental-friendly catalyst. Adv Mater Res 96:183–187
84. Liang X, Liu C, Kuai P (2008) Selective oxidation of glucose to gluconic acid over argon plasma reduced Pd/Al2O3. Green Chem 10:1318–1322. doi:10.1039/b804904a
85. Lin H, Strull J, Liu Y et al (2012) High yield production of levulinic acid by catalytic partial oxidation of cellulose in aqueous media. Energy Environ Sci 5:9773–9777. doi:10.1039/c2ee23225a
86. Liu Z, Li W, Pan C et al (2011) Conversion of biomass-derived carbohydrates to methyl lactate using solid base catalysts. Catal Commun 15:82–87. doi:10.1016/j.catcom.2011.08.019
87. Lourvanij K, Rorrer GL (1993) Reactions of Aqueous Glucose Solutions over Solid-Acid Y-Zeolite Catalyst at 110-160 C. Ind Eng Chem Res 32:11–19
88. Lourvanij K, Rorrer GL (1994) Dehydration of glucose to organic acids in microporous pillared clay catalysts. Appl Catal A 109:147–165
89. Lunt J (1998) Large-scale production, properties and commercial applications of polylactic acid polymers. Polym Degrad Stab 59:145–152. doi:10.1016/S0141-3910(97)00148-1
90. Ma C, Jin F, Cao J, Wu B (2010) Hydrothermal conversion of carbohydrates into lactic acid with alkaline catalysts. In: 2010 4th International Conference on Bioinformatics and Biomedical Engineering, pp 1–4. doi: 10.1109/ICBBE.2010.5516468

91. Maas RHW, Bakker RR, Eggink G, Weusthuis RA (2006) Lactic acid production from xylose by the fungus Rhizopus oryzae. Applied microbiology and biotechnology 72:861–868. doi: 10.1007/s00253-006-0379-5
92. Masingale MP, Alves EF, Korbieh TN et al (2009) An oxidant to replace nitrobenzene in lignin analysis. BioResources 4:1139–1146
93. Mirescu A, Berndt H, Martin A, Prüße U (2007) Long-term stability of a 0.45 % Au/TiO2 catalyst in the selective oxidation of glucose at optimised reaction conditions. Appl Catal A 317:204–209. doi:10.1016/j.apcata.2006.10.016
94. Mirescu A, Prüße U (2007) A new environmental friendly method for the preparation of sugar acids via catalytic oxidation on gold catalysts. Appl Catal B 70:644–652. doi:10.1016/j.apcatb.2006.01.017
95. Möller M, Nilges P, Harnisch F, Schröde U (2011) Subcritical water as reaction environment: fundamentals of hydrothermal biomass transformation. ChemSusChem 4:566–579
96. Mosier NS, Ladisch CM, Ladisch MR (2002) Characterization of acid catalytic domains for cellulose hydrolysis and glucose degradation. Biotechnol Bioeng 79:610–618. doi:10.1002/bit.10316
97. Muangrat R, Onwudili JA, Williams PT (2010a) Reaction products from the subcritical water gasification of food wastes and glucose with NaOH and H2O2. Bioresour Technol 101:6812–6821. doi: 10.1016/j.biortech.2010.03.114
98. Muangrat R, Onwudili JA, Williams PT (2010b) Alkali-promoted hydrothermal gasification of biomass food processing waste: A parametric study. Int J Hydrogen Energy 35:7405–7415. doi: 10.1016/j.ijhydene.2010.04.179
99. Nikov I, Paev K (1995) Palladium on alumina catalyst for glucose oxidation: reaction kinetics and catalyst deactivation. Catal Today 24:41–47. doi:10.1016/0920-5861(95)00011-4
100. Önal Y, Schimpf S, Claus P (2004) Structure sensitivity and kinetics of D-glucose oxidation to D-gluconic acid over carbon-supported gold catalysts. J Catal 223:122–133. doi:10.1016/j.jcat.2004.01.010
101. Onda A, Ochi T, Kajiyoshi K, Yanagisawa K (2008) Lactic acid production from glucose over activated hydrotalcites as solid base catalysts in water. Catal Commun 9:1050–1053. doi:10.1016/j.catcom.2007.10.005
102. Onda A, Ochi T, Kajiyoshi K, Yanagisawa K (2008) A new chemical process for catalytic conversion of d-glucose into lactic acid and gluconic acid. Appl Catal A 343:49–54. doi:10.1016/j.apcata.2008.03.017
103. Onda A, Ochi T, Yanagisawa K (2011) New direct production of gluconic acid from polysaccharides using a bifunctional catalyst in hot water. Catal Commun 12:421–425. doi:10.1016/j.catcom.2010.10.023
104. Onwudili JA, Williams PT (2007) Hydrothermal catalytic gasification of municipal solid waste. Energy Fuels 21:3676–3683. doi: 10.1021/ef700348n
105. Palkovits R (2010) Pentenoic acid pathways for cellulosic biofuels. Angew Chem Int Ed Engl 49:4336–4338. doi:10.1002/anie.201002061
106. Partenheimer W (2009) The aerobic oxidative cleavage of lignin to produce hydroxyaromatic benzehydes and carboxylic acids via metal/bromide catalysts in acetic acid/water mixtures. Adv Synth Catal 351:456–466. doi:10.1002/adsc.200800614
107. Peng L, Lin L, Li H, Yang Q (2011) Conversion of carbohydrates biomass into levulinate esters using heterogeneous catalysts. Appl Energy 88:4590–4596. doi:10.1016/j.apenergy.2011.05.049
108. Peng L, Lin L, Zhang J et al (2010) Catalytic conversion of cellulose to levulinic acid by metal chlorides. Molecules (Basel, Switzerland) 15:5258–5272. doi: 10.3390/molecules15085258
109. Peterson AA, Vogel F, Lachance RP et al (2008) Thermochemical biofuel production in hydrothermal media: A review of sub- and supercritical water technologies. Energy Environ Sci 1:32–65. doi: 10.1039/b810100k

110. Potvin J, Sorlien E, Hegner J et al (2011) Effect of NaCl on the conversion of cellulose to glucose and levulinic acid via solid supported acid catalysis. Tetrahedron Lett 52:5891–5893. doi:10.1016/j.tetlet.2011.09.013
111. Qi X, Watanabe M, Aida TM, Smith RL Jr (2008) Catalytical conversion of fructose and glucose into 5-hydroxymethylfurfural in hot compressed water by microwave heating. Catal Commun 9:2244–2249. doi:10.1016/j.catcom.2008.04.025
112. Rackemann DW, Doherty WOS (2011) The conversion of lignocellulosics to levulinic acid. Biofuels, Bioprod Biorefin 5:198–214
113. Rasrendra CB, Makertihartha IGBN, Adisasmito S, Heeres HJ (2010) Green chemicals from d-glucose: systematic studies on catalytic effects of inorganic salts on the chemo-selectivity and yield in aqueous solutions. Top Catal 53:1241–1247. doi:10.1007/s11244-010-9570-0
114. Rinaldi R, Palkovits R, Schüth F (2008) Depolymerization of cellulose using solid catalysts in ionic liquids. Angew Chem Int Ed Engl 47:8047–8050. doi:10.1002/anie.200802879
115. Rinaldi R, Schüth F (2009) Acid hydrolysis of cellulose as the entry point into biorefinery schemes. ChemSusChem 2:1096–1107. doi:10.1002/cssc.200900188
116. Román-Leshkov Y, Chheda JN, Dumesic JA (2006) Phase modifiers promote efficient production of hydroxymethylfurfural from fructose. Science 312:1933–1937. doi: 10.1126/science.1126337
117. Saeman JF (1945) Industrial and engineering chemistry. Ind Eng Chem 37:43–52
118. Sales FG, Abreu CAM, Pereira JAFR (2004) Catalytic wet-air oxidation of lignin in a three-phase reactor with aromatic aldehyde production. Braz J Chem Eng 21:211–218
119. Sales FG, Maranhão LCA, Lima NM, Abreu CAM (2007) Experimental evaluation and continuous catalytic process for fine aldehyde production from lignin. Science 62:5386–5391. doi:10.1016/j.ces.2007.02.018
120. Savage PE (2009) A perspective on catalysis in sub- and supercritical water. J Supercrit Fluids 47:407–414. doi:10.1016/j.supflu.2008.09.007
121. Scheller HV, Ulvskov P (2010) Hemicelluloses. Annu Rev Plant Biol 61:263–289. doi:10.1146/annurev-arplant-042809-112315
122. Seri K, Inoue Y, Ishida H (2000) Highly efficient catalytic activity of lanthanide(III) ions for conversion of saccharides to 5-hydroxymethyl-2-furfural in organic solvents. Chem Lett 29:22–23. doi: 10.1246/cl.2000.22
123. Serrano-Ruiz JC, Luque R, Sepúlveda-Escribano A (2011) Transformations of biomass-derived platform molecules: from high added-value chemicals to fuels via aqueous-phase processing. Chem Soc Rev 40:5266–5281. doi:10.1039/c1cs15131b
124. Shanableh A (2005) Generalized first-order kinetic model for biosolids decomposition and oxidation during hydrothermal treatment. Environ Sci Technol 39:355–362
125. Simakova OA, Kusema BT, Campo BC et al (2011) Structure sensitivity in L-arabinose oxidation over Au/Al2O3 catalysts. J Phys Chem C 115:1036–1043
126. Smolentseva E, Kusema BT, Beloshapkin S et al (2011) Selective oxidation of arabinose to arabinonic acid over Pd–Au catalysts supported on alumina and ceria. Appl Catal A 392:69–79. doi:10.1016/j.apcata.2010.10.021
127. Takeuchi Y, Jin F, Tohji K, Enomoto H (2007) Acid catalytic hydrothermal conversion of carbohydrate biomass into useful substances. J Mater Sci 43:2472–2475. doi:10.1007/s10853-007-2021-z
128. Tan X, Deng W, Liu M et al (2009) Carbon nanotube-supported gold nanoparticles as efficient catalysts for selective oxidation of cellobiose into gluconic acid in aqueous medium. Chem Commun 46:7179–7181. doi: 10.1039/b917224f
129. Tarabanko VE, Chernyak MY, Aralova SV, Kuznetsov BN (2002) Kinetics of levulinic acid formation from carbohydrates at moderate temperatures. React Kinet Catal Lett 75:117–126
130. Taylor P, Besson M, Descorme C et al (2010) Supported noble metal catalysts in the catalytic wet air oxidation of industrial wastewaters and sewage sludges. Environ Technol. 31:37–41. doi: 10.1080/09593331003628065

131. Tembe SM, Patrick G, Scurrell MS (2009) Acetic acid production by selective oxidation of ethanol using Au catalysts supported on various metal oxide. Gold Bulletin 42:321–327. doi:10.1007/BF03214954
132. Thielecke N, Aytemir M, Prüsse U (2007) Selective oxidation of carbohydrates with gold catalysts: Continuous-flow reactor system for glucose oxidation. Catal Today 121:115–120. doi:10.1016/j.cattod.2006.11.015
133. Thielecke N, Vorlop K-D, Prüße U (2007) Long-term stability of an Au/Al2O3 catalyst prepared by incipient wetness in continuous-flow glucose oxidation. Catal Today 122:266–269. doi:10.1016/j.cattod.2007.02.008
134. Tominaga K, Mori A, Fukushima Y et al (2011) Mixed-acid systems for the catalytic synthesis of methyl levulinate from cellulose. Green Chem 13:810–812. doi:10.1039/c0gc00715c
135. Toor SS, Rosendahl L, Rudolf A (2011) Hydrothermal liquefaction of biomass: a review of subcritical water technologies. Energy 36:2328–2342. doi:10.1016/j.energy.2011.03.013
136. Verendel J, Church T, Andersson P (2011) Catalytic one-pot production of small organics from polysaccharides. Synthesis 2011:1649–1677. doi:10.1055/s-0030-1260008
137. Villa A, Veith GM, Prati L (2010) Selective oxidation of glycerol under acidic conditions using gold catalysts. Angew Chem Int Ed Engl 49:4499–4502. doi:10.1002/anie.201000762
138. Villar JC, Caperos A (2001) Oxidation of hardwood kraft-lignin to phenolic derivatives with oxygen as oxidant. Wood Sci Technol 35(3):245–255
139. Van de Vyver S, Thomas J, Geboers J et al (2011) Catalytic production of levulinic acid from cellulose and other biomass-derived carbohydrates with sulfonated hyperbranched poly(arylene oxindole)s. Energy Environ Sci 4:3601–3610. doi:10.1039/c1ee01418h
140. Watanabe M, Aizawa Y, Iida T et al (2005) Glucose reactions with acid and base catalysts in hot compressed water at 473 K. Carbohydr Res 340:1925–1930. doi:10.1016/j.carres.2005.06.017
141. Wei Y, Liu J, Zhao Z et al (2011) Highly active catalysts of gold nanoparticles supported on three-dimensionally ordered macroporous LaFeO3 for soot oxidation. Angew Chem Int Ed Engl 50:2326–2329. doi:10.1002/anie.201006014
142. Weingarten R, Conner WC, Huber GW (2012) Production of levulinic acid from cellulose by hydrothermal decomposition combined with aqueous phase dehydration with a solid acid catalyst. Energy Environ Sci 5:7559–7574. doi:10.1039/c2ee21593d
143. Wenkin M, Touillaux R, Ruiz P et al (1996) Influence of metallic precursors on the properties of carbon-supported bismuth-promoted palladium catalysts for the selective oxidation of glucose to gluconic acid. Appl Catal A 148:181–199. doi:10.1016/S0926-860X(96)00231-1
144. Wu S, Fogiel AJ, Petrillo KL et al (2008) Protein engineering of nitrilase for chemoenzymatic production of glycolic acid. Biotechnol Bioeng 99:717–720. doi:10.1002/bit.21643
145. Xu B, Madix RJ, Friend CM (2010) Achieving optimum selectivity in oxygen assisted alcohol cross-coupling on gold. J Am Chem Soc 132:16571–16580. doi:10.1021/ja106706v
146. Yan L, Yang N, Pang H, Liao B (2008) Production of levulinic acid from bagasse and paddy straw by liquefaction in the presence of hydrochloride acid. CLEAN–soil, Air, Water 36:158–163. doi:10.1002/clen.200700100
147. Yan X, Jin F, Tohji K et al (2010) Hydrothermal conversion of carbohydrate biomass to lactic acid. AIChE J 56:2727–2733. doi:10.1002/aic
148. Yin H, Zhou C, Xu C et al (2008) Aerobic oxidation of D-glucose on support-free nanoporous gold. J Phys Chem C 112:9673–9678
149. Yin S, Dolan R, Harris M, Tan Z (2010) Subcritical hydrothermal liquefaction of cattle manure to bio-oil: Effects of conversion parameters on bio-oil yield and characterization of bio-oil. Bioresour Technol 101:3657–3664. doi:10.1016/j.biortech.2009.12.058
150. Yokoyama T, Chang H-M, Reiner RS et al (2004) Polyoxometalate oxidation of non-phenolic lignin subunits in water: Effect of substrate structure on reaction kinetics. Holzforschung 58:116–121. doi:10.1515/HF.2004.016

151. Yu Y, Wu H (2010) Understanding the primary liquid products of cellulose hydrolysis in hot-compressed water at various reaction temperatures. Energy Fuels 24:1963–1971. doi:10.1021/ef9013746
152. Yu Y, Wu H (2010) Significant differences in the hydrolysis behavior of amorphous and crystalline portions within microcrystalline cellulose in hot-compressed water. Ind Eng Chem Res 49:3902–3909. doi:10.1021/ie901925g
153. Yu Y, Wu H (2011) Kinetics and mechanism of glucose decomposition in hot-compressed water: effect of initial glucose concentration. Ind Eng Chem Res 50:10500–10508. doi:10.1021/ie2011388
154. Zakzeski J, Agnieszka D, Bruijnincx PCA, Weckhuysen BM (2011) Catalytic oxidation of aromatic oxygenates by the heterogeneous catalyst. Appl Catal A 394:79–85. doi:10.1016/j.apcata.2010.12.026
155. Zeng W, Cheng D, Chen F, Zhan X (2009) Catalytic conversion of glucose on Al–Zr mixed oxides in hot compressed water. Catal Lett 133:221–226. doi:10.1007/s10562-009-0160-3
156. Zhang H, Toshima N (2013) Glucose oxidation using Au-containing bimetallic and trimetallic nanoparticles. Catal Sci Technol 3:268–278. doi:10.1039/c2cy20345f
157. Zhang J, Deng H, Lin L (2009) Wet aerobic oxidation of lignin into aromatic aldehydes catalysed by a perovskite-type oxide: LaFe1-xCuxO3. Molecules 394:2747–2757. doi: 10.3390/molecules14082747
158. Zhang J, Liu X, Hedhili MN et al (2011) Highly selective and complete conversion of cellobiose to gluconic acid over Au-Cs2HPW12O40 nanocomposite catalyst. ChemCatChem 3:1294–1298
159. Zhang J, Liu X, Sun M et al (2012) Direct conversion of cellulose to glycolic acid with a phosphomolybdic acid catalyst in a water medium. ACS Catal 2:1698–1702. doi:10.1021/cs300342k
160. Zhang J, Weitz E (2012) An insitu NMR study of the mechanism for the catalytic conversion of fructose to 5-hydroxymethylfurfural and then to levulinic acid using 13C labeled D-fructose. ACS Catal 2:1211–1218
161. Zhang Q, Chuang KT (1998) Kinetics, catalysis, and reaction engineering alumina-supported noble metal catalysts for destructive oxidation of organic pollutants in effluent from a softwood kraft pulp mill. Ind Eng Chem Res 5885:3343–3349
162. Zhang S, Jin F, Hu J, Huo Z (2011) Improvement of lactic acid production from cellulose with the addition of Zn/Ni/C under alkaline hydrothermal conditions. Bioresour Technol 102:1998–2003. doi:10.1016/j.biortech.2010.09.049
163. Zhao H, Holladay JE, Brown H, Zhang ZC (2007) Metal chlorides in ionic liquid solvents convert sugars to 5-hydroxymethylfurfural. Science 316:1597–1600. doi:10.1126/science.1141199
164. Zope BN, Hibbitts DD, Neurock M, Davis RJ (2010) Reactivity of the gold/water interface during selective oxidation catalysis. Science 330:74–78. doi:10.1126/science.1195055

Chapter 6
Catalytic Hydrothermal Conversion of Biomass-Derived Carbohydrates to High Value-Added Chemicals

Zhibao Huo, Lingli Xu, Xu Zeng, Guodong Yao and Fangming Jin

Abstract The reduction of the dependence on petroleum oil and the development of renewable sources to replace fossil fuels have been a hot topic recently in the field of energy. One of the strategies for the replacement of petroleum oil in the production of fuels is to develop efficient renewable fuels and chemicals from biomass. Efficient conversion of biomass-derived carbohydrates into high value-added chemicals such as alcohols has drawn the attention of many chemists due to their industrial importance. This chapter focuses on our recent research relating to the synthesis of alcohols from biomass-derived esters and their interesting mechanism aspects. New developments including the synthesis of ethylene glycol and 1, 2-propanediol are discussed.

6.1 Introduction

Alcohol is an important intermediate chemicals and solvents, which has been widely used as antifreezes, liquid detergents, biodegradable polyester fibers, cosmetic products, pharmaceuticals, explosives, plasticizers, etc. and also often as a protecting group for carbonyl groups in organic synthesis [1]. Because of its industrial importance, the synthesis of ethylene glycol (EG) has been of considerable interest. Presently industrial production of alcohol is from petroleum-derived compounds [2]. However, over consumption of fossil fuels and their limited supply has strongly requested the reduction of the dependence on petroleum oil and the development of renewable and green sources [3–5]. Over the past several decades, biomass as a source to produce high value-added chemicals has attracted much attention due to its excellent performances such as abundance, renewability, less pollution, etc. [6].

Z. Huo (✉) · L. Xu · X. Zeng · G. Yao · F. Jin
School of Environmental Science and Engineering, Shanghai Jiao Tong University, 800 Dongchuan Road, Shanghai 200240, China
e-mail: hzb410@sjtu.edu.cn

The conversion of biomass-derived carbohydrates into alcohols is currently receiving increasing attention [7, 8]. Figure 6.1 shows some kinds of carbonyl compounds that can be derived from carbohydrates [9].

As we all know, the hydrogenation of esters to alcohols is an important reaction in organic chemistry [10–16]. However, the reaction of ester group, which has a less electrophilic carbonyl group, is still a great challenge with both academic and practical importance [17]. To date very few reports are known on the transformation.

Recently, hydrothermal technology for the transformation of carbohydrate biomass into value-added products has attracted much attention and been considered to be one of the most promising ways for the conversion of biomass because of the unique inherent properties of high-temperature water (HTW), including a high ion product (Kw) and a low dielectric constant [18, 19]. The remarkable properties of HTW allow it to convert a wide range of biomass materials into fuels and other value-added products. More importantly, HTW is an environmentally benign solvent that is preferable to organic solvents and an alternative reaction medium.

Previously, we and other groups have reported a series of experiments on the conversion of CO_2 [20–23] and biomass [24–26] into value-added chemicals in HTW. In this process, water acts not only as an excellent environmentally benign reaction medium but also as the hydrogen source generated by reduction of cheap metal reductants [27]. Recently, some research groups also reported that ZnO can be cycled to Zn through a redox process of a ZnO/Zn by using solar energy [28–30]. On the basis of these findings, we reasoned that esters might produce alcohols in a similar manner. The results obtained showed the feasibility of the hypothesis, and we found that the reaction of esters proceeded effectively in the presence of metal oxides to give the desired alcohols in high yield with high selectivity (Eq. 6.1). The detailed results are reported herein.

$$(CH_2)_n\!\!\underset{C=O}{\overset{O}{\diagup\!\!\diagdown}} + H_2O \xrightarrow[\text{Hydrothermal conditions}]{Cat. / Reductant} HO(CH_2)_{n+1}OH \qquad (6.1)$$

>No need gaseous hydrogen;
>H_2O as hydrogen source and solvent;
>No need base;
>Catalyst and reductant: commercially available;
>high efficiency, high yield.

6.2 Conversion of Biomass-Derived Glycolide to Ethylene Glycol Over CuO

Glycolide as an important biomass-derived compound can be obtained from biomass sources e.g. sugar cane, sugar beet, and immature grape juice [31].

Recently, Milstein et al. reported the hydrogenation of biomass-derived glycolide to the corresponding EG by well-defined electron rich bipyridine-based

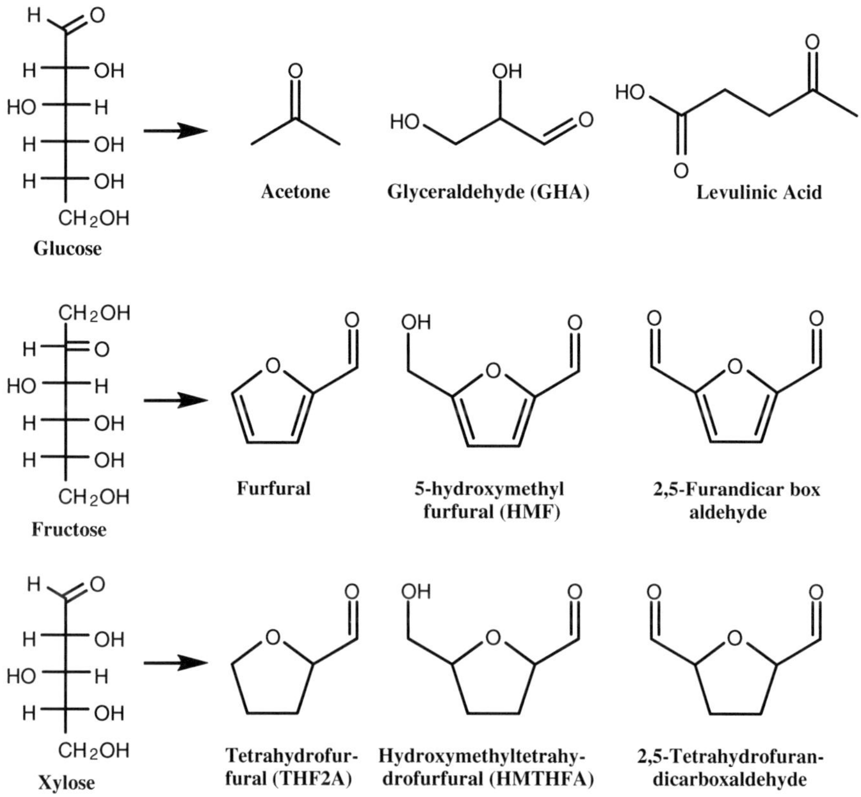

Fig. 6.1 Carbonyl compounds derived from carbohydrates [9]

PNN [2-(di-*tert*-butyl-phosphinomethyl)-6-(diethylaminomethyl) pyridine] Ru (II) pincer complexes under mild conditions [32]. However, the reactions still have considerable drawbacks. For example, use of high-pressure gaseous hydrogen (10–50 atm) which is not easy to obtain with less energy cost, organic solvent (THF or 1,4-dioxane), expensive catalysts (PNN-Ru), and longer reaction time (12–48 h). Therefore, the development of suitable reaction process for the production of EG from glycolide is highly desirable. Efficient conversion of biomass-derived glycolide into EG over CuO in HTW was first investigated (Eq. 6.2).

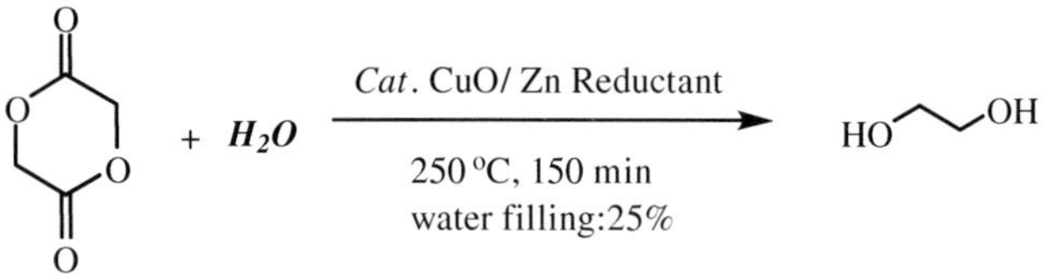

$$\text{glycolide} + H_2O \xrightarrow[\text{water filling:25\%}]{\textit{Cat}.\ \text{CuO/ Zn Reductant} \atop 250\,°C,\ 150\ \text{min}} \text{HO-CH}_2\text{-CH}_2\text{-OH} \quad (6.2)$$

6.2.1 Materials, Analysis, and Experimental Procedure

6.2.1.1 General Information

Glycolide ($\geq$99 %, Sigma-Aldrich) was used as reagent and ethylene glycol ($\geq$99.5 %, GC) was purchased from TCI for the qualitative analysis of the products in the liquid samples. As preliminary tests, various active metals including Zn, Mg, Fe, Al, and Mn (AR, Sinopharm Chemical Reagent Co., Ltd) were used as reductants in powder form, metal oxide including Fe_3O_4, Fe_2O_3, Ni_2O_3, TiO_2, ZrO_2, Cu_2O, and CuO (150–200 mesh, i.e., 0.074–0.1 mm) (AR, Sinopharm Chemical Reagent Co., Ltd) were used as catalysts in powder form.

6.2.1.2 Product Analysis

After reaction, the liquid samples were analyzed with a GC-FID (Shimadzu GC-2010) equipped with an DB-FFAP capillary column with dimensions of 30 m × 250 µm × 0.25 µm for the quantification of ethylene glycol. The liquid samples' analyses were confirmed by GC–MS (Agilent GC7890A–MS5975C) equipped with an HP Innowax polyethylene glycol capillary column with dimensions of 30 m × 250 µm × 0.25 µm. Thin layer chromatography (TLC) was performed on aluminum-precoated plates of silica gel 60 with an HSGF254 indicator and visualized under UV light or developed by immersion in the solution of 0.6 % $KmnO_4$ and 6 % K_2CO_3 in water. The solid samples were characterized by X-ray diffraction (XRD) (Shimadzu XRD-6100) to determine the composition and phase purity.

The yield of ethylene glycol is defined as the carbon mole ratio of produced ethylene glycol to the initial glycolide as below. The yields were obtained from experiments over three times and relative error was less than 5 %.

$$\text{Yield, mmol \%} = \frac{\text{C in ethylene glycols, mmol}}{\text{C in the initial glycolide, mmol}} \times 100\%$$

6.2.1.3 General Procedure for the Synthesis of Ethylene Glycol from Glycolide

Experiments were conducted in a Teflon-lined stainless steel batch reactor with an inner volume of 30 mL. The typical procedure for the synthesis of ethylene glycol from glycolide was as follows. Firstly, glycolide (0.058 g, 0.5 mmol), Zn (1.63 g, 25 mmol), CuO (0.48 g, 6 mmol), and ultrapure water (7.5 mL) were loaded into reactor. Next, nitrogen was charged into reactor in order to exclude the effect of air and then the reactor was sealed and placed into an oven that had been preheated to the desired temperature. After 150 min, the reactor was removed from the oven and cooled to room temperature with an electric fan. The reaction time was defined as the time when the reactor's temperature was increased to 250 °C. Water filling was defined as the ratio of the volume of the ultrapure water (7.5 mL) put into the reactor to the inner volume (30 mL) of the reactor. Finally, after cooling off, liquid sample was collected and filtered with 0.45 μm Syringe Filter, the highest yield of ethylene glycol (94 %) was obtained. Solid sample was collected and washed with deionized water and ethanol several times to remove impurities and dried in the oven at 50 °C for 24 h.

6.2.2 Results and Discussion

6.2.2.1 Effects of Reaction Conditions on the Yields of Ethylene Glycol

Initially, we carried out the experiment to investigate the effect of metal or metal oxides for the synthesis of ethylene glycol from glycolide in the presence of 6 mmol Cu and 20 mmol Zn at 250 °C for 150 min in water (water filling: 25 %). The reaction proceeded efficiently and produced the desired ethylene glycol in 23 % yield analyzed by GC-FID and GC–MS (entry 4). Solid residual after reaction was analyzed by XRD (Figs. 6.2, 6.3, 6.4), it showed Cu still existed in solid samples while Zn was oxidized to ZnO. It is clarified that Zn acts as a reductant, and Cu has catalytic activity for the conversion of glycolide as a catalyst. Next, we screened various catalysts to search the suitable one. The results are summarized in Table 6.1. The reaction did not give the desired ethylene glycol at all in the absence of both reductant and catalyst or reductant only (entries 1 and 3). The reaction in the presence of 20 mmol Zn, without the addition of any catalyst, gave desired ethylene glycol with the yield of 9 % (entry 2), indicating the importance of the combined use of both reductant and catalyst. Among the various catalysts we investigated (entries 5–13), the use of Cu_2O and CuO improved the transformation efficiency and afforded ethylene glycol in good yield with no glycolide recovery monitored by TLC (entries 12 and 13), the latter gave much better result. Other catalysts, such as Fe_2O_3, Fe_3O_4, Ni_2O_3 and Fe, Co, TiO_2, ZrO_2, either gave low yield (entries 7–9) or no reaction took place (entries 5, 6, 10 and 11).

The residual solid sample after reaction with the addition of CuO and Zn was analyzed by XRD. Unexpectedly, the Cu and ZnO were observed in solid sample rather than expected CuO and ZnO. From this result, we reasoned that CuO as a catalyst might be reduced completely to Cu by in situ-formed hydrogen via the oxidation of Zn in HTW [33]. We measured the concentration of Cu ions in the solution by ICP and found few Cu ions (<1 ppm) in liquid sample. It suggested that Cu probably existed in the solid residual. Thus, we performed the experiment under the same reaction conditions without the addition of glycolide in order to investigate the role of in situ-formed hydrogen. The results suggested that hypothesis was completely realistic. CuO was reduced completely to Cu by

Fig. 6.2 GC/MS chart of ethylene glycol

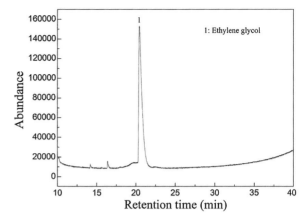

Fig. 6.3 GC-FID chart of ethylene glycol

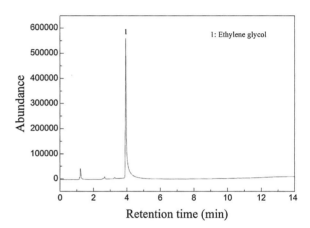

Fig. 6.4 XRD patterns of solid samples after reaction (glycolide: 0.5 mmol; Zn: 20 mmol; Cu (or CuO): 6 mmol; water filling: 25 %; time: 150 min; temp: 250 °C)

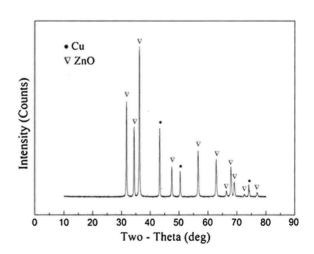

6 Catalytic Hydrothermal Conversion

Table 6.1 Effect of catalysts on the yields of ethylene glycol[a]

Entry	Reductant	Catalyst	Yield (%)[b]
1	—	—	0
2	Zn	—	9
3	—	Cu	0
4	Zn	Cu	23
5	Zn	Fe	0
6	Zn	Co	0
7	Zn	Fe_2O_3	14
8	Zn	Fe_3O_4	10
9	Zn	Ni_2O_3	10
10	Zn	TiO_2	0
11	Zn	ZrO_2	0
12	Zn	CuO	45
13	Zn	Cu_2O	37

[a] *Reaction condition*: glycolide: 0.5 mmol; Zn: 20 mmol; catalyst: 6 mmol; water filling: 25 %; time: 150 min; temp: 250 °C
[b] The yields of ethylene glycol is defined as the ratio of the amount of carbon in produced that of the initial glycolide

in situ-formed hydrogen, Cu and ZnO were found in solid samples by XRD analysis after reaction.

Next, we investigated the effect of the amount of CuO on the yields of ethylene glycol. The reaction was conducted in the presence of 20 mmol Zn with 25 % water filling at 250 °C for 150 min with the amount of CuO varied from 3 to 12 mmol, the results were shown in Fig. 6.5. The yield of ethylene glycol increased significantly at first with the amount of CuO increasing from 0 to 6 mmol, and the best yield of 45 % of ethylene glycol was achieved when the amount of CuO was increased to 6 mmol. The yield was decreased with further increase of the amount of CuO. It might be explained that more CuO would lead to more cracking products such as acetic acid, CO_2 and H_2O (Eq. 6.3), which might result in the decrease in glycolide selectivity [34].

$$CH_2OH - CH_2OH_{(l)} + 5CuO_{(s)} \rightarrow 5Cu^0_{(s)} + 2CO_{2(g)} + 3H_2O_{(g)} \quad (6.3)$$

Finally, a carbon balance was conducted. The result showed that the conversion of glycolide for liquid products was about 98 %. The main product observed for this transformation was ethylene glycol, and the yield of ethylene glycol was 45 %, other products were acetic acid which had a yield of 5 % and unknown compounds. Gas product observed was CO_2 in the transformation.

Experiments were performed to investigate the various active metals, such as Zn, Al, Fe, Mn,and Mg, to search suitable reductant for the conversion of glycolide to ethylene glycol. Figure 6.6a shows the effect of different reductants on the yields of ethylene glycol. Among the various metals tested, Zn had significant effects in the conversion of glycolide and gave the best yield (45 %) of ethylene

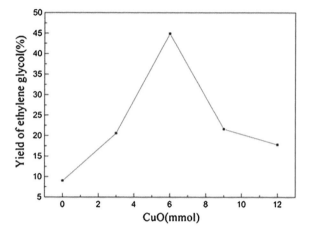

Fig. 6.5 Effect of the amount of CuO on the yields of ethylene glycol (glycolide: 0.5 mmol; Zn: 20 mmol; temp: 250 °C; time: 150 min; water filling: 25 %)

glycol compared to Al, Mg, Mn, and Fe. Thus, Zn was chosen for the following experiments.

Figure 6.6b shows the effect of the amount of Zn on the yields of ethylene glycol at different amounts from 0 to 30 mmol. No desired ethylene glycol was observed without the addition of Zn. The yields of ethylene glycol improved gradually with the increasing amount of Zn from 0 to 15 mmol, and then remarkably increased from 15 to 25 mmol. The yield of ethylene glycol decreased a little when the amount of Zn exceeded 25 mmol. The results indicated that the improvement of hydrogen production with the increase amount of Zn in HTW was favorable for the conversion of glycolide to ethylene glycol. The optimal amount of Zn was 25 mmol and gave the ethylene glycol in the maximum yield of 94 %.

The effect of water filling on the yields of ethylene glycol from glycolide was investigated at different water filling from 15 to 45 % with the addition of 6 mmol CuO and 25 mmol Zn at 250 °C for 150 min. The results were shown in Fig. 6.7. First, the water filling profile had a small change and the ethylene glycol yield gained slight increase from 15 to 20 %. The yield of ethylene glycol significantly increased with the increasing water filling from 15 to 25 % and then decreased after 25 %. The optimum water filling was 25 % to obtain the highest yield of 94 % ethylene glycol from glycolide. The decrease in conversion of glycolide at a higher water filling was most likely due to the decreasing concentration of the initial starting materials when water filling increased.

The effect on the yields of ethylene glycol at different temperatures from 100 to 250 °C was examined. The results were shown in Fig. 6.8a. The reaction did not take place at 100 °C and no ethylene glycol was observed. It indicated that lower reaction temperature was completely ineffective in the conversion of glycolide. The yield of ethylene glycol improved gradually with the increase of reaction temperature from 100 to 200 °C, and then remarkably increased as the temperature increased from 200 to 250 °C. The maximum value of 94 % ethylene glycol was obtained at 250 °C. However, the temperature was limited to 250 °C due to the maximum durable temperature of the Teflon material.

Fig. 6.6 **a** Effect of reductants (glycolide: 0.5 mmol; Metal: 20 mmol; CuO: 6 mmol; water filling: 25 %; time: 150 min; temp: 250 °C). **b** Effect of the amount of Zn (glycolide: 0.5 mmol; CuO: 6 mmol; time: 150 min; temp: 250 °C; water filling: 25 %)

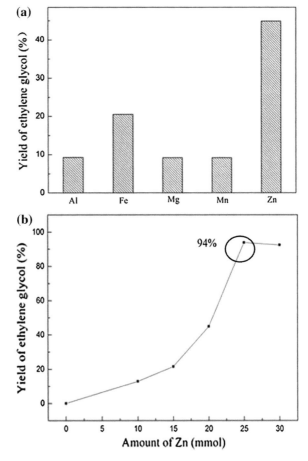

Fig. 6.7 Effect of water filling on the yields of ethylene glycol (glycolide: 0.5 mmol; Zn: 25 mmol; CuO: 6 mmol; time: 150 min; temp: 250 °C)

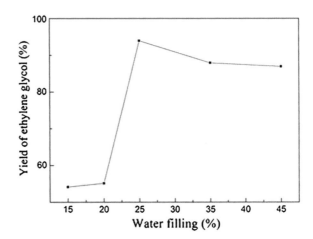

Fig. 6.8 Effect of reaction temperature and reaction time on the yields of ethylene glycol (**a** glycolide: 0.5 mmol; Zn: 25 mmol; CuO: 6 mmol; time: 150 min; water filling: 25 %; **b** glycolide: 0.5 mmol; Zn: 25 mmol; CuO: 6 mmol; temp: 250 °C; water filling: 25 %)

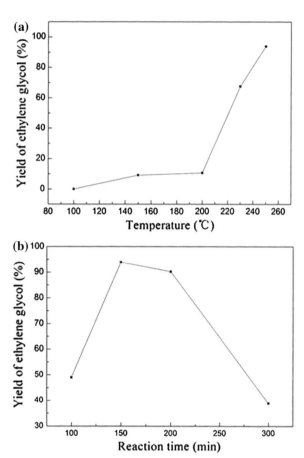

Experiments were performed by changing reaction time from 100 to 300 min at 250 °C to examine the effect of reaction time on the yields of ethylene glycol. Figure 6.8b indicated that the yield increased drastically with reaction time increased in the first 150 min, when the time was increased to 150 min, the yield was obtained in the maximum value of 94 %. However, the yield of ethylene glycol decreased with further increase of the time. The decreasing yield of ethylene glycol after 150 min can be attributed to the decomposition of produced ethylene glycol. To confirm this assumption, the decomposition reaction of ethylene glycol was performed in the presence of 6 mmol Cu and 25 mmol ZnO with 25 % water filling at 250 °C for 300 min. Experiment result indicated that ethylene glycol was decomposed and the concentration of which was decreased from 120 to 52 mmol/L by GC-FID, XRD analysis also showed that Cu and ZnO did not change and still existed in solid residual. From the result above, it can be suggested that long reaction time has an influence on the yields of ethylene glycol in the presence of Cu and ZnO obtained in the reaction [34].

6.2.2.2 Possible Mechanism for the Production of Ethylene Glycol

Based on previous reports [31] and findings in this work, the pathway for the conversion of glycolide to ethylene glycol over CuO in HTW was proposed in Scheme 6.1. First, hydrogen was formed via the oxidation of Zn in HTW [35, 36]. The in situ-formed hydrogen was adsorbed on the surface of Cu/CuO though the formation of hydrogen bonds between hydrogen and Cu/CuO. Also glycolide was adsorbed on the surface of Cu/CuO by the combination of Cu/CuO with carbonylic C and O atoms [37, 38]. When the hydrogen reacted with glycolide molecule, an intermediate 2-hydroxyacetaldehyde was obtained. Next, both the intermediate 2-hydroxyacetaldehyde and hydrogen formed were adsorbed on the surface of Cu/CuO, first H was added to C=O and an intermediate linked by an α-bond was formed, the intermediate acquired one more H from the surface of Cu/CuO, and desired ethylene glycol was obtained.

6.3 Conversion of DL-Lactide into 1, 2-Propanediol Over CuO

Lactides as an important biomass-derived compound can be produced from biomass sources, e.g., lactic acid (from fermentation of glucose) via self-esterification [31]. With success in developing the conversion of glycolide to ethylene glycol in HTW, we next examined the scope of reaction with respect to DL-lactide in HTW.

6.3.1 General Procedure for the Formation of 1, 2-Propanediol

Experiments were conducted in a Teflon-lined stainless steel batch reactor with an inner volume of 30 mL. The typical procedure for the synthesis of ethylene glycol from glycolide was as follows. First, DL-lactide (0.072 g, 0.5 mmol), Zn (1.63 g, 25 mmol), CuO (0.48 g, 6 mmol), and ultrapure water (7.5 mL) were loaded into reactor. Next, nitrogen was charged into reactor in order to exclude the effect of air and then the reactor was sealed and placed into an oven that had been preheated to the desired temperature. After 150 min, the reactor was removed from the oven and cooled to room temperature with an electric fan. The reaction time was defined as the time when the reactor's temperature was increased to 250 °C. Water filling was defined as the ratio of the volume of the ultrapure water (7.5 mL) put into the reactor to the inner volume (30 mL) of the reactor. Finally, after cooling off, liquid sample was collected and filtered with 0.45 μm Syringe Filter, the highest yield of 1,2-propanediol (84 %) was obtained. Solid sample was collected and washed with deionized water and ethanol several times to remove impurities and dried in the oven at 50 °C for 24 h.

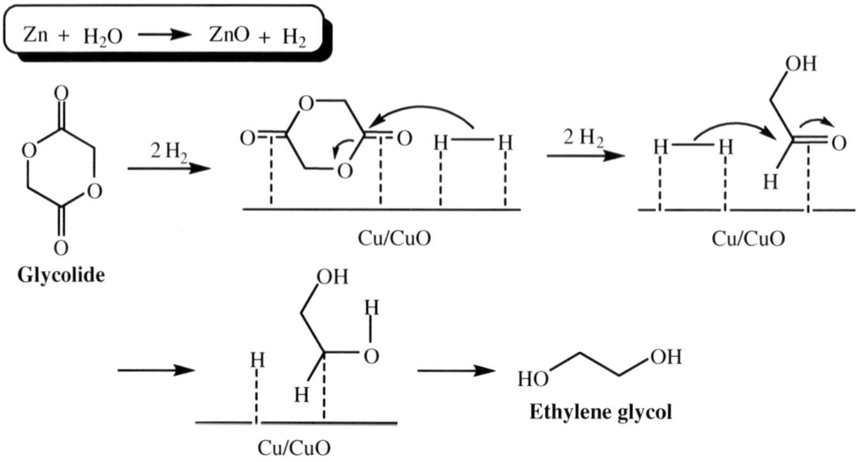

Scheme 6.1 Proposed mechanism for the formation of ethylene glycol

6.3.2 The Formation of 1, 2-Propanediol from DL-Lactide

We investigated the scope of reaction of DL-lactide in HTW and the result was shown in Eq. (6.4). The reaction of DL-lactide was carried out in the presence of 25 mmol Zn and 6 mmol Cu with 25 % water filling at 250 °C for 150 min and gave desired 1,2-propanediol in 84 % yield. The result indicated that substituted cyclic di-ester can also proceed efficiently and give the product in high yield in HTW.

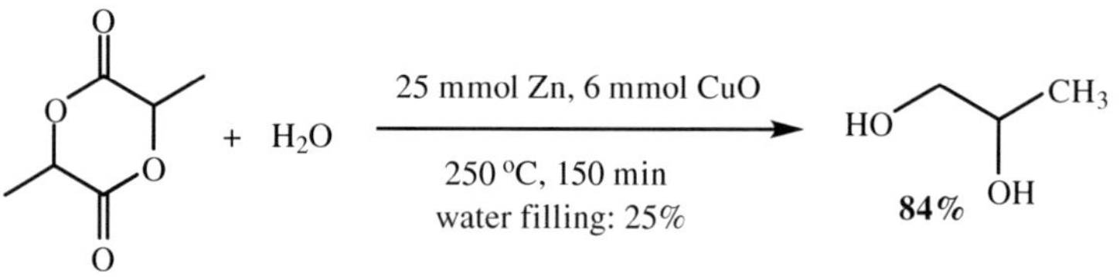

(6.4)

6.4 Conclusions and Prospects

In summary, our results show that it is technically possible to efficiently and directly convert biomass-derived carbohydrates into various types of alcohols over CuO in HTW. This study provided a significant process for the conversion of biomass-derived cyclic di-esters to 1, 2-diols with high selectivity and high yield. Further research is undergoing to develop the efficient methods for the conversion of biomass-derived carbohydrates to value-added chemicals in high temperature water.

References

1. Xu G, Li Y, Li Z, Wang H (1995) Kinetics of hydrogenation of diethyl oxalate to ethylene glycol. Ind Eng Chem Res 34:2371–2378
2. Haas T, Jaeger B, Weber R, Mitchell SF, King CF (2005) New diol processes: 1,3-propanediol and 1,4-butanediol. Appl Catal A 280:83–88
3. Hermann BG, Block K, Patel MK (2007) Producing bio-based bulk chemicals using industrial biotechnology saves energy and combats climate change. Environ Sci Technol 41:7915–7921
4. Chheda JN, Huber GW, Dumesic JA (2007) Liquid-phase catalytic processing of biomass-derived oxygenated hydrocarbons to fuels and chemicals. Angew Chem 46:7164–7183
5. Huber GW, Corma A (2007) Synergies between bio- and oil refineries for the production of fuels from biomass. Angew Chem Int Ed 46:7184–7201
6. Jin FM, Enomoto H (2011) Rapid and highly selective conversion of biomass into value-added products in hydrothermal conditions: chemistry of acid/base-catalysed and oxidation reactions. Energy Environ Sci 4:382–397
7. Yin AY, Guo XY, Dai WL, Fan KN (2009) The synthesis of propylene glycol and ethylene glycol from glycerol using raney Ni as a versatile catalyst. Green Chem 11:1514–1516
8. Zhou JX, Guo LY, Guo XW, Mao JB, Zhang SG (2010) Selective hydrogenolysis of glycerol to propanediols on supported Cu-containing bimetallic catalysts. Green Chem 12:1835–1843
9. Chheda JN, Dumesic JA (2007) An overview of dehydration, aldol-condensation and hydrogenation processes for production of liquid alkanes from biomass-derived carbohydrates. Catal Today 123:59–70
10. Kuriyama W, Ino Y, Ogata O, Sayo N, Saito T (2010) A homogeneous catalyst for reduction of optically active esters to corresponding chiral alcohols without loss of optical purities. Adv Synth Catal 352:92–96
11. Fogler E, Balaraman E, Ben-David Y, Leitus G, Shimon LJW, Milstein D (2011) New CNN-type ruthenium pincer NHC complexes. Mild efficient catalytic hydrogenation of esters. Organometallics 30:3826–3833
12. Ino Y, Kuriyama W, Ogata O, Matsumoto T (2010) An efficient synthesis of chiral alcohols via catalytic hydrogenation of esters. Top Catal 53:1019–1024
13. Saudan LA, Saudan CM, Debieux C, Wyss P (2007) Dihydrogen reduction of carboxylic esters to alcohols under the catalysis of homogeneous ruthenium complexes: high efficiency and unprecedented Chemoselectivity. Angew Chem Int Ed 46:7473–7476
14. Wu Y, Perez M, Scalone M, Ayad T, Ratovelomanana-Vidal V (2013) Ruthenium-catalyzed asymmetric transfer hydrogenation of 1-Aryl-sunstituted dihydroisoquinolines: access to valuable chiral 1-aryl-tetrahydroisoquinoline scaffolds. Angew Chem Int Ed 52:4925–4928
15. Yakebayashi S, Bergens SH (2009) Facile bifunctional additional of lactones and esters at Low temperature. The first intermediates in lactone/ester hydrogenations. J Am Chem Soc 133:4240–4242
16. Ito M, Ikariya T (2007) Catalytic hydrogenation of polar organic functionalities based on Ru-mediated heterolytic dihydrogen cleavage. Chem Commun 43:5134–5142
17. Ito M, Ootsuka T, Watari R, Shiibashi A, Himizu A, Ikariya T (2011) Catalytic hydrogenation of carboxamides and esters by well-defined Cp*Ru complexes bearing a protic amine ligand. J Am Chem Soc 133:4240–4242
18. Shaw RW, Brill YB, Clifford AA, Eckert CA, Franck EU (1991) Supercritical water a medium for chemistry. Chem Eng News 69:26–39
19. Akiya N, Savage PE (2002) The roles of water for chemical reactions in high-temperatures water. Chem Rev 102:2725–2750
20. Huo ZB, Hu MB, Zeng X, Yun J, Jin FM (2012) Catalytic reduction of carbon dioxide into methanol over copper under hydrothermal conditions. Catal Today 194:25–29

21. Jin FM, Gao Y, Jin YJ, Zhang YL, Cao JL, Wei Z, Smith RL (2011) High-yield reduction of carbon dioxide into formic acid by zero-valent metal/metal oxide redox cycles. Energy Environ Sci 4:881–884
22. Tian G, Yuan HM, Mu Y, He C, Feng SH (2007) Hydrothermal reactions from sodium hydrogen carbonate to phenol. Org Lett 9:2019–2021
23. He C, Tian G, Liu ZW, Feng SH (2010) A mild hydrothermal route to fix carbon dioxide to simple carboxylic acids. Org Lett 12:649–651
24. Huber GW, Cortright RD, Dumesic JA (2004) Renewable alkanes by aqueous-phase reforming of biomass-derived oxygenates. Angew Chem Int Ed 43:1549–1551
25. Huber GW, Chheda JN, Barrett CJ, Dumesic JA (2005) Production of liquid alkanes by aqueous-phase processing of biomass-derived carbohydrates. Science 308:1446–1450
26. Chheda JN, Huber GW, Dumesic JA (2007) Liquid-phase catalytic processing of biomass-derived oxygenated hydrocarbons to fuels and chemicals. Angew Chem Int Ed 46:7164–7183
27. Kruse A, Dinjus E (2007) Hot compressed water as reaction medium and reactant: 2. degradation reactions. J Supercrit Fluids 41:361–379
28. Service RF (2009) Sunlight in your tank. Science 326(5959):1472–1475
29. Steinfeld A (2005) Solar thermochemical production of hydrogen-a review. Sol Energy 78:603–615
30. Stamatiou A, Steinfeld A, Jovanovic Z (2013) On the effect of the presence of solid diluents during Zn oxidation by CO_2. Ind Eng Chem Res 52:1859–1869
31. Davachi SM, Kaffashi B, Roushandeh JM (2012) Synthesis and characterization of a novel terpolymer based on *t*-lactide, glycolide and trimethylene carbonate for specific medical applications. Polym Adv Technol 23:565–573
32. Balaraman E, Fogler E, Milstein D (2012) Efficient hydrogenation of biomass-derived cyclic di-esters to 1,2-diols. Chem Comm 48:1111–1113
33. Li Y, Xu G, Liu C, Eliasson B, Xue B (2001) Co-generation of syngas and higher hydrocarbons from CO_2 and CH_4 using dielectric-barrier discharge: effect of electrode materials. Energy Fuels 15:299–302
34. Larcher D, Patrice R (2000) Preparation of metallic powders and alloys in polyol media: a thermodynamic approach. J Solid State Chem 154:405–411
35. Resende FLP, Savage PE (2010) Effect of metals on supercritical water gasification of cellulose and lignin. Ind Eng Chem Res 49:2694–2700
36. Elliott DC (2008) Catalytic hydrothermal gastification of biomass. Biofuels Bioprod Bioref 2:254–265
37. Yan ZP, Lin L, Liu S (2009) Synthesis of γ-valerolactone by hydrogenation of biomass-derived levulinic acid over Ru/C catalyst. Energy Fuels 23:3853–3858
38. Kieboom APG, van Rantwiik F (1981) Hydrogenation and hydrogenolysis in synthetic organic chemistry. Science Press, Beijing

Part III
Hydrothermal Conversion of Biomass into Fuels

Chapter 7
Effective Utilization of Moso-Bamboo (*Phyllostachys heterocycla*) with Hot-Compressed Water

Satoshi Kumagai and Tsuyoshi Hirajima

Abstract In this study, the hydrothermal carbonization behavior of bamboo in hot-compressed water (HCW) using a batch-type reactor at 180–300 °C was observed to investigate the effective utilization of bamboo as a biomass resource. Polysaccharides (hemicellulose and cellulose) in the bamboo were changed to water soluble products. At 180–220 °C, hemicellulose (arabinoxylan) was first hydrolyzed to xylooligosaccharides and then to xylose that was further decomposed to various organic acids and furfural. However, most of the cellulose was not decomposed and was recovered as a solid residue at this temperature range. Cellulose began hydrolyzing to glucose at temperatures above 240 °C. The glucose was further decomposed to various organic acids and 5-HMF. The recovered oligosaccharides and monosaccharides can be used as functional food, food additives, and feedstocks for ethanol and lactic fermentation. Furthermore, organic acids and furans can be used as various chemicals. More hemicellulose and cellulose, which have relatively low carbon content in the bamboo, were decomposed and dissolved in water. As a result, the solid residue consisted mainly of lignin, which has higher carbon content compared to cellulose and hemicellulose. Hence, the heating value of the solid residue increased at higher temperature during treatment and the residue could be considered as a solid fuel.

S. Kumagai (✉)
Research and Education Center of Carbon Resource, Kyushu University, Nishi-ku, Fukuuoka 819-0395, Japan
e-mail: f9229@cc.saga-u.ac.jp; kumagais@hotmail.co.jp

T. Hirajima
Faculty of Engineering, Kyushu University, Nishi-ku, Fukuuoka 819-0395, Japan

S. Kumagai
Organization for Cooperation with Industry and Regional Community, Honjyo, Saga 840-8502, Japan

7.1 Introduction

From viewpoints of global environmental problems and diminishing fossil-fuel resources, renewable, and carbon neutral lignocellulosic biomass have been drawing attention as environmentally friendly resource. However, conversion processes for its effective utilization have not been established. For that reason, we have been studying the application of hydrothermal reactions for utilization of lignocellulosic biomass.

On the other hand, hydrothermal treatment involving the conversion of lignocellulosic biomass in hot-compressed water (HCW; ~300 °C, ~10 MPa) has received significant attention in recent years [1–6] (Fig. 7.1). Characteristics of HCW are represented by the high ionic product. The ionic product of water changes as the temperature changes. It reaches the maximum value, 6.34×10^{-12}, at around 250 °C under saturated vapor pressure condition and decreases to 1.88×10^{-16} at critical point (374.15 °C, 22.12 MPa).

This type of treatment is environmentally friendly, since typically only water is used in the process. Additionally, the advantage of this treatment is that it can convert wet input low-grade material into higher grade substance at relatively high yields without the need for an energy-intensive drying before or during process. For example, this treatment is used for a saccharification method. Bobleter et al. are pioneers in this research area and have studied this process using cellobiose as a model component of cellulose [1, 2]. They demonstrated that the reaction mechanism of hydrothermolysis is different from that of acidic hydrolysis. In the former, cellulose is hydrolyzed to water solubles by using water alone [3]. Previous experiments by our group have shown the decomposition behavior of some real lignocellulosic biomass (rice hull [5] and barley straw [6]) in HCW. However, the purpose of their experiments was limited to sugar production from cellulose and hemicellulose. Distribution of other water soluble products and characteristics of the treatment residue have not been clarified in detail.

On the other hand, bamboo is considered as a promising biomass resource and an alternative to fossil fuel resources recently.

Bamboo has been used in agriculture, architecture, and the production of bamboo shoots. However, in recent years, alternatives have been developed and cheaper bamboo shoots are imported from other countries. Bamboo forest areas have thus expanded rapidly in Japan, and bamboo plantations are invading natural or plantation forest areas [7]. Bamboos grow relatively fast and attain stand maturity within 5 years. Stands of tall species may attain a height of 15–20 m. The largest known bamboos grow to a height of 40 m and have a stem diameter of 30 cm [8].

Hence, in this study, moso-bamboo was used as a raw material and treated in HCW using a batch-type reactor at 180–300 °C for the effective utilization of lignocellulosic biomass.

Then, the hydrothermal decomposition behavior of chemical components, water soluble product distribution, and characteristics of the treatment residue were

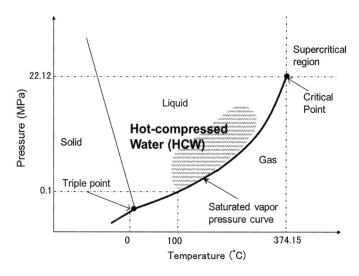

Fig. 7.1 Phage diagram of pure water

examined. Finally, we conducted the hydrothermal treatment of other lignocellulosic biomass to use water insoluble as a solid fuel, and then investigated their characteristics.

7.2 Experimental

7.2.1 Raw Materials

Moso-bamboo (*Phyllostachys pubescens*) was used in this study. It was pulverized to under 0.5 mm and then dried in a vacuum-drying oven at 60 °C for 24 h before treatment. The components were analyzed making reference to a report by the U.S. National Renewable Energy Laboratory [9].

In brief, the wax content in the sample was measured by the soxhlet extraction of 5.0 g of the sample with ethanol. The amount of dewaxed carbohydrates and lignin were measured as follows: 150 mg of the dewaxed dried sample was treated with 1.5 ml of 72 w/w% H_2SO_4 at 30 °C for 1 h. Subsequently, 42 ml of H_2O was added and treated in an autoclave at 121 °C for 1 h. After cooling, the reaction products were washed and filtered using a GP 16 glass filter in vacuum. The solid residue on the filter was dried to a constant weight at 105 °C. The weight of the solid residue was calculated as Klason lignin. Before performing high performance liquid chromatography (HPLC) on the sugars, the filtrate was treated in a column (OnGuard IIA, Dionex) to remove H_2SO_4 and then filtered through a 0.2 μm membrane filter. Glucose, xylose, and arabinose were analyzed by HPLC where the apparatus was equipped with a KC-811 column (Shodex) and refractive index

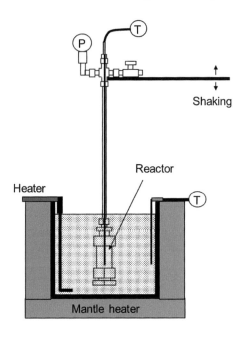

Fig. 7.2 Schematic diagram of the batch type apparatus

(RI) detector (RI-2031, JASCO). HPLC was conducted at 50 °C with 2 mM HClO$_4$ flowing at a rate of 0.7 ml/min. The obtained concentrations of each sugar were transformed to their polysaccharide concentrations by the following formula:

$$\text{cellulose (wt\%)} = \text{glucose (wt\%)} \times 0.9,$$

$$\text{hemicellulose (wt\%)} = (\text{xylose} + \text{arabinose})(\text{wt\%}) \times 0.88$$

Furthermore, the ash content was determined by measuring the weight of the residue after heating 1.0 g of the sample at 575 °C for 5 h. Summative analyses of moso bamboo showed 39.4 wt% cellulose, 20.2 wt% hemicellulose (19.3 wt% xylan, 0.9 wt% arabinan), 24.7 wt% lignin, 1.7 wt% ash, and 7.5 wt% wax.

7.2.2 Hydrothermal Treatment

The experiments were performed in a batch-type reactor (SUS 316, 14 ml volume) with a valve, as shown in Fig. 7.2. The reaction temperature was measured using a K-thermocouple in the reactor. A sample of 1.5 g of dried moso bamboo powder and 12 g of pure water was charged in the reactor and connected to hydrothermal equipment. Air in the equipment was displaced with nitrogen gas pressurized at 0.5 MPa before sealing the equipment. Hydrothermal treatment was conducted in a fused salt bath (KNO$_3$:NaNO$_2$:NaNO$_3$ = 53:40:7); that is, the reactor was immersed and shaken vigorously in the bath heated at various temperatures. It took 3 mins for

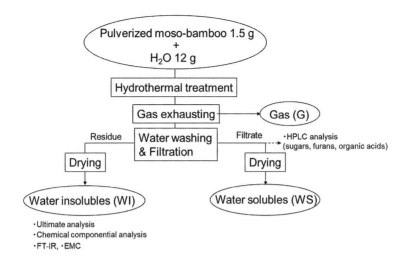

Fig. 7.3 Experimental procedure

the temperature to increase from room temperature to the target temperature. After the reactor had been immersed in the salt bath for 10 mins, it was immediately transferred into a water bath and cooled to end the reaction.

7.2.3 Analysis Procedure

The product separation and analysis procedure is illustrated in Fig. 7.3. After cooling the reactor, the gas was exhausted. The reaction mixture was recovered and filtered using a GP 5 glass filter in vacuum for solid–liquid separation after washing with enough water. Next, the water insoluble solid residue (WI) and water soluble components (WS) were dried at 105 °C. WS and WI were weighed, and the product yields were calculated based on the dry feed weight. The weight loss included gaseous products (G).

G was obtained by subtracting the yields of WS and WI from 100.

The Monosaccharides and organic acids of WS were analyzed by HPLC as described above. Then, xylooligosaccharides (xylobiose and xyrotriose) were analyzed by HPAE-PAD (Dionex, DX-500) [5]. After drying, the WI residue was subject to chemical composition described above and ultimate analyses (Yanagimoto, MT-5). The higher heating value (HHV) was calculated according to Dulong's formula [10]:

$$\text{HHV (MJ/kg)} = 0.3383\text{C} + 1.442(\text{H} - 1/8\text{O})$$

where C, H, and O are the weight percentage of carbon, hydrogen, and oxygen, respectively.

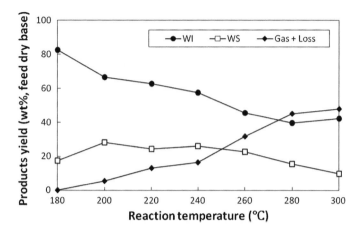

Fig. 7.4 Yields of WI, WS, and G + Loss as a function of reaction temperature

For equilibrium moisture content (EMC) determined according to JIS M811, an aliquot of the sample was placed in a desiccator containing saturated salt solution. In this study, the relative humidity inside desiccator was maintained 75 % using a saturated NaCl solution at 20 °C. After equilibrium was reached, the moisture content was measured by the drying at 105 °C. Identification of the chemical functional groups were performed on the Fourier transform infrared (FTIR, JASCO 670 Plus) using a KBr disk method.

7.3 Results and Discussion

7.3.1 Mass Balance

Figure 7.4 shows the yields of WI, WS, and G as a function of reaction temperature for 10 mins. WI yield decreased gradually with increasing reaction temperature. Finally, WI yield reached 39.5 wt% at 280 °C. WS yield was comparatively steady though two weak peaks were observed at around 200 and 240 °C. This behavior is presumably due to WS generated by the decomposition of bamboo, and then the WS was quickly converted to G by excessive decomposition. In addition, as described later, the two peaks are attributed to the difference in decomposition characteristics of hemicellulose and cellulose. The former decomposes at around 200 °C and the latter at temperatures over 240 °C. Hence, it is assumed that the WS peak at 200 °C is derived from hemicellulose, and the peak at 240 °C is derived from cellulose. At lower treatment temperatures, the yield of

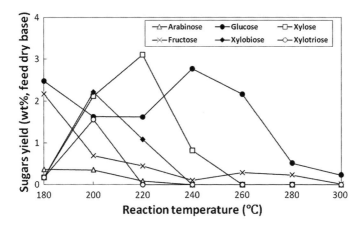

Fig. 7.5 Relationship between reaction temperature and sugars yield

G was very low. However, the yield increased with treatment temperature. Finally, the yield attained approximately 47.9 wt% at 300 °C.

7.3.2 Sugar Production

Figure 7.5 shows the relationship between reaction temperature and sugars yield. The main compounds in WS at 180 °C were glucose and fructose. It was assumed that these sugars were not derived from the decomposition products of cellulose but derived from free sugars because the decomposition ratio of cellulose is very low in this temperature range, as described later. Kozukue et al. reported that bamboo contained some free sugars such as glucose and fructose [11].

A small amount of arabinose, xylose, xylobiose, and xylotriose derived from hemicellulose were observed at this temperature. When the temperature reached 200 °C, the yields of glucose, fructose, and arabinose decreased while those of xylose, xylobiose, and xylotriose, derived from xylan, increased up to 2.1, 2.2, and 1.6 wt%, respectively. Ando et al. investigated xylooligosaccharides production from moso-bamboo with this region HCW using a percolator type reactor [12]. After that at 220 °C, only the yield of xylose increased up to 3.1 wt%, while that of the xylooligosaccharide (Xylobiose, Xylotriose) decreased. At 240 °C, only the yield of glucose increased to a maximum of 2.8 wt% and then decreased. The yield of all sugars was very low. In particular, hemicellulose-derived sugars completely disappeared.

It is thought that the decrease in product yields was caused by decomposition of the generated sugars to organic acids and furan compounds [13–15]. Hence, we investigated the generation of these compounds.

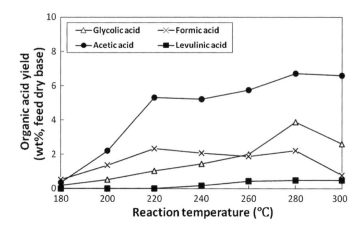

Fig. 7.6 Organic acids yield as a function of reaction temperature

7.3.3 Organic Acid Production

Four organic acids (acetic acid, formic acid, glycolic acid, and levulinic acid) were mainly obtained by the hydrothermal treatment of bamboo in HCW. Organic acids are mainly derived from the degradation of cellulose and hemicellulose and not from that of lignin during the hydrothermal treatment of lignocellulosic biomass. As shown in Fig. 7.6, the yield of acetic acid was notably higher than that of the others under these conditions. It is assumed that acetic acid is formed both by the degradation of sugars and by the detachment of the acetyl group from hemicellulose. The yield increased rapidly with treatment temperature until the temperature reached 220 °C and then the yield increased gradually. The yield reached 6.7 wt% at 280 °C. In the case of formic acid, the yield increased with temperature until 220 °C. After that, it decreased gradually as temperature increased. In addition, the yields of glycolic acid and levulinic acid tended to increase with temperature; however, their yields were relatively low.

7.3.4 Furan Compound Production

Figure 7.7 shows the relationship between reaction temperature and furan compounds. Furfural was generated at a relatively lower temperature compared to 5-HMF. In other words, the yield of furfural increased rapidly to 4.3 wt% until the temperature reached 220 °C and then decreased gradually. Generally, furfural is formed by the degradation of pentose (arabinose, xylose), a constituent sugar of hemicelluloses [14]. On the other hand, 5-HMF is mainly formed by the degradation of hexose (glucose), a constituent sugar of cellulose [15]. The temperature at which decomposition of hemicellulose started was lower than that of cellulose [5, 6].

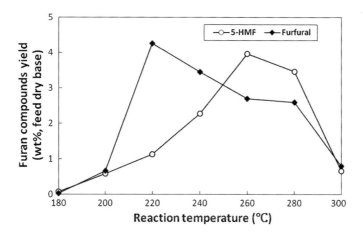

Fig. 7.7 Furan compounds yield as a function of reaction temperature

The maximum yield was 4.0 wt% at 260 °C. Hence, the generation behaviors of furfural and 5-HMF reflected the decomposition of hemicellulose and cellulose, respectively.

7.3.5 Chemical Composition of Water Insolubles

So far, we have described WS compounds obtained by HCW treatment of bamboo. However, the treatment residue must be used effectively. Because, the WI yield was higher than the others' yields. Hence, we investigated the chemical and ultimate compositions of the WI residue as follows. First, the chemical composition of the residue was analyzed, as shown in Fig. 7.8. The chemical composition was represented as the remaining ratio of each component in the WI. In addition, we examined the constituents of hemicellulose: xylan and arabinan.

At 180 °C, most of the cellulose, lignin, and xylan remained in the WI. However, the recovery ratio of arabinan was only 29.1 wt%. Then, arabinan disappeared from the WI at 200 °C. In addition, 79.6 wt% of xylan was decomposed and subsequently disappeared at 220 °C. These observations suggest that the decomposition of hemicellulose (arabinoxylan) occurred in two steps. Arabinan decomposed first followed by xylan. During this period, cellulose was recovered as the WI at a yield of over 90 wt%. However, at temperatures above 220 °C, the recovery ratio gradually decreased to 73.0 wt% at 240 °C; above this temperature, the ratio rapidly decreased. At 280 °C, no sugars were observed in the WI. The recovery ratio of lignin increased gradually. The yield exceeded 100 % when the temperature reached 220 °C. The final yield was approximately 150 wt%.

This can be explained as follows: Minowa et al. reported that WS generated by the hydrothermal treatment of cellulose polymerized and changed to a WI, such as

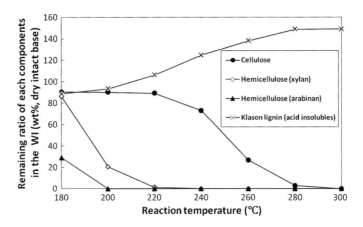

Fig. 7.8 Remaining ratio of each component in the WI

a char, in a prolonged reaction [16]. In this study, we believe similar reactions occurred. It is assumed that sulfuric acid insoluble char-like products were formed from the components of except lignin which were found because lignin was determined using sulfuric acid insolubles (Klason lignin) in this method.

7.3.6 FTIR Spectra

To understand changes in chemical functional groups during the hydrothermal treatment. FTIR analysis of the WI was performed. Peak assignment was made based on literature data [17–20]. Figure 7.9 shows spectras of the intact sample and the residues samples. The intensity of the peak around 3500 cm^{-1} attributed to hydroxyl groups decreased at elevated temperature, indicating that water molecules within the WI were gradually expelled. The peak at 2900 cm^{-1} region shows the CH_n stretch in methyl and methylene group also weakened. The peak observed at around 1700 cm^{-1} represents C=O stretching vibrations. The peaks 1,600, 1,510, and 1,400 cm^{-1} reveal to aromatic skeletal vibration derived from lignin. Intensity of these peaks was hardly changed during the treatment. The peak at around 1,050 cm^{-1} attributed to cyclic C–O–C indicating the presence of cellulose. These results were agreed with the chemical composition of the WI showed in Fig. 7.8.

7.3.7 Ultimate Analysis of Water Insolubles

Many researchers (Inoue et al. [17, 21–23], Kobayashi et al. [19, 24], Hirajima et al. [20, 25, 26]) have already reported that the WI, obtained by hydrothermal treatment of lignocellulosic biomass, could be used for solid fuel. This is why the

7 Effective Utilization of Moso-Bamboo   165

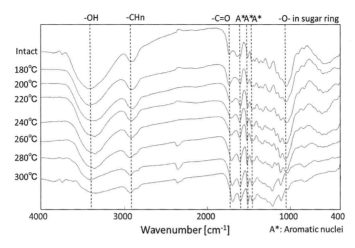

Fig. 7.9 FTIR spectra of WI obtained by HCW treatment of moso-bamboo at various reaction temperatures

Table 7.1 Ultimate composition of intact bamboo and the WI

	C (wt%)	H (wt%)	O (wt%)	N (wt%)	HHV (MJ/kg-dry)
Intact bamboo	48.1	6.0	45.7	0.2	16.7
180 °C-treated	50.1	6.0	43.8	0.2	17.7
200 °C-treated	51.4	6.0	42.4	0.2	18.4
220 °C-treated	54.1	5.9	39.7	0.3	19.6
240 °C-treated	55.4	5.9	38.5	0.3	20.2
260 °C-treated	59.2	5.2	35.3	0.3	21.2
280 °C-treated	70.2	5.3	24.1	0.4	27.0
300 °C-treated	72.1	5.3	22.3	0.3	28.0

O*: calculated by difference

WI has high heating value. The ultimate composition of the raw bamboo and residue are shown in Table 7.1. The carbon content of the residue increased gradually when the temperature increased to 240 °C. After that, the carbon content increased rapidly as more hemicellulose and cellulose, which have relatively low carbon content in bamboo, were decomposed and dissolved in water at higher reaction temperatures, as shown in Fig. 7.7. As a result, the solid residue consisted mainly of lignin, which has higher carbon content compared to cellulose and hemicellulose.

Under hydrothermal treatment, biomass undergoes a coalification-like process, as demonstrated in the van Klrevelen diagram in Fig. 7.10. The intact sample has high atomic H/C and O/C ratios, which both gradually decreased during the treatment. The slope of the trajectories suggests that the O content decreased in proportion to the H content, due to dehydration. It is clear that decrease in O and H

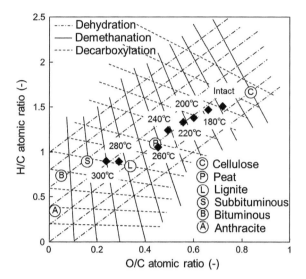

Fig. 7.10 van Klevelen diagram for products obtained at various reaction temperatures and other solid fuel

content occurred mainly in this treatment temperature range. The WI resulting from higher temperature treatment approached to sub-bituminous coal.

The heating value of the residue, which was calculated using Dulong's formula, is also shown in Table 7.1. The heating value of the intact raw bamboo was only 16.7 MJ/kg. However, the heating value of the residue increased with the reaction temperature, rising up to 28.0 MJ/kg. Hence, the solid residue could be utilized as a solid fuel.

7.3.8 Equilibrium Moisture Content

Figure 7.11 shows the relationship between treatment temperature and EMC. The value of EMC was reduced during the treatment. That is, the treatment at 180 °C reduced the intact sample value of 10.1 to 8.4 wt%. Further treatment at 300 °C led to the value as low as 3.4 wt%. These behaviors were in agreement with the change of chemical components shown in Fig. 7.8. In the ability of water adsorption, polysaccharides (hemicellulose and cellulose) are superior to lignin. Since polysaccharides could be removed from water insolubles by the treatment up to 280 °C. EMC and heating value are two important properties of solid fuel. When the fuel is combusted, a part of energy is consumed for water vaporization. Hence, a solid fuel with higher EMC will consume more energy to evaporate the moisture. Thus a good solid fuel should have a high-heating value, and a low EMC. Our results show that both properties could be adequately improved by the hydrothermal carbonization process.

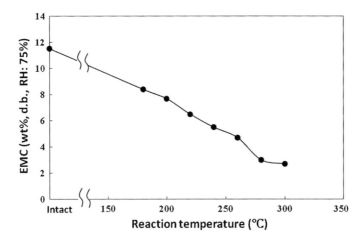

Fig. 7.11 Relationship between reaction temperature and EMC of WI (RH: 75 %)

Table 7.2 Chemical component, elemental and HHV of raw sample

	Chemical composition (wt%)					Elemental composition (wt%, daf)				HHV (MJ/kg-dry)
	Cellulose	Hemicellulose	Klason lignin	Ethanol extracts	Ash	C	H	O	N	
Cypress	43.6	17.5	33.4	1.4	0.6	50.7	6.2	43.1	0.1	18.3
Japanese cedar	42.3	17.1	28.9	3.6	0.4	51.1	6.1	42.8	0.1	18.3
Chinquapin	40.9	22	23.7	10.1	0.2	50.2	6.2	43.5	0.1	18.1
Eucalyptus	39.1	14.5	26.8	1.3	0.7	50.1	5.9	43.8	0.2	17.6
EFB	32.5	21.8	20.1	6.9	2.3	47.1	6.2	43.5	0.1	16.5
Kenaf core	38.2	18.6	19.8	3.6	2.9	47.2	5.9	46.6	0.3	16.1
Bamboo	39.4	20.2	22.8	7.5	1.7	48.1	6.0	45.7	0.2	16.7
Oil palm shell	14.7	17.8	46.2	3.9	2.4	52.8	5.8	41.0	0.4	18.8

7.3.9 Solid Fuel Production from Various Lignocellulosic Biomass by Hydrothermal Treatment

We investigated the production of carbonaceous material through hydrothermal treatment of eight species of lignocellulosic biomass. Table 7.2 shows the chemical composition, elemental composition, and HHV of each raw material and Table 7.3 summarizes the value of water insoluble (WI) obtained by hydrothermal treatment at 300 °C for 30 min and the corresponding WI yield. Their chemical composition was depended on their species. The carbon content in all the raw

Table 7.3 WI yield, elemental composition and HHV of WI obtained by HCW treatment at 300 °C for 30 min

	WI yield (wt%)	Elemental composition (wt%, daf)				HHV (MJ/kg-dry)
		C	H	O	N	
Cypress	46.0	72.6	5.5	22.0	0.4	28.6
Japanese cedar	47.7	73.8	5.3	21.9	0.2	28.8
Chinquapin	47.8	72.3	5.0	22.8	0.1	27.5
Eucalyptus	46.1	72.0	5.2	22.8	0.2	27.7
EFB	38.7	71.3	6.0	22.7	0.3	28.7
Kenaf core	46.1	72.7	5.3	22.0	0.4	28.3
Bamboo	41.7	72.1	4.8	23.1	0.2	27.1
Oil palm shell	46.9	72.2	5.0	22.8	0.5	27.6

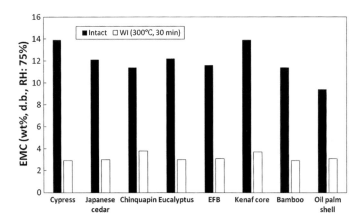

Fig. 7.12 Change in EMC before and after hydrothermal treatment

materials was estimated to be approximately 50 %. After the hydrothermal treatment, the value of WI increased up to approximately 72 % regardless of the species. Furthermore, the variation in WI value among the samples became smaller and increased up to 1.6 times compared with the raw material. On the other hand, WI yield varied depending on the nature of the species. For instance, herbaceous species like kenaf, EFB and bamboo exhibited lower WI yield than that the woody species. Figure 7.12 shows the change in EMC before and after treatment. The EMC value of the corresponding raw material was approximately 10 % (RH: 75 %). Moreover, the variation in EMC value among the raw materials was considerably larger. However, the EMC value of each raw material dropped to 2–3 % and the corresponding variation became smaller after the hydrothermal treatment. This implies that homogenization of the samples also simultaneously progressed during the hydrothermal treatment.

7.4 Conclusion

Bamboo was treated in HCW at a temperature range of 180–300 °C to examine properties that are important in understanding its maximum utilization as a biomass resource.

We found that hemicellulose (arabinoxylan) was first decomposed to produce arabinose and then xylan was decomposed. In addition, xylan was hydrolyzed to xylooligosaccharides and then to xylose at 180–220 °C. Xylose was further decomposed to various organic acids and furfural. However, most of the cellulose were not decomposed and was recovered as a solid residue. Cellulose began hydrolyzing to glucose at temperatures above 240 °C. Glucose was further decomposed to various organic acids and 5-HMF.

These products can be used for various applications such as food additives and feedstocks for plastics and fuels. Furthermore, the heating value of the residue increased with the reaction temperature, reaching 28.0 MJ/kg. Additionally, the moisture absorbency dramatically became lower. Hence, the solid residue could be used as a solid fuel.

Acknowledgments The authors are grateful for the support of this research by a Grant-in-Aid for Scientific Research No. 24246149 from the Japan Society for the Promotion of Science (JSPS), the Global COE Program (Novel Carbon Resources Sciences, Kyushu University) and the Japan Science and Technology Agency (JST).

References

1. Bobleter O, Bonn G (1983) The hydrothermolysis of cellobiose and its reaction product d-glucose. Carbohydr Res 124:185–193
2. Bobleter O (1994) Hydrothermal degradation of polymers derived from plants. Prog Polym Sci 19:797–841
3. Ando H, Sakaki T, Kokusho T, Shibata M, Uemura Y, Hatate Y (2000) Decomposition behavior of plant biomass in hot-compressed water. Ind Eng Chem Res 39(10):3688–3693
4. Sakaki T, Shibata M, Sumi T, Yasuda S (2002) Saccharification of cellulose using a hot-compressed water-flow reactor. Ind Eng Chem Res 41:661–665
5. Kumagai S, Hayashi N, Sakaki T, Nakada M, Shibata M (2004) Fractionation and saccharification of cellulose and hemicellulose in rice hull by hot-compressed-water treatment with two-step heating. J Jpn Inst Energy 83:776–781
6. Kumagai S, Ota M, Nakano S, Hayashi N, Sakaki T (2008) Solubilization and saccharification of barley straw by hot-compressed water treatment. J Food Eng 9(2):115–119
7. Fujii Y, Shigematsu T, Nishiura C (2005) Reproductive dynamics on cleared bamboo forest stands in Northern Kyushu. LRJ 68(5):689
8. Scurlock JMO, Dayton DC, Hames B (2000) Bamboo: an overlooked biomass resource? Biomass Bioenergy 19(4):229–244
9. Sluiter A, Hames B, Ruiz R, Scarlata C, Sluiter J, Templeton D, Crocker D (2005) Determination of structural carbohydrates and lignin in biomass, Technical Report NREL
10. Selvig WA, Gibson FH (1945) Chemistry of coal utilization. John Wiley, New York, p 132

11. Yoshimoto T, Morita S (1985) Studies on hot-water extractives of bamboo stem (*Phyllostachys pubescens* Mazel): seasonal variation of the content of free sugars. Bull Tokyo Univ Forests 74:9–15
12. Ando H, Morita S, Furukawa I, Kamino Y, Sakaki T, Hirosue H (2003) Generation of xylooligosaccharides from moso bamboo (*Phyllostachy pubescens*) using hot compressed water. Mokuzai Gakkaishi 49:293–300
13. Yoshida K, Kusaki J, Ehara K, Saka S (2005) Characterization of low molecular weight organic acids from beech wood treated in supercritical water. Appl Biochem Biotechnol 121:795–806
14. Antal MJ, Leesomboon T, Mok WS, Richards GN (1991) Mechanism of formation of 2-furaldehyde from D-xylose. Carbohydr Res 217:71–85
15. Antal MJ, Mok WS, Richards GN (1990) Mechanism of formation of 5-(hydroxymethyl)-2-furaldehyde from d-fructose and sucrose. Carbohydr Res 199:91–109
16. Minowa T, Fang Z, Ogi T, Varhegyi G (1998) Decomposition of cellulose and glucose in hot-compressed water under catalyst free conditions. J Chem Eng Jpn 31(1):131–134
17. Inoue S (2010) Hydrothermal carbonization of empty fruit bunches. J Chem Eng Jpn 43:972–976
18. Liu Z, Zhang F-S, Wu J (2010) Characterization and application of chars produced from pinewood pyrolysis and hydrothermal treatment. Fuel 89:510–514
19. Kobayashi N, Okada N, Hirakawa A, Sato T, Kobayashi J, Hatano S, Itaya Y, Mori S (2009) Characteristics of solid residues obtained from hot-compressed-water treatment of woody biomass. Ind Eng Chem Res 48(1):373–379
20. Yuliansyah AT, Hirajima T, Kumagai S, Sakaki K (2010) Production of solid biofuel from agricultural wastes of the palm oil industry by hydrothermal treatment. Waste Biomass Valor 1:395–405
21. Inoue S, Hanaoka H, Minowa T (2002) Hot compressed water treatment for production of charcoal from wood. J Chem Eng Jpn 35(4):1020–1023
22. Inoue S, Uno S, Minowa T (2008) Carbonization of cellulose using the hydrothermal method. J Chem Eng Jpn 41(3):210–215
23. Inoue S, Yoshimura T (2010) Behavior of wood compositions during hydrothermal carbonization treatment of *Eucalyptus*. Wood Carbonization Res 6(2):47–52
24. Kobayashi N, Okada N, Tanabe Y, Itaya Y (2011) Fluid behavior of woody biomass slurry during hydrothermal treatment. Ind Eng Chem Res 50(7):4133–4139
25. Hirajima T, Kobayashi H, Yukawa K, Tsunekawa M, Fukushima M, Sasaki K (2003) Fundamental study on the production of woody biomass fuel using hydrothermal treatment. Shigen-to-Sozai 119:118–124
26. Yuliansyah AT, Hirajima T, Kumagai S, Sasaki K (2010) Solid fuel production from oil palm shell by hydrothermal carbonization. Wood Carbonization Res 7(1):19–26

Chapter 8
Hydrothermal Liquefaction of Biomass in Hot-Compressed Water, Alcohols, and Alcohol-Water Co-solvents for Biocrude Production

Chunbao Charles Xu, Yuanyuan Shao, Zhongshun Yuan,
Shuna Cheng, Shanghuang Feng, Laleh Nazari
and Matthew Tymchyshyn

Abstract Hydrothermal liquefaction (HTL) is a technology for directly converting biomass into clean liquid fuels (biocrude) in the presence of water or water-containing solvent/co-solvent and more commonly a suitable catalyst at a temperature of 200–400 °C and under moderate to high pressure (5–25 MPa). Two key operating parameters, solvent type/composition and catalysts play significant roles in the performance of a HTL process including biomass conversion, biocrude oil yield and oil quality, etc., which are closely related to the economic feasibility of the process for industrial/commercial applications. A HTL process with properly designed solvent and catalysts would lead to a high yield of biocrude oil (up to 65 %) with a high quality (lower oxygen content). This chapter provides an overview on the effects of solvents (focusing on water, alcohols, and alcohol-water co-solvents) and catalysts on the HTL processes, and their industrial applications.

8.1 Introduction

There are increasing concerns worldwide over declining nonrenewable fossil resources, energy security, climate change, and sustainability. It is thus of strategic significance to explore alternatives to fossil resources for both energy and chemical production. Among all the potential alternatives to fossil resources, biomass such as wood and wood waste, forestry and agricultural residues are promising because biomass is immense and renewable, and hence sustainable sources, for both energy and chemical production (the potential energy to be

---

C. C. Xu (✉) · Y. Shao · Z. Yuan · S. Cheng · S. Feng · L. Nazari · M. Tymchyshyn
Institute for Chemical and Fuels from Alternative Resources (ICFAR), The University of Western Ontario, London, ON N6A 5B9, Canada
e-mail: cxu6@uwo.ca

derived from the global annual production of biomass is about 7–8 times that of the world energy consumption) [1, 2].

There are tremendous opportunities for utilizing forestry and agricultural residues. For instance, forest trimmings (e.g., leaves and branches) account for 15 % of the harvested biomass, and 40 % of felled timber ends as sawmill co-products and wastes (e.g., wood chips, saw dust, and bark) [3]. A fraction of forestry residues and by-products, such as lignin, sawdust, and bark are currently burned to provide heat for sawmills and paper mills. Pulp and paper plants and sawmills are the largest users of bioenergy in Canada, generating 513 × 1015 J, equivalent to 16.2 billion m^3 of natural gas, to supply 50 % of their own energy needs. Although 70 % of Canadian sawmill residues are combusted for energy, the remaining 30 % (more than 5 million dry tons per annum) are not used [4]. In addition, it is estimated that 18 Mt of biomass per year would be available for energy and chemical production from Canadian agricultural residues [5]. In the USA, the US DOE, and the US Department of Agriculture estimated that 1.3 billion metric tons of biomass can be produced exclusively for biofuel production in the US each year, which could supply 21 % of the U.S. energy requirement or 33 % of the U.S. transport fuels.

Although biomass resources are renewable, carbon-neutral, and remarkably massive in amount, they are very bulky and difficult to transport, handle, and store. Therefore, appropriate biomass conversion technologies are required to densify and upgrade them into more convenient forms, whether liquid (ethanol, biodiesel, bio-oils) or solid (pellets, biochar), through a variety of biochemical and thermochemical conversion technologies. Current, first-generation processes convert food crops into bioethanol or biodiesel: they inflate the cost of the crops, which in turn increases the biofuels cost and puts food out of reach of a significant fraction of the World's population. Cellulosic ethanol production from residues is not a viable solution due to the costly pretreatment and saccharification (hydrolysis) processes [6]. Moreover, ethanol is not an ideal transportation fuel because of its low energy density (ethanol-fueled vehicles have only about 66 % of efficiency compared with those fueled with gasoline), its high volatility and its readiness for absorption of water from the atmosphere.

Targeting to produce liquid biofuels, other thermochemical conversion processes include gasification of biomass for syngas combined with various catalytic processes for production of synthetic fuels (e.g., high quality Fischer–Tropsch diesel), and pyrolysis and other direct liquefaction processes (e.g., hydrothermal liquefaction) for the production of biooil or biocrude. The technical route via biocrude oil production is much more flexible than biochemical processes and can handle a wider range of feedstocks. More importantly, biocrude oil after proper upgrading (to reduce its oxygen content and improve its compatibility with petroleum) can be coprocessed with petroleum crude oil in existing refineries or used as a drop-in fuels. Not limited to biofuel production, biocrude oils can also be used for producing various bioproducts such as phenolic resins, expoxy resins, and polyols and polyurethane foam materials, etc. In the authors' group, a novel biooil

8 Hydrothermal Liquefaction of Biomass

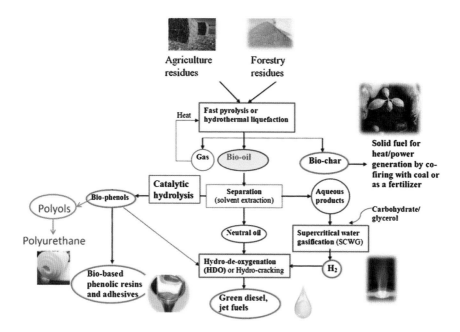

Fig. 8.1 Biooil platform biorefinery

platform biorefinery (as illustrated in Fig. 8.1) was proposed and currently under investigation.

Pyrolysis of biomass is operated in an inert atmosphere at a temperature of 400–800 °C under low pressure (0.1–0.5 MPa), usually without catalyst. At high temperature, solid lignocellulosic materials thermally decompose into fragments react in the vapor phase, yielding about 50–75 wt% liquid bio-oil products and 15–30 wt% of solid bio-char; the balance consists of noncondensable gases that are combusted to provide energy for the process. Fast or flash pyrolysis maximizes the liquid yield to 75 wt%, by operating at a 500–650 °C, with a high heating rate of 1000 °C/s (or even 10000 °C/s) and a short vapor residence time of less than 2 s [7–9]. Bio-oil from fast pyrolysis is a complex mixture composed of acids, alcohols, aldehydes, esters, ketones, sugars, guaiacols, syringols, furans, lignin-derived phenols, and extractible terpene with multi functional groups [10].

An alternative thermochemical route for biocrude production is biomass hydrothermal liquefaction (HTL), which is performed under an inert or reducing atmosphere at moderate temperature (200–400 °C) and high pressure (5–25 MPa), in the presence of a suitable solvent. Under such conditions, the feedstock macromolecules decompose into fragments of light molecules directly in the presence of a hot-compressed solvent (water, alcohol, alkanes, phenols, and tetralin, etc.) and with a suitable catalyst. A pioneering work of biomass HTL technology was reported by Appell et al. [11] at the Pittsburgh Energy Technology Center (PETC), where a variety of lignocellulosic materials were efficiently converted to oily products in water at an elevated temperature in the presence of CO and Na_2CO_3 as the catalyst. The PETC's biomass direct liquefaction technology was further advanced by the research group led by Elliott at Pacific Northwest Laboratory in USA [12, 13]. Recent developments of this HTL technology include efficient liquefaction of woody biomass in near-/super-critical water and organic solvents (alcohol, alkanes, phenols, and tetralin, etc.) with a suitable catalyst (e.g., K_2CO_3, KOH, $FeSO_4$, FeS), obtaining a total liquid yield of 40–60 wt% [14, 15]. In a recent study by the authors' group [16], alcohol (methanol or ethanol) and water showed synergistic effects on biomass direct liquefaction. A 50/50 mix of either methanol/water or ethanol/water was the most effective solvent for the liquefaction of white pine sawdust at 300 °C for 15 min, resulting in 65 wt% bio-oil yield and a 95 wt% biomass conversion. An advantage of using alcohols for biomass liquefaction is that they can also be renewable chemicals derived from biomass, and they can be recycled and reused after the liquefaction operation. Compared with the fast pyrolysis technology, biomass HTL technology produces higher quality bio-oil (biocrude) with better chemical and physical properties, an oxygen content that is reduced by about 50 % and a greater HHV (30–35 MJ/kg as compared with 15–20 MJ/kg for pyrolytic oil) [15]. HTL technology is particularly promising for converting wet biomass resources such as microalgae, agro waste streams (e.g., manures), municipal/industrial wastewater sludge and fresh/green forest biomass/residues, into biocrude. After further treatment and upgrading in a petroleum refinery, the obtained biocrude (with a low oxygen content and a high calorific value up to 35 MJ/kg) could be a potential substitute for petroleum for the production of liquid transportation fuels and a range of chemicals such as phenols, aldehydes, and organic acids.

This chapter provides a critical review of the HTL technologies for the production of biocrude oils, focusing on the effects of solvents and catalysts on the HTL processes.

8.2 HTL Technologies for Biocrude Production

Hydrothermal liquefaction (HTL) is performed under an inert or preferable reducing atmosphere at a moderate temperature (200–400 °C) in water or water containing medium. Macromolecule compounds contained in a biomass feedstock can decompose into fragments of light molecules in the presence of a solvent (water, alcohol, alkanes, phenols, tetralin, or their mixtures, etc.) and a suitable catalyst. The biomass-derived fragments are unstable and reactive, and would repolymerize into oily compounds having various molecular weights. The detailed classification, reaction conditions, and reaction pathways for HTL processes of biomass were extensively described and summarized in the literatures [17–20]. Compared with fast pyrolysis, an HTL process can produce higher quality bio-oils (heavy liquid oils or bio-crude oils) with beneficial chemical and physical properties (e.g., a lower oxygen content and smaller molecular weights, and increased heating values). HTL technologies also have the potential for producing a range of chemicals including vanillin, phenols, aldehydes, and acetic acids, etc., as well documented in many literature papers [17–19, 21, 22]. Behrendt et al. [17] critically evaluated various technical implementations of historic and current HTL processes for biomass liquefaction Peterson et al. [21] and Toor et al. [18] reported HTL processes involving sub- and supercritical water. Moreover, Zhang [19] published a book chapter which gives a comprehensive introduction about fundamentals of the biomass HTL processes for various feedstocks. The biocrude oil yield is strongly dependent on the various operating parameters such as type of feedstock, type of solvent, temperature, substrate concentration, reaction time, and catalyst, etc., as reviewed by Akhtar et al. [22]. However, comprehensive discussion is not available by far in the literature on the effects of two important chemical parameters (i.e., type of solvents and catalyst) on a biomass HTL process. This section is focused on such discussion.

8.2.1 Effects of Type of Solvents

In a biomass liquefaction process, the presence of solvent promotes such reactions as solvolysis, hydration, and pyrolysis, which helps achieving better fragmentation of biomass and enhancing dissolution of reaction intermediates. The type of solvents is thus one of key parameters that determine the yield and composition of the biocrude. For example, Mazheri et al. observed that subcritical 1, 4-dioxane was a more effective solvent for conversion of fruit press fiber into biocrude oil than other tested solvents including methanol, ethanol and acetone at the same condition [23]. Liu and Zhang produced biocrude with different major components via HTLs using acetone and ethanol as the solvent, respectively [24]. Fan et al. demonstrated that water, ethanol, and toluene yielded biocrude phenolic compounds as the major compounds, while ethylene glycol favors the formation of alcohol compounds, and acetone yields a high concentration of ketone and aldehyde compounds in the biocrude [25]. The type of solvents can be generally divided into two classes, water [18, 21] and organic solvents (alcohols, tetralin, glycerin, etc.) [26, 27]. Mixtures of these two types of solvents in the form of water-organic or organic-organic co-solvent were used in biomass HTL and achieved satisfactory results recently [16–31]. The following sections discuss the performance of different solvents (water, alcohols, water-alcohol co-solvents) in HTL of biomass.

8.2.1.1 Water

Water is the cheapest and greenest solvent with abundant supply, so water is regarded as the most promising solvent, especially for HTL processes. A Sub/super-critical fluid has unique ability to dissolve materials not normally soluble in either liquid or vapor phase and has complete miscibility with the liquid/ vapor products from the processes, providing a single-phase environment for reactions that would otherwise occur in a multiphase system under conventional conditions [32]. Sub/near-critical water (below or near 374 °C and 22 MPa) has widely been used for biomass liquefaction as opposed to supercritical water mainly for gasification of aqueous biomass [33–38].

Xu's group has achieved HTL of various feedstocks such as paper sludge and woody biomass, lignocellulosic wastes and model biomass compounds using sub/near-critical water under various operational conditions (reaction time, reaction temperature, and type of atmosphere) [34–37]. As concluded from their experimental results, a reaction temperature around 300 °C with an initial pressure of 2 MPa in hydrogen atmosphere are the optimum conditions for producing biocrude oil when using water as the liquefaction solvent. Moreover, the products (bio-oil/biocrude) from the HTL experiments have higher caloric values (>30 MJ/kg) and less oxygen contents (10–20 %), when comparing with fast pyrolysis oils. [39–44]. The GC–MS measurements for the biocrude products revealed that such components as carboxylic acids, phenolic compounds, and derivative, as well as long-chain alkanes are major components in the biocrude, which may account for its higher caloric values.

Biomass consists of three major components: lignin, cellulose, and hemicellulose, which have quite different properties and thus each component could undergo different degradation pathways during HTL in sub/near-critical water, as summarized in Table 8.1 [21, 33, 35, 45–50]. For instance, hemicelluloses (biopolymer of glucose/xylose with a lower degree of crystallinity) is easily thermal decomposed at a low temperature (>180 °C), while high-molecular weight lignin with three cross-linked structure cannot be degraded until the temperature reached 280 °C and alkaline medium was found to catalyze the liquefaction (depolymerization) process [46–49]. Based on the different decomposition conditions and mechanisms for different components of biomass, Ramsurn and Gupta [38] developed a two-step biomass HTL (using acidic subcritical water followed by alkaline supercritical water) to maximize the biocrude production while minimize char formation.

8.2.1.2 Alcoholic Solvents

Although water is regarded as the most environmentally friendly solvent, a drawback of utilizing water as the biomass liquefaction solvent lies in the fact that the yield of water-insoluble oil product (with a greater heating value) was generally lower than that of aqueous (water-soluble) products, and the gas formation became significant in supercritical water [51, 52]. Solvolysis liquefaction of biomass in the presence of an organic solvent proved effective for lowering the viscosity of biocrude oil produced in biomass liquefaction [49]. For example, phenolysis liquefaction of wood biomass (i.e., biomass liquefaction in phenol) is a commonly-used method to prepare bio-based phenolic feedstock for the

8 Hydrothermal Liquefaction of Biomass 177

Table 8.1 HTL conversion of biomass components during sub-/near-critical water liquefaction

Component	Properties	Share in biomass (wt%)	Degradation mechanisms	Favorite reaction conditions	
				Temperature (°C)	Medium
Lignin	Highly stable biopolymer of three highly cross-linked phenylpropane (C_6–C_3) units bonded mainly by ether bonds (C–O–C)	~20–30	Hydrolysis via cleavage of the ether bonds	280–400	Alkaline
Cellulose	Biopolymer of glucose with a high degree of crystallinity	~40–50	Hydrolysis, Saccharification	240–350	Acidic/Alkaline
Hemicellulose	Biopolymer of glucose/xylose with a lower degree of crystallinity	~20–30	Hydrolysis, Thermal decomposition	180–290	Acidic

production of phenolic resins and adhesives [53, 54]. However, the use of phenol as the liquefaction solvent does not appear to be feasible for biofuel production, simply because phenol is expensive, and recycling of phenol from the liquefied products is very challenging, as well as the use of phenol is associated with some environmental concerns. As a result, other organic solvents including alcohols and cyclic carbonates were employed in low-temperature liquefaction of biomass [55, 56]. These solvents could be recycled by evaporation/distillation after the liquefaction operation and they are much cheaper than phenol.

Sub-/supercritical alcohols (methanol, ethanol, propanol, and butanol, etc.) have been widely tested for liquefaction of lignocellulosic materials [14, 15, 57–61] to improve the yields of liquid oil products of a greater calorific value. Since these alcohols have critical temperatures and pressures lower than water, much moderate reaction conditions can be employed. Another advantage of using alcohols as the solvent for biomass liquefaction is that these alcohols are expected to readily dissolve relatively high-molecularweight polar products derived from cellulose, hemicelluloses, and lignin due to their lower dielectric constants when compared with that of water [14]. Among all the alcohols tested, ethanol and methanol have been more commonly employed.

Miller et al. [62] studied the depolymerization of Kraft and organosolv–derived lignins in supercritical methanol or ethanol in the presence of KOH, when high biomass conversions were realized. The ether insoluble material, or solid residues, remaining after treating an organosolv lignin ethanol at 290 °C in the presence of KOH was as low as 7 wt%. Cemek and Kucuk [63] reported the liquid yields of 44.4 and 43.3 wt% in liquefaction of Verbascum stalk at 300 °C with supercritical methanol and ethanol, respectively. The conversion was rapid, reaching the maximum value within 10–15 min.

Minami and Saka [57–59] reported that 90 % of beech wood was successfully converted in supercritical methanol and the optimal conditions for the process were 350 °C/43 MPa. To enhance liquid yields further and to obtain oil products of a lower oxygen content, supply of hydrogen to the liquefaction process, or called hydro-liquefaction, proved to be effective [15, 64].

8.2.1.3 Alcohol-Water Co-solvents

Organic-water co-solvents were introduced in biomass HTL and initially tested with a phenol-water mixture [65–68]. Okuda et al. [69] treated lignin by combining depolymerisation and phenolation in a water-phenol mixture at 400 °C, in which phenol acted as a capping agent to react with the lignin-degraded intermediates to prevent repolymerization. However, similar to the phenolysis liquefaction of biomass for biofuel production, the water-phenol mixture could suffer from the same disadvantages. In contrast, alcohol-water co-solvents were adopted in many studies [16, 31, 70–73]. In a well-known organosolv delignification process—the Alcell process, lignin in woody biomass could be effectively extracted using a co-solvent of ethanol-water at 180–210°C and 2–3.5 MPa [71].

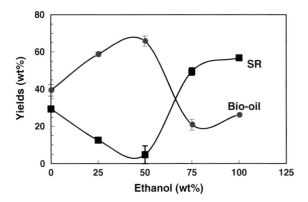

Fig. 8.2 Dependency of yields of Bio-oil and SR on ethanol water co-solvent composition (liquefaction conditions: 300 °C and 15 min). Modified with the permission from Ref. [16]. Copyright 2010 American Chemical Society

Yuan et al. [31] investigated the HTL of rice straw in a mixture of ethanol and water, where the experimental results showed that increasing ethanol share in the mixture from 10 to 50 vol.% not only promoted the conversion of rice straw but reduced the oxygen content in the bio-oil.

The authors' group [16] used methanol water and ethanol water as co-solvent for the liquefaction of eastern white pine sawdust and concluded that 50 wt% aqueous alcohol at 300 °C for 15 min produced a very high bio-oil yield of close to 65 wt% and almost complete biomass conversion (<5 % solid residue yield), as illustrated in Fig. 8.2.

The authors' group [72] also achieved nearly complete degradation of Kraft lignin (a very high-molecular weight type of lignin with a weight-average molecular weight, Mw, of over 10,000 g/mol) in a hot-compressed water ethanol medium with NaOH as the catalyst and phenol as the capping agent at 260 °C for 1 h, with the lignin/phenol ration of 1:1 (w/w). In another study by the authors' group [73], we performed HTL of alkali lignin in sub/supercritical ethanol and water-ethanol co-solvent and observed that the optimal temperature for the process in 50 vol.% ethanol in water-ethanol co-solvent is approximately 100 °C lower than that with pure ethanol.

8.2.2 Effects of Catalysts

Catalysts play an important role in HTL processes to suppress char formation and improve oil yield and quality. Table 8.2 summarized various catalysts commonly used in biomass HTL processes and their effects on bio-oil production.

Alkaline solutions including Na_2CO_3, K_2CO_3, KOH, $Ca(OH)_2$, $Ba(OH)_2$, RbOH and CsOH were widely employed as catalysts in biomass HTL [36, 74–77]. Apell et al. [76, 77] conducted pioneering studies on sodium carbonate-catalyzed liquefaction of carbohydrate in the presence of carbon monoxide and proposed the reaction pathways involving deoxygenation occurs through decarboxylation from

Table 8.2 Summary of catalytic conversion of biomass and model compounds in HTL processes

Catalyst	Feedstock	T, (°C)	Effects	Ref.
K_2CO_3	Woody biomass	280	Reduced solid residue	[78, 79]
Rb_2CO_3, Cs_2CO_3	Woody biomass	280	Increased oil yield	[74]
Na_2CO_3	Corn stalk	276–376	Increased oil yield	[80]
K_2CO_3	Woody biomasses	280–360	Reduced solid residue	[81]
Inorganic acids	Woody powder	150	Less solid residue	[82]
$ZnCl_2$, Na_2CO_3, NaOH	Oil palm fruit fiber	210–330	Promoted gas yields	[83]
FeS, $FeSO_4$	Woody biomass	350	Increased oil yield	[15]
KOH, $FeCl_3$, $FeSO_4$	Peat	350–500	Promoted H_2 yield	[84]
Iron ore	Peat	350–500	Increase oil yield	[84]
Na_2CO_3, Ni	Cellulose	200–350	Char reduction	[85]
K_2CO_3, Ni	Glucose	350	Enhanced water-gas shift	[86]
NaOH, H_2SO_4, TiO_2, ZrO_2	Glucose	200	Increased isomerization of glucose	[87]
$KHCO_3$, HCO_2H, Co_3O_4	Glucose	300	Increased oil yield	[88]

ester formed by the hydroxyl group and formate ion derived from the carbonate, as detailed below.

Reaction of sodium carbonate and water with carbon monoxide, to yield sodium formate:

$$Na_2CO_3 + 2CO + H_2O \rightarrow 2HCOONa + CO_2$$

Dehydration of vicinal hydroxyl groups in a carbohydrate to an enol, followed by isomerization to ketone:

$$CH(OH) - CH(OH) - \rightarrow -CH = C(OH) - + H_2O \rightarrow -CH_2 - CO - + H_2O$$

Reduction of newly formed carbonyl group to the corresponding alcohol with formate ion and water:

$$HCOO - + - CH_2 - CO - \rightarrow -CH_2 - CH - (O-) - + CO_2$$
$$-CH_2 - CH - (O^-) - + H_2O \rightarrow -CH_2 - CH - (OH) - + OH^-$$

The hydroxyl ion reacts with additional carbon monoxide to regenerate the formate ion:

$$OH - + CO \rightarrow HCOO-$$

Karagoz et al. [78] performed hydrothermal treatment of woody biomass in hot compressed water at 280 °C for 15 min in the presence of an alkaline solution (NaOH, Na_2CO_3, KOH and K_2CO_3). Based on the biomass conversion and the yields of liquid products, the catalytic activity showed the following sequence: K_2CO_3 > KOH > Na_2CO_3 > NaOH. Generally, the use of alkaline catalysts hinders the formation of char but favors the formation of oil products. However, the activity of an alkaline catalyst appeared to depend on the properties and type of biomass feedstock. Zhong et al. [81] carried out direct liquefaction of different types of woody biomass in hot-compressed water at 280–360 °C, and a heavy oil yield of over 30 % coupled with less than 10 % yield of solid residue was obtained for all the wood samples tested. However, although the addition of K_2CO_3 catalyst reduced the solid residue yield for all the woods tested, the catalyst was less effective for the production of oil products for a biomass feedstock of a lower lignin content.

Mazaheri et al. [83] conducted HTL treatment of oil palm fruit press fiber in subcritical water with and without a catalyst such as $ZnCl_2$, Na_2CO_3, or NaOH. When 10 % $ZnCl_2$ was added as a catalyst to the reaction system, the biomass conversion increased slightly but the gaseous products increased significantly. However, when 10 % Na_2CO_3 or NaOH was added, the solid conversion increased to as high as 90 %, accompanied by increased yields of liquid bio-oil products.

Acids were not regarded as effective catalysts for HTL of biomass in water or an alcohol medium, except for the processes employing phenol as the liquefaction solvent. A strong acid, such as concentrated sulfuric acid, has been used widely as an active catalyst in the phenolysis liquefaction of woody biomass, and the resulting liquefied products could be used for the synthesis of phenolic resins with satisfactory mechanical properties [89]. However, disadvantages of the biomass liquefaction processes using strong acids include (1) partial carbonization of biomass during liquefaction and (2) corrosion of the equipment. Thus, weaker acids, such as hydrochloric and phosphoric acids, were employed in the low-temperature biomass liquefaction via phenolysis. Zhang et al. [69] investigated liquefaction of wood powder of Chinese fir and poplar in phenol with the presence of a variety of inorganic acids: 85 % phosphoric acid, 36 % sulfuric acid, 37 % hydrochloric acid and 99.5 % oxalic acid. The results showed that 85 % phosphoric acid and 36 % sulfuric acid were very effective for enhancing the liquefaction (phenolysis) efficiencies. It was found that an extremely low yield (<5 wt%) of solid residue was obtained from the operation at 150 °C for 2 h when phosphoric or sulfuric acid was used.

In the work by the authors' group [15] hydro-liquefaction of a woody biomass (Jack pine powder) was studied in sub-/supercritical ethanol without and with iron-based catalysts (5 wt% FeS or $FeSO_4$). With the catalyst, the oil yields significantly increased while the yields of solid residue and gases and water decreased. A very high liquid yield at 63 % was obtained in the operation at 350 °C for 40 min with the presence of $FeSO_4$ and H_2 of a cold pressure of 5.0 MPa, as illustrated in Fig. 8.3. The mechanism concerning the high activity for $FeSO_4$ is yet to be elucidated. The authors' group also performed HTL tests with different Fe-based catalysts (iron ore, $FeCl_3$, FeOOH, Fe_2O_3) on HTL of peat in supercritical water [84]. The supercritical water treatment of peat with the iron ore generally resulted in 19–40 % heavy oil yield that has a higher heating value of 30–37 MJ/kg.

Watanabe et al. [88] tested three different-type catalysts ($KHCO_3$, HCO_2H and Co_3O_4) in glucose liquefaction in hot-compressed water at 300 °C for 30 or 60 min. In their experiments, $KHCO_3$ and HCO_2H showed a positive effect on oil

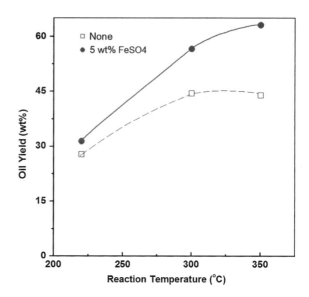

Fig. 8.3 Biocrude oil yields in hydro-liquefaction of pine wood in supercritical ethanol for 40 min under H2 of an initial pressure of 5.0 MPa. Modified from Ref. [15], Copyright 2008, with permission from Elsevier

formation. Co_3O_4 was found to be an advantageous additive as well, increasing the oil formation from glucose, but its stability under reaction conditions was quite low.

8.3 Industrial Applications of HTL Technology

HTL technology is particularly promising for converting wet biomass resources such as microalgae, agro waste streams (e.g., manures), municipal/industrial wastewater sludge and fresh/green forest biomass/residues, etc., into liquid biocrude oil. After further treatment and upgrading in a petroleum refinery, the obtained biocrude (with oxygen content lower than 10 wt% and a high calorific value higher than 35 MJ/kg) can be a potential substitute for petroleum for the production of liquid transportation fuels or a range of chemicals such as phenols, aldehydes, and organic acids, etc. Thus, the HTL technology offers industrial application potential as an economically viable solution to densification of the bulky and green/wet biomass feedstock into industrial bioproducts (liquid biofuels, bio-based chemicals and materials). Although main technical challenge which the HTL technology is facing for industrialization and commercialization is in feeding the biomass-water slurry feedstock into a high-pressure process, usually operating at 100–200 atm. This challenge has been addressed nicely with the advance in the high-pressure syringe/piston-type feeder (pumps) technologies. There are a growing number of companies for HTL technologies for biomass, peat, or low rank coal conversions, e.g., Changing World Technologies (West Hampstead, NY), EnerTech Environmental Inc (Atlanta, GA), and Bio-fuel B.V. (Heemskerk, Netherlands). Changing World Technologies (CWT) has developed an HTL technology that turns waste into oil - the Thermal Conversion Process, or TCP. The company is currently demonstrating its HTL (i.e., TCP) technology through its affiliate company (Renewable Environmental Solutions) in Carthage, Missouri to convert animal wastes into biocrude oils (http://www.changingworldtech.com/who/index.asp). Steeper Energy (a company based in Demark) developed its proprietary technology Hydrofaction^{TM} using supercritical chemistry to transform low-energy density input streams (biomass and lignite) into valuable high-energy density products. As per the company's website (http://steeperenergy.com/), with 50 % moisture lignite and 65 % moisture biomass, about 1 tonne of syncrude oil can be produced from 5 tonnes of input materials at 50 % moisture. The Hydrofaction^{TM} process produces a relatively low-oxygenated syncrude oil (containing <10 wt% oxygen) that can be easily hydrodeoxygenated to transport diesel in a modern refinery. The company has recently commissioned a 15 + kg/hr Continuous Bench Scale unit in University of Aalborg (Denmark). By late 2013 or early 2014 Steeper Energy Canada Ltd is also looking toward building a 10–25 barrel per day Pilot Scale unit in cooperation with the Province of Alberta and the Government of Canada. Moreover, the company is currently working towards building a large-scale bioenergy and combination biofossil to synthetic crude oil facility with 1,000 + barrel per day capacity in or around 2016 in Alberta.

8.4 Summary

Hydrothermal liquefaction (HTL) technology operates, at a moderate temperature <400 °C but high pressure up to 10–20 MPa, usually in the presence of a suitable solvent and catalyst. It has the potential for producing biocrude oils with much higher caloric values (25–35 MJ/kg) than fast pyrolysis oils. Hot-compressed or sub-/supercritical water treatment with alkaline catalysts has been widely employed for biomass liquefaction mainly due to the fact that water is the most environmentally benign and safer solvent and the process does not require predrying of the feedstock which will yield much better energy efficiency. Sub-/super-critical alcohols and alcohol-water co-solvents proved to be very effective solvents for HTL of biomass.

Biocrude oils produced from biomass HTL have received increasing interest owing to their great potential to substitute petroleum as fuels and resources for chemical and material production. The major challenges that hinder the industrialization and commercialization of HTL technology are: (1) biocrude oils are complex and chemically unstable mixtures, so more efforts are needed in stabilizing and upgrading of bio-oils; (2) it is a big technical challenge to feed the biomass water slurry feedstock into a high-pressure process, usually operating at 10–20 MPa.

Acknowledgments The authors are grateful for the financial support from the Natural Science and Engineering Research Council of Canada (NSERC) through the Discovery Grant. We also acknowledge the funding from the NSERC/FPInnovations Industrial Research Chair Program and

Ontario Research Fund—Research Excellence Program in Forest Biorefinery, partnered with FPInnovations, Arclin and Bioindustrial Innovation Center. "In addition, the financial support from Lignoworks, BioFuelNet, MITACS and CENNATEK is gratefully acknowledged."

References

1. Kucuk M, Demirbas A (1997) Biomass conversion processes. Energy Convers Manage 38:151–165
2. Mckendry P (2002) Energy production from biomass (Part 1): overview of biomass. Bioresour Technol 83:37–46
3. Harms H (1998) Wood, a versatile chemical material. In: European conference on renewable raw materials, Gmunden, 1998
4. (S&T)^2 Consultants Inc (2000) Liquid fuels from biomass: North America impact of non-technical barriers on implementation, prepared for IEA Bioenergy Task 27. http://intohouse.p2ric.org/ref/38/37723.pdf
5. Wood S, Layzell DB (2003) A Canadian biomass inventory: feedstocks for a bio-based economy. BIOCAP Canada Foundation. www.biocap.ca/images/pdfs/BIOCAP_Biomass_Inventory.pdf
6. Yang B, Wyman CE (2008) Pretreatment: the key to unlocking low-cost cellulosic ethanol. Biofuels Bioprod Biorefin 2:26–40
7. Bridgwater AV, Bridge SA (1991) Review of biomass pyrolysis processes. In biomass pyrolysis liquids upgrading and utilisation. Elsevier, New York, pp 11–92
8. Demirbas A (2005) Pyrolysis of ground beech wood in irregular heating rate conditions. J Anal Appl Pyrolysis 73:39–43
9. Bridgwater AV, Cottam ML (1992) Opportunities for biomass pyrolysis liquids production and upgrading. Energy Fuels 6(2):113–120
10. Guo Y, Wang Y, Wei F (2001) Research progress in biomass flash pyrolysis technology for liquids production. Chem Ind Eng Progr 8:13–17
11. Appell HR, Fu YC, Friedman S, Yavorsky PM, Wender I (1971) Converting organic wastes to oil. US Bureau of Mines, Report of Investigation, No. 7560
12. Elliott DC (1980) Bench-scale research in biomass liquefaction by the CO-steam process. Can J Chem Eng 58:730–734
13. Schirmer RE, Pahl TR, Elliott DC (1984) Analysis of a thermochemically-derived wood oil. Fuel 63:368–372
14. Yamazaki J, Minami E, Saka S (2006) Liquefaction of beech wood in various supercritical alcohols. J Wood Sci 52:527–532
15. Xu C, Etcheverry T (2008) Hydro-liquefaction of woody biomass in sub- and super-critical ethanol with iron-based catalysts. Fuel 87:335–345
16. Cheng S, Dcruz I, Wang M, Leitch M, Xu C (2010) Highly efficient liquefaction of woody biomass in hot-compressed alcohol-water co-solvents. Energy Fuels 24:4659–4667
17. Behrendt F, Neubauer Y, Oevermann M, Wilmes B, Zobel N (2008) Direct liquefaction of biomass. Chem Eng Technol 31:667–677
18. Toor SS, Rosendahl L, Rudolf A (2011) Hydrothermal liquefaction of biomass: a review of subcritical water technologies. Energy 36:2328–2342
19. Zhang Y (2010) Hydrothermal liquefaction to convert biomass into crude oil. In: Ezeji TC, Jürgen S, Blaschek HP (eds) Biofuels from agricultural wastes and byproducts. Blackwell, London
20. Chornet E, Overend RP (1985) Biomass liquefaction: an overview. In: Overend RP, Milne TA, Mudge LK (eds) Fundamentals of thermochemical biomass conversion. Elsevier, London

21. Peterson AA, Vogel F, Lachance RP, Froling M, Antal MJ et al (2008) Thermochemical biofuel production in hydrothermal media: A review of sub- and supercritical water technologies. Energy Environ Sci 1:32–65
22. Akhtar J, Amin NAS (2011) A review on process conditions for optimum bio-oil yield in hydrothermal liquefaction of biomass. Renew Sustain Energy Rev 15:1615–1624
23. Mazaheri H, Lee KT, Bhatia S, Mohamed AR (2010) Sub/supercritical liquefaction of oil palm fruit press fiber for the production of bio-oil: effect of solvents. Bioresour Technol 101:7641–7647
24. Liu Z, Zhang F (2008) Effects of various solvents on the liquefaction of biomass to produce fuels and chemical feedstocks. Energy Convers Manag 49:3498–3504
25. Fan S, Zakaria S, Chia C, Jamaluddin F, Nabihah S et al (2011) Comparative studies of products obtained from solvolysis liquefaction of oil palm empty fruit bunch fibres using different solvents. Bioresour Technol 102:3521–3526
26. Pan H (2011) Synthesis of polymers from organic solvent liquefied biomass: a review. Renew Sustain Energy Rev 15:3454–3463
27. Zou X, Qin T, Wang Y, Huang L (2011) Mechanisms and product specialties of the alcoholysis processes of poplar components. Energy Fuels 25:3786–3792
28. Li H, Yuan X, Zeng G, Tong J, Yan Y et al (2009) Liquefaction of rice straw in sub- and supercritical 1,4-dioxane-water mixture. Fuel Process Technol 90:657–663
29. Wang Y, Wu L, Wang C, Yu J, Yang Z (2011) Investigating the influence of extractives on the oil yield and alkane production obtained from three kinds of biomass via deoxy-liquefaction. Bioresour Technol 102:7190–7195
30. Wang M (2011) Reductive degradation of lignin in supercritical solvent and application in phenolic resin synthesis. Acta Polym Sin 12:1433–1438
31. Yuan X, Li H, Zeng G, Tong J, Xie W (2007) Sub- and supercritical liquefaction of rice straw in the presence of ethanol-water and 2-prop anol-water mixture. Energy 32:2081–2088
32. Savage PE (1999) Organic chemical reactions in supercritical water. Chem Rev 99:603–621
33. Savage PE (1996) Reaction model of cellulose decomposition in near-critical water and fermentation of products. Bioresour Technol 58:197–202
34. Zhang L, Champagne P, Xu C (2011) Bio-crude production from secondary pulp/paper-mill sludge and waste newspaper via co-liquefaction in hot-compressed water. Energy 36:2142–2150
35. Tymchyshyn M, Xu C (2010) Liquefaction of bio-mass in hot-compressed water for the production of phenolic compounds. Bioresour Technol 101:2483–2490
36. Xu C, Lad N (2008) Production of heavy oils with high caloric values by direct liquefaction of woody biomass in sub/near-critical water. Energy Fuels 22:635–642
37. Xu C, Lancaster J (2008) Conversion of secondary pulp/paper sludge powder to liquid oil products for energy recovery by direct liquefaction in hot-compressed water. Water Res 42:1571–1582
38. Ramsurn H, Gupta RB (2012) Production of bio-crude from biomass by acidic subcritical water followed by alkaline supercritical water two-step liquefaction. Energy Fuels 26:2365–2375
39. Goudnaan F, van de Beld B, Boerefijn FR, Bos GM, Naber JE et al (2008) Thermal efficiency of the HTU® process for biomass liquefaction. In: Bridgwater AV (ed) Progress in thermochemical biomass conversion. Blackwell Science Ltd, Oxford
40. He B, Zhang Y, Yin Y, Funk TL, Riskowski GL (2001) Preliminary characterization of raw oil products from the thermochemical conversion of swine manure. Trans Asae 44:1865–1871
41. Theegala CS, Midgett JS (2012) Hydrothermal liquefaction of separated dairy manure for production of bio-oils with simultaneous waste treatment. Bioresour Technol 107:456–463
42. Alba LG, Torri C, Samori C, van der Spek J, Fabbri D et al (2012) Hydrothermal treatment (HIT) of microalgae: evaluation of the process as conversion method in an algae bio-refinery concept. Energy Fuels 26:642–657
43. Akalin MK, Tekin K, Karagoz S (2012) Hydrothermal liquefaction of cornelian cherry stones for bio-oil production. Bioresour Technol 110:682–687

44. Liu H, Xie X, Li M, Sun R (2012) Hydrothermal liquefaction of cypress: effects of reaction conditions on 5-lump distribution and composition. J Anal Appl Pyrol 94:177–183
45. Amen-Chen C, Pakdel H, Roy C (2001) Production of monomeric phenols by thermochemical conversion of biomass: a review. Bioresour Technol 79:277–299
46. Demirbas A (2000) Effect of lignin content on aqueous liquefaction products of biomass. Energy Convers Manag 41:1601–1607
47. Demirbas A (2010) Sub- and super-critical water depolymerization of biomass. Energy Sources Part A 32:1100–1110
48. Durot N, Gaudard F, Kurek B (2003) The unmasking of lignin structures in wheat straw by alkali. Phytochemistry 63:617–623
49. Demirbas A (2000) Mechanisms of liquefaction and pyrolysis reactions of biomass. Energy Convers Manag 41:633–646
50. Bobleter O (1994) Hydrothermal degradation of polymers derived from plants. Prog Polym Sci 19:797–841
51. Matsumura Y, Sasaki M, Okuda K, Takami S, Ohara S et al (2006) Supercritical water treatment of biomass for energy and material recovery. Combust Sci Technol 178:509–536
52. Xu C, Donald J (2012) Upgrading peat to gas and liquid fuels in supercritical water with catalysts. Fuel 102:16–25
53. Shiraishi N, Kishi H (1986) Wood-phenol adhesives prepared from carboxymet hylated wood. J Appl Polym Sci 32:3189–3209
54. Alma M, Basturk M (2006) Liquefaction of grapevine cane (*Vitis vinisera* L.) waste and its application to phenol-formaldehyde type adhesive. Ind Crops Prod 24:171–176
55. Hassan ME, Mun SH (2002) Liquefaction of pine bark using phenol and lower alcohols with methane sulfonic acid catalyst. J Ind Chem 8:359–364
56. Yamada T, Ono H (1999) Rapid liquefaction of lignocellulosic waste by using ethylene carbonate. Bioresour Technol 70:61–67
57. Minami E, Kawamoto H, Saka S (2003) Reaction behaviors of lignin in supercritical methanol as studied with lignin model compounds. Jpn Wood Res Soc 49:158–165
58. Minami E, Saka S (2003) Comparison of decomposition behaviors of hard wood and soft wood in supercritical methanol. Jpn Wood Res Soc 49:73–78
59. Minami E, Saka S (2005) Decomposition behavior of woody biomass in water-added supercritical methanol. Jpn Wood Res Soc 51:395–400
60. Ishikawa Y, Saka S (2001) Chemical conversion of cellulose as treated in supercritical methanol. Cellulose 8:189–195
61. Tsujino J, Kawamoto H, Saka S (2003) Reactivity of lignin in supercritical methanol studied with various lignin model compounds. Wood Sci Technol 37:299–307
62. Miller JE, Evans L, Littlewolf A, Trudell DE (1999) Batch micro-reactor studies of lignin and lignin model compound depolymerization by bases in alcohol solvents. Fuel 78:1363–1366
63. Cemek M, Kucuk MM (2001) Liquid products from verbascum stalk by supercritical fluid extraction. Energy Convers Manag 42:125–130
64. Georget DMR, Cairns P, Smith AC, Waldron KW (1999) Crystallinity of lyophilised carrot cell wall components. Int J Biol Macromol 26:325–331
65. Maldas D, Shiraishi N (1997) Liquefaction of biomass in the presence of phenol and H_2O using alkalies and salts as the catalyst. Biomass Bioenergy 12:273–279
66. Saisu M, Sato T, Watanabe M, Adschiri T, Arai K (2003) Conversion of lignin with supercritical water-phenol mixtures. Energy Fuels 17:922–928
67. Wang M, Leitch M, Xu C (2009) Synthesis of phenolic resol resins using cornstalk-derived bio-oil produced by direct liquefaction in hot-compressed phenol-water. J Ind Eng Chem 15:870–875
68. Wang M, Xu C, Leitch M (2009) Liquefaction of cornstalk in hot-compressed phenol-water medium to phenolic feedstock for the synthesis of phenol-formaldehyde resin. Bioresour Technol 100:2305–2307

69. Okuda K, Umetsu M, Takami S, Adschiri T (2004) Disassembly of lignin and chemical recovery-rapid depolymerizatin of lignin without char formation in water-phenol mixtures. Fuel Process Technol 85:803–813
70. Pasquini D, Pimenta MTB, Ferreira LH, Curvelo AAS (2005) Extraction of lignin from sugar cane bagasse and *Pinus taeda* wood chips using ethanol-water mixtures and carbon dioxide at high pressures. J Supercrit Fluids 36:31–39
71. Kleinert M, Barth T (2008) Towards a lignocellulosic bio-refinery: direct one-step conversion of lignin to hydrogen-enriched bio-fuel. Energy Fuels 22:1371–1379
72. Yuan Z, Cheng S, Leitch M, Xu C (2010) Hydrolytic degradation of alkaline lignin in hot-compressed water and ethanol. Bioresour Technol 101:9308–9313
73. Cheng S, Wilks C, Yuan Z, Leitch M, Xu C (2012) Hydrothermal degradation of alkali lignin to bio-phenolic compounds in sub/supercritical ethanol and water–ethanol co-solvent. Polym Degrad Stab 97:839–848
74. Karagoz S, Bhaskar T, Muto A, Sakata Y (2004) Effect of Rb and Cs carbonates for production of phenols from liquefaction of wood biomass. Fuel 83:2293–2299
75. Qian Y, Zuo C, Tan H, He J (2007) Structural analysis of bio-oils from sub- and supercritical water liquefaction of woody biomass. Energy 32:196–202
76. Appell HR (1967) Fuels from waste. Academic Press, New York
77. Appell HR, Wender I, Miller RD (1969) Conversion of urban refues to oil. US Bureau of Mines
78. Karagoz S, Bhaskar T, Muto A, Sakata Y, Oshiki T et al (2005) Low-temperature catalytic hydrothermal treatment of wood biomass: analysis of liquid products. Chem Eng J 108:127–137
79. Karagoz S, Bhaskar T, Muto A, Sakata Y (2006) Hydrothermal upgrading of biomass: Effect of K_2CO_3 concentration and biomass/water ratio on products distribution. Bioresour Technol 97:90–98
80. Song C, Hu H, Zhu S, Wang G, Chen G (2004) Non-isothermal catalytic liquefaction of corn stalk in subcritical and supercritical water. Energy Fuels 18:90–96
81. Zhong C, Wei X (2004) A comparative experimental study on the liquefaction of wood. Energy 29:1731–1741
82. Zhang Q, Zhao G, Chen J (2006) Effects of inorganic acid catalysts on liquefaction of wood in phenol. Front For China 2:214–218
83. Mazaheri H, Lee KT, Bhatia S, Mohamed AR (2010) Subcritical water liquefaction of oil palm fruit press fiber in the presence of sodium hydroxide: an optimisation study using response surface methodology. Bioresour Technol 101:9335–9341
84. Li H, Hurley S, Xu C (2011) Liquefactions of peat in supercritical water with a novel iron catalyst. Fuel 90:412–420
85. Minowa T, Zhen F, Ogi T (1998) Cellulose decomposition in hot-compressed water with alkali or nickel catalyst. J Supercrit Fluids 13:253–259
86. Sinag A, Kruse A, Rathert J (2004) Influence of the heating rate and the type of catalyst on the formation of key intermediates and on the generation of gases during hydropyrolysis of glucose in supercritical water in a batch reactor. Ind Eng Chem Res 43:502–508
87. Watanabe M, Aizawa Y, Lida T, Aida TM, Levy C et al (2005) Glucose reactions with acid and base catalysts in hot compressed water at 473 K. Carbohydr Res 340:1925–1930
88. Watanabe M, Bayer F, Kruse A (2006) Oil formation from glucose with formic acid and cobalt catalyst in hot-compressed water. Carbohydr Res 341:2891–2900
89. Alma MH, Yoshioka M, Yao Y, Shiraishi N (1998) Preparation of sulfuric acid-catalyzed phenolated wood resin. Wood Sci Technol 32:297–308

Chapter 9
Hydrothermal Liquefaction of Biomass

Saqib Sohail Toor, Lasse Aistrup Rosendahl, Jessica Hoffmann,
Thomas Helmer Pedersen, Rudi Pankratz Nielsen
and Erik Gydesen Søgaard

Abstract Biomass is one of the most abundant sources of renewable energy, and will be an important part of a more sustainable future energy system. In addition to direct combustion, there is growing attention on conversion of biomass into liquid energy carriers. These conversion methods are divided into biochemical/biotechnical methods and thermochemical methods, such as direct combustion, pyrolysis, gasification, liquefaction, etc. This chapter focuses on hydrothermal liquefaction, where high pressures and intermediate temperatures together with the presence of water are used to convert biomass into liquid biofuels, with the aim of describing the current status and development challenges of the technology. During the hydrothermal liquefaction process, the biomass macromolecules are first hydrolyzed and/or degraded into smaller molecules. Many of the produced molecules are unstable and reactive and can recombine into larger ones. During this process, a substantial part of the oxygen in the biomass is removed by dehydration or decarboxylation. The chemical properties of the product are mostly dependent of the biomass substrate composition. Biomass consists of various components such as carbohydrates, lignin, protein, and fat, and each of them produce distinct groups of compounds when processed individually. When processed together in different ratios, they will most likely cross-influence each other and thus the composition of the product. Processing conditions including temperature, pressure, residence time, catalyst, and type of solvent are important for the bio-oil yield and product quality.

S. S. Toor (✉) · L. A. Rosendahl · J. Hoffmann · T. H. Pedersen
Department of Energy Technology, Aalborg University, Pontoppidanstræde 101,
Aalborg Ø 9220, Denmark
e-mail: sst@et.aau.dk

R. P. Nielsen · E. G. Søgaard
Department of Biotechnology, Chemistry and Environmental Engineering,
Section of Chemical Engineering, Aalborg University, Esbjerg Campus, Niels Bohrs Vej 8,
Aalborg Ø 6700, Denmark

9.1 Introduction

One of the grand challenges of society is to ensure an ample supply of sustainably produced liquid energy carriers, to fill the void as fossil reserves run out and to reduce greenhouse gas emissions due to human activity. Significant effort is put into all aspects of meeting this challenge: biomass availability, conversion technologies, and product range. In a resource-constrained market scenario, which is probably the most likely one, successful conversion processes must demonstrate the ability either to convert cheap, low quality feedstocks, to be extremely resource and energy efficient, or even both. Hydrothermal liquefaction (HTL) has the potential to fulfill both criteria, and to do so sustainably. Furthermore, with appropriate upgrading, the bio-crude produced from the process as well as transport grade biofuels will have drop-in properties, allowing for complete miscibility in the existing fuel stream and thus allowing for a gradual but complete replacement of the fossil components. Providing, of course, ample feedstock is available to secure the necessary volume. To fully realize the potential of HTL, an in-depth understanding of the complex chemical interaction between real biomasses and their components as well as simplified model compounds is necessary, in order to design HTL systems operating at temperatures and pressures ensuring maximum oil yields and quality.

9.2 Biomass as a HTL Feedstock

Biomass is a broad definition and includes a wide range of materials with varying compositions. The main biomass components are: carbohydrates, lignin, protein, and lipids.

9.2.1 Carbohydrates

The most abundant carbohydrates in biomass are the polysaccharides cellulose, hemicellulose, and starch. Under hydrothermal conditions, carbohydrates undergo rapid hydrolysis to form glucose and other saccharides, which are then further degraded. The rate of hydrolysis varies between different carbohydrates. Hemicelluloses and starch are hydrolyzed much faster than cellulose, which in contrast to the former two has a mainly crystalline structure. The degradation of carbohydrates in subcritical and supercritical water has been thoroughly reviewed by several authors. [1–3].

9.2.1.1 Cellulose

Cellulose consists of glucose units, linked by β-(1→4)-glycosidic bonds. In contrast to starch. it consists of straight chains, which enables formation of strong intra-and inter-molecule hydrogen bonds [4]. Thus cellulose has a high degree of crystallinity, which renders it insoluble in water and resistant to attack by enzymes. However, at subcritical conditions cellulose is rapidly solubilised and hydrolyzed to its constituents. A proposed reaction mechanism for the conversion of microcrystalline cellulose in subcritical and supercritical water at 25 MPa can be seen in Fig. 9.1.

Rogalinski et al. carried out an excellent kinetic study on hydrolysis of different biopolymers. The authors determined rate constants for the degradation of cellulose, starch, and protein. The hydrolysis rates of the three different biopolymers differed quite significantly. These results imply that rapid heating is important to avoid that some biopolymers depolymerise and start degrade before the right reaction temperature is reached. It was observed that cellulose hydrolysis rate in water at 25 MPa increased tenfold between 240 and 310 °C, and that cellulose hydrolysis was considerably slower than starch hydrolysis. At 280 °C, 100 % cellulose conversion was achieved within 2 min. The glucose decomposition rate increased rapidly with temperature and became higher than the glucose release rate already between 250 and 270 °C. The addition of carbon dioxide yielded a significant increase of the hydrolysis rate. This effect was most likely due to the formation of carbonic acid, acting as an acid catalyst. However, the effect got less pronounced above 260 °C [5].

In another kinetic study [6], the relationship between cellulose hydrolysis rate and glucose decomposition rate was also investigated, and in addition, the spectrum of hydrolysis and degradation products was carefully analyzed. Microcrystalline cellulose was subjected to subcritical and supercritical water (320–400 °C, 25 MPa, and 0.05–10.0 s). At 400 °C, hydrolysis products were mainly obtained, whereas at 320–350 °C, aqueous decomposition products of glucose, such C3–C6 sugars, aldehydes and furans, were the main products. The author's explanation for these rather unexpected results was that below 350 °C the cellulose hydrolysis rate is slower than the glucose decomposition rates, whereas above 350 °C, the cellulose hydrolysis rate drastically increases and becomes higher than the glucose decomposition rate [6].

Kamio et al. reported that cellulose hydrolysis drastically increased above 240 °C and that the hydrolysis proceeded by initial formation of oligosaccharides, which were subsequently converted into monosaccharides and pyrolysis products. This study, in contrast to the two previous trials, was carried out batch-wise and the moderate heating rate (6 K/min) might have affected the accuracy of the results, especially when the results show degradation also at medium temperatures [8].

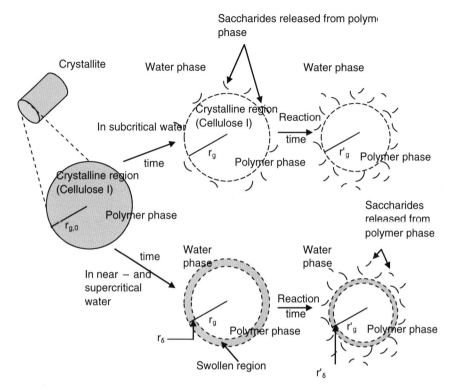

Fig. 9.1 Estimated reaction mechanism for the conversion of microcrystalline cellulose in subcritical and supercritical water at 25 MPa. Reprinted with the permission from Ref. [7]. Copyright 2004 American Institute of Chemical Engineers

9.2.1.2 Hemicelluloses

Hemicelluloses make up 20–40 % of plant biomass. It is a heteropolymer composed of various monosaccharides, including xylose, mannose, glucose, and galactose [2]. The composition varies significantly between plant types; grass hemicellulose is mainly composed of xylan, whereas wood hemicelluloses are rich in mannan, glucan, and galactan. Due to the abundance of side-groups and the less uniform structure of the hemicellulose, it has a much lower degree of crystallinity than cellulose [2, 4]. Hemicelluloses are easily solubilised and hydrolyzed in water at temperatures above about 180 °C, and the hydrolysis is both acid- and base-catalyzed [2]. Mok and Antal found that close to 100 % of the hemicellulose of various wood and herbaceous biomass materials was hydrolyzed at 230 °C, 2 min, and 34.5 MPa. Rapid heating made the experiments very accurate and the authors demonstrated how hemicellulose is selectively removed from the biomass at these conditions [9].

Just as in the case of cellulose, the saccharides released during subcritical hemicellulose hydrolysis will also be degraded in a similar way. Sasaki et al. conducted well-controlled decomposition experiments of D-xylose in subcritical and supercritical water at temperatures 360–420 °C, pressures of 25–40 MPa, and residence times of 0.02–1 s. The results showed that retro-aldol condensation of D-xylose was the dominant reaction and that the contribution of dehydration was small in subcritical water. The main degradation products were glycolaldehyde, glyceraldehydes, and dihydroxyacetone [10].

9.2.1.3 Starch

Starch, another main biomass component, is a polysaccharide consisting of glucose monomers connected with β-(1→4) and α-(1→6) bonds [11]. There are two different forms of starch; amylose with a linear structure, and amylopectin with a more branched structure. Compared to cellulose, starch is relatively readily hydrolyzed. Below some selected work on starch hydrolysis at subcritical conditions is summarized.

Starch (from sweet potato) was decomposed in a batch reactor at hydrothermal conditions (180–240 °C and unspecified pressures) in the absence of catalyst. The reaction temperature was reached within a minute. Rapid heating time is crucial when studying hydrolysis and degradation at these conditions. The starch was completely solubilised already after 10 min in 180 °C, however glucose yields were negligible. The maximum glucose yield was about 60 % and was obtained both at 200 °C and a residence time of 30 min and at 220 °C and 10 min. At more severe conditions (240 °C and 10 min), the yield was considerably lower due to glucose degradation. The main degradation product was 5-hydroxymethylfurfural (HMF) [12]. Here it was obvious that the relatively low temperatures favoured dehydration instead of defragmentation to short-chained acids and aldehydes.

In a similar study on hydrolysis of sweet potato, Miyazawa and Funazukuri reported a glucose yield of just 4 % after 15 min at 200 °C and unspecified pressures [13]. This seems to be low compared to the previous study. It was observed that the hydrolysis rate increased drastically if the medium was acidified with CO_2, and the amount of glucose released, increased linearly with increasing CO_2 concentration, in the range of 0–0.1 g CO_2 per g H_2O. This effect has been mentioned before and was reported to decrease with temperature and thus the importance for subcritical processes is limited [5]. In a third study, a maximum glucose yield of 43.8 % was attained during hydrolysis of starch from sweet potato (240 °C, 3.64 min, and no addition of catalyst) again dehydration of the produced glucose was observed producing 1,6-anhydroglucose and HMF [14].

9.2.2 Lignin

Lignin is together with cellulose and hemicellulose a major component of plant materials. It is an aromatic heteropolymer consisting of p-hydroxyphenylpropanoid units held together by C–C or C–O–C bonds. The three basic building blocks are transp- coumaryl alcohol, coniferyl alcohol, and sinapyl alcohol as can be seen in Fig. 9.2.

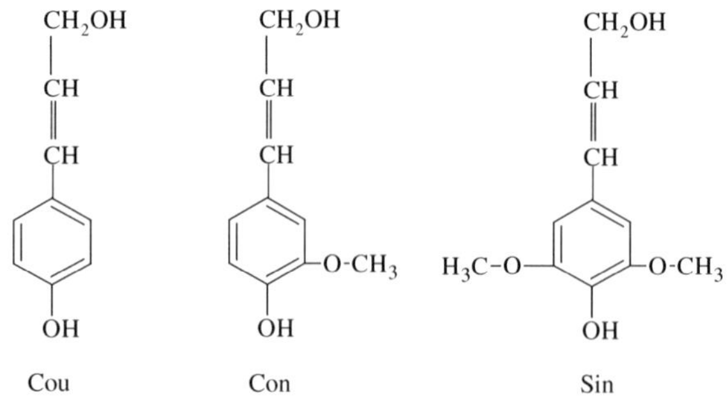

Fig. 9.2 Monomeric lignin building units: Cou, Con, Sin (p-coumaryl, coniferyl, sinapyl alcohols with CH$_2$OH–CH=CH– side chains on phenolic rings with 0, 1, and 2 methoxy (O-CH$_3$) substituents respectively)

Lignin is relatively resistant to chemical or enzymatic degradation [2]. However, during hydrothermal degradation various phenols and methoxy phenols are formed by hydrolysis of ether-bonds. These products can also degrade further by hydrolysis of methoxy groups, however, the benzene ring is stable at these conditions. Lignin hydrolysis is catalyzed by alkaline pH. Hydrothermal liquefaction of lignin produces significant amounts of solid residue and thus the amount of lignin in the substrate should be carefully balanced. The hydrothermal degradation pattern of lignin can be seen in Fig. 9.3.

9.2.3 Lipids

Fats and oils are nonpolar compounds with mainly aliphatic character. Chemically, they are referred to as triacylglycerides (TAGs), i.e., triesters of fatty acids and glycerol. Fats are insoluble in water at ambient conditions. However, the dielectric constant of water is significantly lower at subcritical conditions, allowing greater miscibility [11, 16]. TAGs are readily hydrolyzed in hot compressed water and catalysts are normally not required. On the other hand, the produced free fatty acids are relatively stable in subcritical water [17, 18]. King et al. achieved rapid hydrolysis of soybean oil in subcritical water at 330–340 °C and 13.1 MPa. Free fatty acid yields of 90–100 % were attained in 10–15 min. By using a reactor with transparent windows, the phase behavior at subcritical conditions could be studied. The authors noted that the reaction quickly went to completion when the mixture became a single phase at 339 °C [18].

Glycerol is one of the products of triglyceride hydrolysis and therefore a major co-product in bio-diesel production. It can be used to synthesize specialty chemicals, but is also considered as an important source for producing energy or fuels. Glycerol is not converted to an oily phase during hydrothermal liquefaction, but rather to water-soluble compounds. Thus, glycerol alone is not a suitable substrate for hydrothermal production of bio-oil.

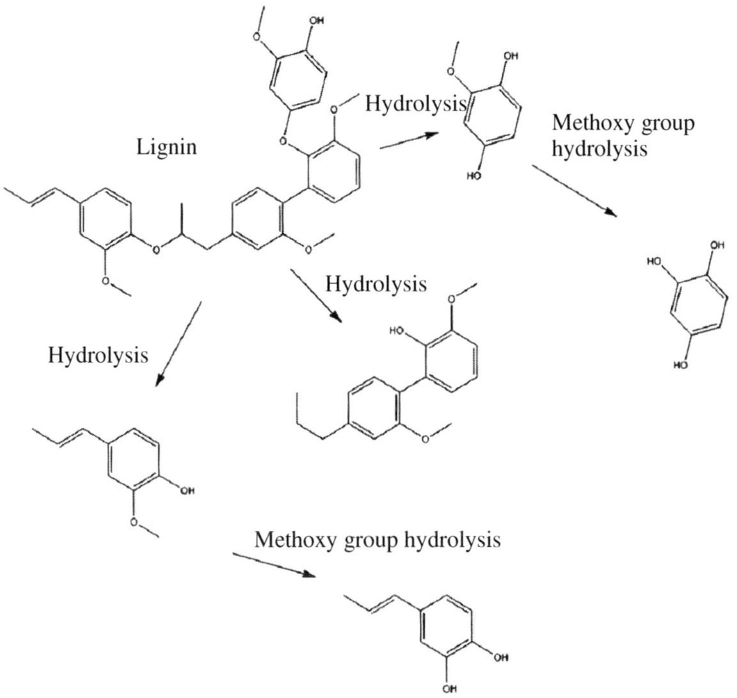

Fig. 9.3 Simplified scheme of lignin degradation. Reprinted with the permission from Ref. [15]. Copyright 2006 American Chemical Society

9.2.4 Proteins

Proteins are major biomass components, found particularly in animal and microbial biomass. Proteins consist of one or several peptide-chains, which in their turn are polymers of amino acids. The structural bond that links amino acids together is the peptide bond and amide bond between carboxyl and amine groups of the amino acids. A considerable fraction of the nitrogen in proteins will be incorporated in the bio-oil in a hydrothermal liquefaction process, affecting smell, combustion, and various other properties, and thus understanding the degradation of proteins is important.

The peptide bonds of proteins are more stable than the glycosidic bonds in cellulose and starch, and thus only slow hydrolysis occurs below 230 °C [5, 19]. The amino acid yield during hydrothermal liquefaction is generally significantly lower than achieved during conventional low-temperature acid hydrolysis, since the degradation rate of amino acids at hydrothermal conditions is relatively rapid compared to other biomass monomers [11, 20]. In one study on hydrolysis of bovine serum albumin (BSA), the subsequent amino acid degradation was carefully studied. The highest amino acid yield was obtained at 290 °C and 65 s, but still the total hydrolysis yields were low due to degradation. It was also observed that above 250 °C, the amino acid decomposition rate exceeded the hydrolysis rate. However, there were differences between the individual amino acids [21]. Some researchers have reported that the hydrolysis can be accelerated by acid catalysts. Rogalinski et al. observed that the amino acid yield increased from 3.7 to 15 % by acidification with CO_2 in hydrothermal liquefaction of BSA at 250 °C, 25 MPa, and 300 s. It was however also seen that the catalytic influence of carbon dioxide decreased with increasing temperature [5].

9.3 Reaction Pathways and Products from Primary Compounds

Biomass consists of a range of different macromolecules such as polysaccharides, lignin, protein, and lipids mixed in a variety of ratios, which determine the nature of each individual biomass species. When these macromolecules are hydrothermally processed they break down into their backbone units, which then further degrade by, e.g., fragmentation and dehydration. Fragments may then re-polymerize or recombine to form long-chain liquids or solids. The degradation mechanisms and final recombination products depend completely on the macromolecule in question and on the process conditions. Bio-oil, produced by hydrothermal liquefaction, is commonly defined as the final fraction immiscible in water but miscible in organic solvents such as acetone, chloroform, dichloromethane, and others. Insight into the breakdown and recombination mechanisms is vital in the effort to optimize the bio-oil both in terms of yield and quality by the biomass composition and process conditions.

Woody biomass also known as lignocellulosic biomass is a major renewable carbon resource with the potential to be turned into biofuels. Lignocellulose consists primarily of three different macromolecules, namely: cellulose, hemicellulose and lignin. The relative amount of the three different compounds is highly biomass dependent, but generally the distribution is approximately 35–50 % cellulose, 20–35 % hemicellulose, and 10–25 % lignin [22]. Cellulose and hemicellulose are polysaccharides, carbohydrate polymers of sugar units, whereas lignin is an amorphous phenolic polymer.

Physically, the rigid cellulose is embedded in a matrix of hemicellulose and lignin crosslinked in a complex structure [23]. Pretreatment of the complex structure toward accessibility of all macromolecules has been recognized for a long time, see Fig. 9.4. Normally, these pretreatment methods serve the purpose of delignification and removal of hemicellulose to cover cellulose fibers or glucose units. For hydrothermal liquefaction the purpose is to break down and liquefy all macromolecules and further concentrate the energy content by removal of contained oxygen.

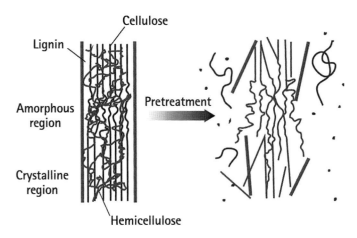

Fig. 9.4 Pretreatment of lignocellulose to fracture the structure. Adapted from Ref. [24], Copyright 2005, with permission for Elsevier

9.3.1 Degradation of Cellulose

The crystalline structure together with its insolubility characteristics in water hampers the degradation of cellulose due to its resistance to both chemical and enzymatic hydrolysis [25]. The characteristics arise from the equatorially positioned functional groups enhancing intra- and intermolecular hydrogen bonds to form between cellulose chains.

Attributed to the abundance of cellulose and its many uses, e.g., in paper production and as a precursor for renewable fuels, the degradation of cellulose has naturally by far been the most scientifically studied biomass compound.

The first degradation mechanism of cellulose polymers is scission of glycosidic bonds by hydrolysis for depolymerization. To obtain reasonable depolymerization rates, temperatures above 220 °C are required, which possibly also explains why temperature studies below this temperature are infrequent [11, 26]. The depolymerization of cellulose in a hydrothermal environment can occur homogeneously or heterogeneously depending on the solubility, which can be seen in Fig. 9.5. Factors influencing the solubility are degree of polymerization (DP) of oligomers and whether the degradation proceeds sub- or supercritically determined by the thermodynamic state.

In subcritical water low DP oligomers are soluble in water and therefore degrade homogeneously [2, 27]. For larger oligomers and polymers, degradation will proceed heterogeneously.

The cellulose solubility was indirectly studied by [25] in a clear view reactor by observing the loss of crystallinity as a function of temperature. The results were based on the loss of cellulose birefringence as temperature increased. It is clearly observed that the transition from a crystalline to an amorphous structure initiates at around 320 °C and determinates at 340 °C, see Fig. 9.6. Since it has been argued

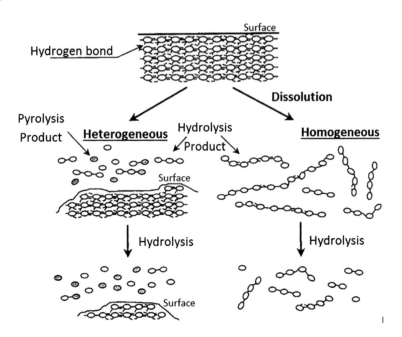

Fig. 9.5 Heterogeneous and homogeneous degradation pathway. Reprinted with the permission from Ref. [6]. Copyright 2000 American Chemical Society

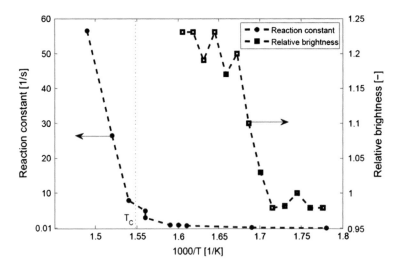

Fig. 9.6 Reaction constant and relative brightness of cellulose hydrolysis at 25 MPa as a function of temperature. ● Reprinted with the permission from ref. [6]. Copyright 2000 American Chemical Society, □ Reproduced from Ref. 25 with permission from the Centre National de la Recherche Scientifique (CNRS) and The Royal Society of Chemistry

9 Hydrothermal Liquefaction of Biomass

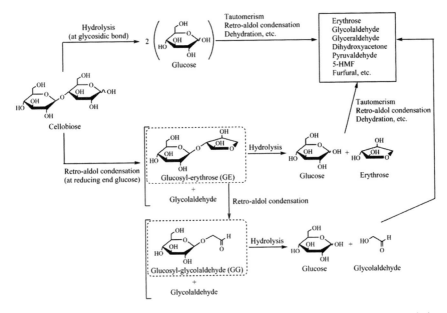

Fig. 9.7 Hydrothermal degradation mechanism of cellobiose. Reprinted with the permission from Ref. [32]. Copyright 2002 American Chemical Society

that the crystalline structure is causing the resistance to hydrolysis, one could expect that a structural transition would increase the hydrolysis rate. However, when plotting the temperature dependent reaction constant against the loss in birefringence no such conclusion can be drawn. Based on this observation, it is more likely that kinetics hamper the global reaction rate, not mass transfer or the lack of reaction sites due to crystallinity.

The sudden increase in reaction constant is more likely offset by entering the supercritical region. It has been suggested that in the subcritical region ionic reactions are dominating and that radical reactions dominate in the supercritical region [28].

The degradation rates are naturally also affected by present catalysts. The depolymerization of cellulose is both acidic and alkaline aided, where the latter demonstrates the highest rates. However, acidic conditions exhibit the highest rate for further degradation reactions of fragmented cellulose [2].

The degradation products of cellulose have been studied using cellobiose as a model compound; the dimer of cellulose [29–31]. Like cellulose, cellobiose hydrolyses into glucose but may also condense through Retro-aldol condensation into pyrolysis products at any reducing end, see Fig. 9.7.

Further degradation compounds can then be lumped into hydrolysis or Retro-aldol condensation products. It has been shown that kinetics of both lumps is nearly independent of pressure in the subcritical region. However, when entering the supercritical region, pressure effects can be quite significant, why selectivity in the supercritical region can be controlled by changing both temperature and pressure [32].

9.3.1.1 Glucose Degradation

The rate of glucose degradation displays an Arrhenius relation through the critical region. At subcritical conditions, the degradation rate has been shown to be much faster than the formation of glucose units from the cellulose scissoring. However, at the critical point, the degradation rate of cellulose discontinuously increases by approximately an order, thereby the glucose formation rate exceeds the degradation rate. As for cellobiose, the conversion rate of glucose shows hardly any affection on pressure in the subcritical region [33].

The selectivity of the different glucose degradation components is highly affected by whether an acidic or alkaline catalyst is present. In acidic solutions, high concentrations of 5-HMF and furfural are found. At alkaline conditions mainly lactic acid or other carboxylic acids are formed, see Fig. 9.8 Lactic acid is formed from pyruvaldehyde.

The fact that selectivity of different product components can be affected by the pH value, yield of bio-oil obtained from glucose is therefore also affected by the pH value. The high concentration of short chained carboxylic acids (C2–C5) which are highly water soluble, does not contribute to oil formation [34]. On the contrary, 5-HMF tends to form polymers and oil products, see Fig. 9.9; hence oil yields increase at low pH values [35].

Moreover, temperature and reaction time have been proven important parameters in the formation of bio-oil from cellulose. Yin and Tan concluded that in their setup maximum bio-oil was obtained at 300 °C and short reaction times with an oil yield of more than 30 %.

In addition to its relation to reaction kinetics, temperature also plays another important role in glucose conversion into bio-oil. At high temperatures, glucose tends to epimerize into fructose, see Fig. 9.10.

It has been found that the selectivity of fructose conversion into 5-HMF over fragmentation products is higher than for glucose. This may also explain why hydrothermal processed fructose produces more oily products than glucose.

9.3.2 Degradation of Hemicellulose

The conversion of different monosaccharides has been studied by Lü and Saka [37] in the temperature window of 160–250 °C. The conversion was expressed in terms of the process severity factor, $\log(R_0)$, taking the time-dependent process temperature profile into account. The severity factor is expressed in reaction (9.1). More about process severity can be found in [38].

$$\log(R_0) = \log\left(\int_0^t \exp\left(\frac{T - 100}{14.75}\right) dt\right) \quad (9.1)$$

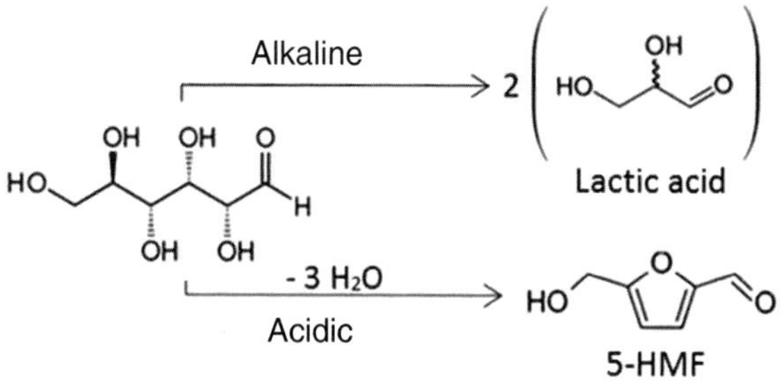

Fig. 9.8 Products formation in acidic and alkaline solutions

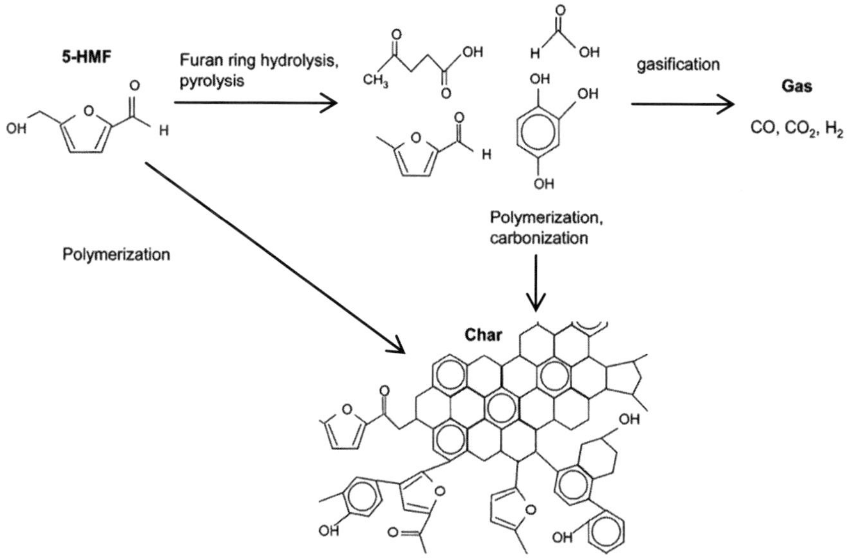

Fig. 9.9 Degradation mechanism of 5-HMF as suggested by Chuntanapum and Matsumura. Adapted with the permission from Ref. [36]. Copyright 2009 American Chemical Society

In Fig. 9.11, it is observed that the conversion of the represented monosaccharides clearly increases as process temperature and reaction time increases. The monosaccharides investigated both represent pentoses and hexoses for which the distributions of degradation products were found dissimilar. For pentoses, furfural was the major degradation product, whereas 5-HMF was mainly found in the products of hexose. The amount of 5-HMF and furfural increases with increased reaction time and temperature, but the formation cannot be tended as a function of severity factor.

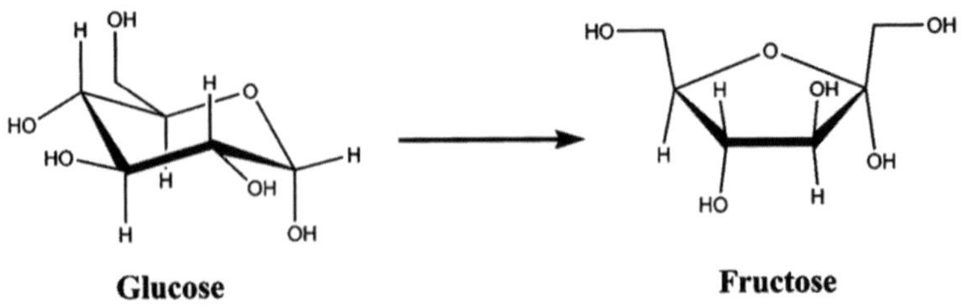

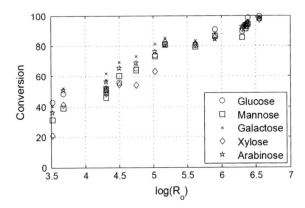

Fig. 9.10 Glucose-fructose epimerization

Fig. 9.11 Conversion of monosaccharides as a function of process severity. Adapted from Ref. [41], Copyright 2007, with permission for Elsevier

9.3.2.1 Xylose Degradation

Xylan is commonly the major component of hemicellulose. This also explains why xylan or more common xylose, the monomer of xylan, is often used as a representative model compound for hemicellulose [39–41].

Like glucose, xylose degrades through two main pathways: The Retro-aldol condensation and dehydration [40], see Fig. 9.12.

In connection with previous statements, in the low temperature range 160–250 °C it has been found that the main decomposition products of D-xylose are furfural and formic acid [37, 41]. Here it was also shown that the xylose hydrolysis activation energy was not changed by increasing the acidity of the aqueous solution. The activation energy ranges from 119 to 130 KJ/mol. On the other hand, an alkaline solution showed a significant reduction and demonstrated an activation energy of only 63 KJ/mol, which also changed the reaction pathway from furfural to increased acid formation [42, 43]. In the temperature range of 160–250 °C the furfural formation from xylose increases with increased reaction time and temperature. However, as temperature increases to near critical values it is observed that dehydration reactions selectively decrease, hence the conversion into furfural also decreases [10, 44].

9 Hydrothermal Liquefaction of Biomass

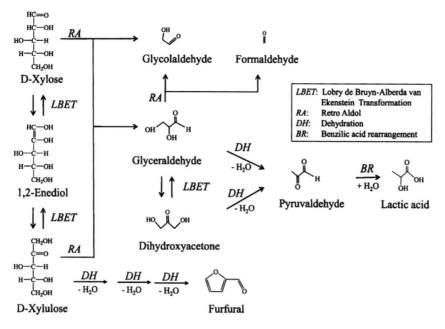

Fig. 9.12 Proposed degradation mechanism of xylose. Reprinted from Ref. [40], Copyright 2010, with permission for Elsevier

Since furfural is also a precursor for polymerization, the same observations of glucose may also be applicable for xylose or pentoses in general.

9.3.3 Degradation of Lignin

In general, lignin is the minor compound of lignocellulose, it has a greater heating value than both cellulose and hemicellulose. Hence, 40–50 % of the total heating value of lignocellulose is attributed to lignin, that is why lignin is a very important compound in the BTL conversion [23].

Isolated lignin is mainly produced as a by-product from large industrial processes like pulp production. During these processes, lignin undergoes chemical structure modifications and must therefore be distinct from native lignin. It has been suggested that cleavage of ether bonds in pretreatment processes lead to a higher fraction of C–C bonds, hence a more crosslinked and stable lignin structure is obtained after pretreatment processes.

The degradation of lignin for bio-oil production is very important, in order to gain accessibility to other macromolecules. When lignin breaks down mainly by weak ether bond hydrolysis, mostly phenolic monomers are formed. Subsequent dealkylation forms highly reactive and unstable fragments in water, as a result

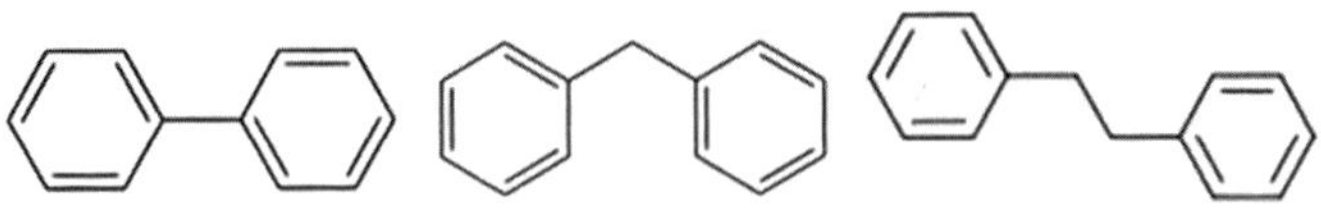

Fig. 9.13 Unreactive lignin model compound in ethanol at 290 °C. *Left* Biphenyl, *Middle* Diphenyl methane, *Right* Bibenzyl [46]

re-polymerization, oligomerization, condensation, and char formation occurs. This is an undesirable reaction pathway of lignin, when bio-oil production is the purpose.

In [45], it was visually observed during hydrothermal treatment of willow, that lignin and hemicellulose dissolute around 200 °C. As temperature increased to 250 °C lignin precipitated and formed capping fragments preventing cellulose dissolution. As a consequence, when temperature was further increased to 350 °C cellulose underwent pyrolysis instead of hydrolysis.

In the pursuance to prevent char formation, radical scavengers like phenols and alcohols have been used to cap reactive fragments.

Base catalyzed alcoholysis of Kraft and Organosolv derived lignin and lignin model compounds in methanol and ethanol at 290 °C was studied by Miller et al. [46]. Diethyl ether insoluble fraction levels as low as 5 % were achievable with an ethanol solvent. From the model compound study it is clear that at these conditions alkylation and dealkylation of benzene rings occur to interact with ethanol. In addition it was found that C–C bonds are not cleaved at these conditions, hence compounds like biphenyl, diphenyl methane, and bibenzyl are unreactive (see Fig. 9.13). A similar fact has been confirmed in a hydrous environment. It has concluded that aryl-aryl and methylene bonds can only be cleaved below 400 °C with the addition of a proper base catalyst.

The alcoholysis is somewhat different from hydrolysis. In hydrothermal medium, phenols are stable compounds, whereas in alcoholic solvents they undergo alkylation. The difference between degradation of alkali lignin in a pure ethanol, pure water or a co-ethanol-water solution was investigated by Cheng et al. [47]. The following sequence was concluded for the yield of degraded lignin; co-solution > water > ethanol. The higher yield of degraded lignin was accompanied by the lowest yield of solid residues. Solid production can almost be neglected in a 1:1 co-solution.

Although the depolymerization/repolymerization mechanisms have been widely studied, previous work dedicated for turning isolated lignin into a bio-oil is scarce. Roberts et al. reports the yield of a dark-brown product, formed by the utilization of base catalyzed depolymerization and boric acid as a repolymerization inhibition agent. A maximum of 52 % oil yield was obtained at a NaOH/boric acid weight ratio of 0.75. It was found that a multitude of process parameters influenced the yield like temperature, pressure, residence time and weight ratios between lignin, catalyst and boric acid.

9.4 Reaction Pathways and Products from Real Biomass

Compared to the conversion of model components, the conversion of real biomasses poses a series of challenges. First of all, the complexity of the real biomass compared to a model component such as cellulose or lignin is increased significantly, and second, the effect of process parameters may not be directly transferable from model studies due to various synergetic effects of the various components of the biomass. A large amount of work has been put into studies of the conversion of actual biomass and model substances in HTL and HTL-like processes [48–79], which forms a foundation of knowledge for controlling the process.

To be able to understand the processes that occur when processing actual biomass in an HTL process it is necessary to have a basic understanding of the reaction types and reaction pathways that are likely to occur in such an environment. This part will initially go over the basic reaction types that are likely to occur during processing and the implications of these reactions on the products, followed by a discussion of the effect of processing parameters and addition of catalysts to the process.

9.4.1 Overall Reaction Pathways

Since many types of biomass contain significant quantities of cellulose, hemicellulose and lignin, these are of main interest with regards to degradation. At the high pressures and temperatures, the HTL process operates the reaction media has to be considered. In this case, water constitutes a majority of the process stream, and due to the proximity to the critical point the auto-dissociation of water (reaction 9.2) must be considered.

$$H_2O \rightarrow H^+ + OH^- \tag{9.2}$$

Near the critical point the auto-dissociation is approx. 3 orders of magnitude larger than that at ambient conditions [80]. This means that the increased amount of ions from the water will make acid and base catalyzed reactions far more likely to occur at these conditions than at ambient.

Also the high processing temperature of the process makes thermal cracking a risk that needs to be considered, since the formation of char is an unwanted side effect. Char may be formed by either thermal cracking (reaction 9.3) or Boudouard reactions (reactions 9.4 and 9.5)

$$C_nH_{2n} \rightarrow C_n + nH_2 \tag{9.3}$$

$$2CO \rightarrow C + CO_2 \tag{9.4}$$

$$CH_4 \rightarrow C + 2H_2 \tag{9.5}$$

These reactions relate to the steam reforming reactions for hydrogen production from hydrocarbons, as given in reactions (9.6) through (9.8), with the water-gas shift as reaction (9.7). These reactions normally operate at high temperatures [81, 82], which is why they to some extent should be considered in the HTL process.

$$C_nH_m + nH_2O \rightarrow nCO + (n + m/2)H_2 \tag{9.6}$$

$$CO + H_2O \rightarrow CO_2 + H_2 \tag{9.7}$$

$$CO + 3H_2 \rightarrow CH_4 + H_2O \tag{9.8}$$

If processing larger hydrocarbons then the amount of CO formed will increase as pr. reaction (9.6), this could lead to increased char formation through the Boudouard reactions (reaction 9.4), which is in accordance with observations in steam reforming [82].

Also reactions to consider are hydrolysis, hydrogenation and decarboxylation (reactions 9.9 through 9.11), where especially hydrolysis is of interest when processing lipids since this is part of the transesterfication of lipids into carboxylic acids.

$$R_1 - O - R_2 + H_2O \rightarrow R_1 - OH + R_2 - OH \tag{9.9}$$

$$H_2C = CH_2 + H_2 \rightarrow H_3C - CH_3 \tag{9.10}$$

$$R - COOH \rightarrow RH + CO_2 \tag{9.11}$$

These reactions may also lead to the formation of alcohols and CO_2 as well as the saturation of unsaturated bonds. The hydrogenation is likely to occur due to the production of hydrogen from the steam reforming reactions and any thermal cracking.

With the formation of hydrogen and CO from the various reaction sets, the Fischer-Tropsch synthesis of hydrocarbons cannot be disregarded. Through this reaction (reaction 9.12) CO and H_2 reacts to form alkanes and water.

$$nCO + (2n + 1)H_2 \rightarrow C_nH_{(2n+2)} + nH_2O \tag{9.12}$$

Fischer-Tropsch synthesis of biodiesel is a known and used process, so it must be noted that the HTL process is not a FT synthesis but rather that the reactions of the FT synthesis are likely to occur during HTL processing.

Another aspect that should be considered as well is the formation of furfurals from the sugars of hydrolyzed cellulose. It has been proposed [70, 83] that below the critical temperature of water, it is likely that furfurals are formed which are subsequently converted into phenols and higher molecular weight components. This is unlike supercritical conditions where radical reactions dominate [70], which promotes gas production. This follows the general idea that liquefaction reactions are occurring below the critical temperature while gasification is running at supercritical temperatures [51]. Also lignin itself is likely to produce a large number of aromatics, since the lignin structure contains significant amount of benzene rings.

Having gone over some of the main reactions that are likely to occur during processing these may be applied to model compounds such as cellulose, lignin and lipids.

It must be assumed that any lignocellulosic material would be subject to hydrolysis of the ether bonds in the cellulose and hemicellulose. This hydrolysis would yield significant amount of sugars which would either remain or be converted into oil through thermal decomposition and cracking or through formation of furfurals and phenols. It is also possible that some of these sugars would go into the steam reforming reactions, i.e. be broken down into small volatile components, which subsequently could participate in a Fischer-Tropsch synthesis of alkanes and water, thus yielding the desired oil phase. This is illustrated in Fig. 9.14 including the formation of char.

Besides lignocellulosic material also the conversion of lipids should be considered. Figure 9.15 illustrates the possibilities when fatty acids are treated at high pressures and temperatures and the possible reactions.

As it may be seen, there are several possibilities for fatty acids and lipids when looking at a wider context. In the case of triglycerides hydrolysis of the ether bonds is likely to occur thus forming glycerol and free fatty acids [48]. Decarboxylation reactions would yield ketones and CO_2 while hydrogenation would saturate any unsaturated bonds in the molecule. Finally there is a chance of simple thermal decomposition which would lead to production of methane, CO_2, and possibly some ketones.

In general, the reactions happening during HTL processing makes up a complex network and pathway system where the extent of reactions and their tendency to occur depends on both process parameters and the concentration of the various components in the reaction mixture. This means, that identifying every single reaction for each component is a very demanding task. Still it is important to have estimates of the conversion of model components before venturing into the conversion of actual biomass, since this knowledge will help give overall estimates of the product from the conversion process based on the composition of the biomass.

9.4.2 Effect of Process Parameters

Having established a basis of possible reactions that may occur in the HTL process the effect of process parameters on the reactions and products is a major area of interest. Since equilibrium constants will vary with temperature, and pressure will affect equilibriums in the system the process parameters cannot be neglected when investigating a high temperature, high pressure process. At process conditions, molecules will have a high kinetic energy due to the elevated temperature as well as a narrow spatial distribution due to the high pressure. This means that molecular

Fig. 9.14 Reaction pathways for lignocellulosic material [48]

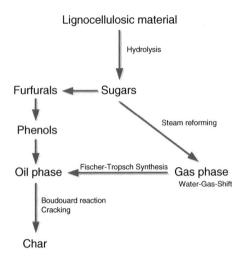

Fig. 9.15 Reaction pathways for fatty acids [48]

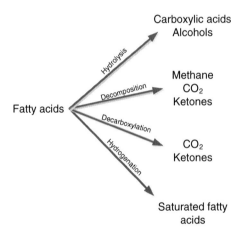

collisions and interactions are far more likely to occur at process conditions for the HTL process than at ambient conditions, i.e., the system is very reactive due to process conditions.

It is generally accepted that the transition from near-critical water to supercritical water as a reaction medium changes biomass conversion processes from liquefaction to gasification processes [19, 49, 51, 54]. Still, even though the focus is on liquefaction valuable, information may be gained from gasification processes.

When increasing temperature reaction rates must be assumed to increase, i.e., the conversion will proceed at a higher pace. However, increasing temperatures also bring other advantages. It has been reported [84] that at high temperatures the formation of tar from lignocellulosic material is reduced in gasification processes. Since tar is a high-molecular weight aromatic compound from thermal degradation of lignin without much practical use when focusing on biofuels, the formation of this is preferably avoided. As high temperatures tends to reduce tar formation it must be assumed that tar will form in a temperature range and that a high heating rate may reduce tar formation significantly, once the "tar-band" is determined. Although increasing temperature generally increases reaction rates, it must be noted that not all reactions may be promoted by increasing temperature. Any exothermic reactions in the system will inhibited due to the high temperature of the process. Furthermore, the temperature may also affect the distribution of nitrogen in the product phases [85]. As the critical temperature is approached, more and more nitrogen will be found in the oil phase, while it at lower processing temperatures will be mainly in the aqueous phase. This shift in nitrogen-distribution must be considered carefully, and it may be difficult to avoid nitrogen as well as tar formation due to mismatched temperature effects.

With regard to pressure it must of course be noted that reactions must be assumed to behave according to le' Chatelier's principle, i.e., equilibriums will shift toward the fewest molecules formed. However, most effects on reactions must be attributed to the high processing temperatures rather than the pressure. Pressure is still needed though, to get to the near-critical region and thus achieve desired properties of water as a reaction media at these conditions.

The use of catalysts in HTL processing has been investigated as well over the years [48–50, 86–90]. In processing there are the options of heterogeneous and homogeneous catalysis. In general, the homogeneous catalysts utilized are a number of alkali salts, while the heterogeneous are metal oxide catalysts. These two options will be treated in turn.

Using a homogeneous alkali catalyst such as KOH or NaOH has been shown to increase the gas, especially hydrogen, formation in supercritical water [55, 73, 86]. It was shown that the increased formation of hydrogen and CO_2 was at the expense of CO production. A reason for this may be that the alkali will promote the water gas shift (reaction 4.6) through the formation of methanoate (reactions 9.13 through 9.16) [48, 58, 71, 72].

$$K_2CO_3 + H_2O \rightarrow KHCO_3 + KOH \quad (9.13)$$

$$KOH + CO \rightarrow K(HCOO) \quad (9.14)$$

$$K(HCOO) + H_2O \rightarrow KHCO_3 + H_2 \quad (9.15)$$

$$2KHCO_3 \rightarrow CO_2 + K_2CO_3 + H_2O \quad (9.16)$$

Evidently in this case potassium carbonate catalyzes the formation of CO_2 and H_2 at the expense of CO. In a liquefaction process, the added gas production is not an objective in itself, however the produced hydrogen could be utilized in any Fischer-Tropsch reactions occurring, and thus actually promote the production of higher alkanes.

However, when using salts as homogeneous catalysts there is a risk of precipitation as the critical point of water is approached. As this happens the dielectric constant of water decreases significantly thus rendering salts and any ionic species almost insoluble in water while nonpolar substances such as alkanes become fully miscible with water. If precipitation of the homogeneous catalyst occurs, it reduces the contact area with the catalyst and thus the efficiency of this. Also with precipitate in the system there is a risk of plugs forming as well as abrasion in bends and small orifices such as valves.

With regard to heterogeneous catalysts, generally metal oxides have been attempted. Results show that a small increase in gas production is observed, however this is very limited to the increase observed for the homogeneous alkali catalysts [86]. When using metal oxides some of these contain active sites that may facilitate both acid- and base catalyzed reactions. With the increased amount of ions due to the near-critical condition of the water, these sites will promote any acid or base catalyzed reaction in the process. There is, however, a downside to using heterogeneous catalysts, since any tar and coke formation may deposit on the catalysts thus fouling/poisoning them and rendering it at a lower efficiency or fully deactivated.

It has also been proposed that the reactor wall itself has an effect on the conversion reactions [91]. Due to process conditions, reactor walls need to be made of highly alloyed steels, and it cannot be ruled out that some of the components in the alloys may actually react with material in the process stream. This must be assumed to have the functionality of a heterogeneous catalyst where minor constituents in the steel may have this functionality.

There are a lot of possibilities for tuning the process using processing parameters as well as adding catalysts to the process. This means that this type of liquefaction may be potentially designed to achieve a specified product by tuning the process conditions and composition.

9.5 Properties and Upgrading of HTL Bio-oil

Hydrothermal liquefaction of biomass delivers high valuable bio crude which can be used for burner firing. Upgrading of the HTL crude is required to reach conventional diesel or jet-fuel standards. Most research done on bio-oils upgrading focuses on pyrolysis bio-oils, upgrading of bio-oils from hydrothermal liquefaction (HTL) has not yet been extensively studied.

The liquefaction process produces a crude oil replacement, which has an important key difference from conventional crude oil: the oxygen and water content in the crude are significantly higher, typically 9–25 and 6–25 wt. %, respectively in HTL bio-oils versus <1 % in conventional petroleum [92]. In Table 9.1 unwanted characteristics of bio-oil and their effects are listed.

In regard to other thermochemical biomass conversion processes HTL delivers raw oil, with a comparable low oxygen and water content. In Table 9.2 property ranges of bio-crude obtained by HTL and in comparison to pyrolysis of wood are shown.

Table 9.1 Unwanted characteristics of bio-oil and their effects collected from [93, 94]

Characteristic	Effect
Low pH value	Corrosion problems
High viscosity	Handling and pumping problems
Instability and temperature sensitivity	Storage problems
	Phase separation
	Decomposition and gum formation
	Viscosity increase
Char and solids content	Combustions problems
	Equipment blockage
	Erosion
Alkali metals	Depositions of solids in boilers, engines, and turbines
Water content	Complex effect on heating value, viscosity, pH, homogeneity and other characteristics
Oxygen content	Higher viscosity and lower heating value

Table 9.2 Property ranges of bio-crudes

	HTL	Pyrolysis	Fossil oil
Carbon (wt. %)	68–81	56–66	83.0–87.0
Sulfur + Nitrogen (wt. %)	0.1	0.1	0.01–5
Oxygen (wt. %)	9–25	27–38	0.05–1.5
Water (wt. %)	6–25	24–52	<1
Density (kg L^{-1})	1.10–1.14	1.11–1.23	0.75–1.0

Reprinted from Ref. [95], Copyright 2013, with permission for Elsevier

Compared to pyrolysis oil though HTL oils have almost half of the oxygen content, which makes upgrading of HTL much easier and less expensive.

When considering upgrading of HTL bio-crude, it is advantageous to look into standard conventional refinery procedures, to be able to use existing refinery infrastructure and knowledge.

Hydroprocessing is one way of upgrading crude oils. As it can be seen in the Van-Krevelen diagram in Fig. 9.16, the oxygen content in the oil has to be decreased and the hydrogen content increased (red arrow). Hydroprocessing is a generic term that describes the use of hydrogen and an appropriate catalyst to remove undesired components form refinery streams. In refinery, hydroprocessing is used to remove reactive compounds like olefins, sulfur, and nitrogen compounds. At higher temperatures and pressures, aromatic rings can be saturated and all sulfur and nitrogen are removed. Hydroprocessing at higher temperatures and pressures is referred to as hydrocracking. Large molecules are broken down and saturated with hydrogen. In standard refinery, this process is used for cracking of vacuum gas oil (VGO). Hydrocracking in standard refineries yields a high percentage of products in the diesel and kerosene boiling range.

With regard to bio-crude, the high oxygen content is the major issue. Hydroprocessing can be used to deoxygenate the crude and subsequently improve the

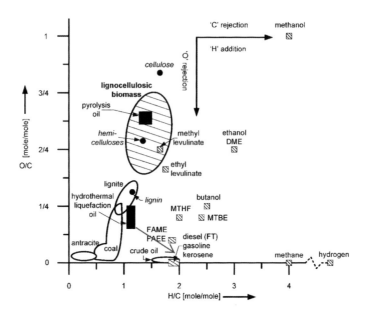

Fig. 9.16 Van Krevelen diagram [96]

properties of the bio-crude. This deoxygenating process is called hydrodeoxygenation (HDO). The oil is treated at temperatures of 300–400 °C under a hydrogen atmosphere of 150–300 bar. The presence of water is crucial since it prevents from char formation during the reaction. Therefore, high pressures of 150–300 bar are necessary to prevent from char formation [93].

During the reaction oxygen from oxygen containing compounds (like phenols, ketones, and fatty acids) is partially or totally eliminated in the form of water in the presence of hydrogen and a suitable catalyst such as Ruthenium on Carbon, Cobalt–Molybdenum (CoMo /γ Al$_2$O$_3$) or Nickel-Molybdenum (NiMo /γ Al$_2$O$_3$) sulphides catalyst [93–95, 97]. Studies on upgrading of HTL bio-oil crude have not yet been published and upgrading work mainly includes upgrading of fast pyrolysis bio-oils [95] Since water is formed during HDO standard hydrodesulphirisation (HDS), and/or hydrodenitrification (HDN) catalyst can't be used because water affects catalyst performance negatively [96].

During hydrodeoxygenation, oxygen is removed from the biocrude through the formation of water. The overall reaction stochiometrie is shown in reaction (9.17).

$$-(CH2O) - + H2 \rightarrow - (CH2) - + H2O \qquad (9.17)$$

During HDO dehydration, through which oxygen is released in the form of water, decarboxylation reaction, through which oxygen is released in form of CO$_2$, hydrogenation reaction, in which hydrogen reacts with unsaturated carbon bounds of bio-oil compounds. Through hydrogenolysis, C–O bonds are broken up and oxygen is removed in the form of water. Also hydrocracking reactions may occur in which big molecules are broken down into smaller ones and saturated with hydrogen.

For upgrading of the HTL crude fuel, it could be advantageous to upgrade before separating the oil and aqueous phase. By this most polar components can be converted to hydrocarbons and the aqueous phase can be easily separated from the hydrocarbon phase. Research is being done on upgrading of biocrudes in the aqueous phase in either sub-critical or super critical conditions and a recent review by Furimsky has been published addressing also upgrading of biocrudes in aqueous phases [95].

9.6 Conclusion

In this chapter, hydrothermal liquefaction has been discussed from the perspective of achieving a fundamental understanding of the reactions involved, and the influence on these from feedstock or biomass composition and process conditions. Through the main components of biomass, carbohydrates, lipids, protein, and lignin, and model compounds for these, reaction mechanisms have been established. The most abundant resource, lignocellulosic biomass, has been broken down into its major constituents, lignin, cellulose, and hemicellulose, and pathways for the conversion of these and factors either inhibiting or enhancing their conversion to bio-crude have been described. In all, it is clear that HTL is an attractive technology for conversion of a wide range of biomasses to bio-crude, and for some biomasses, technology implementation is possible already at this point. However, it is also clear that although an overall understanding of HTL is in place, and several authors bring forward data on conversion of specific biomasses or model compounds, there are still significant gaps in our knowledge of all parameters affecting this process that will only be filled by focussed research efforts.

References

1. Behrendt F, Neubauer Y, Oevermann M, Wilmes B, Zobel N (2008) Direct liquefaction of biomass-review. Chem Eng Technol 31:667–677
2. Bobleter O (1994) Hydrothermal degradation of polymers derived from plants. Polym Sci 19:797–841
3. Yu Y, Lou X, Wu H (2008) Some recent advances in hydrolysis of biomass in hot-compressed water and its comparisons with other hydrolysis methods. Energy Fuels 22:46–60
4. Delmer DP, Amor Y (1995) Cellulose biosynthesis. Plant Cell 7:987–1000
5. Rogalinski T, Liu K, Albrecht T, Brunner G (2008) Hydrolysis kinetics of biopolymers in subcritical water. J Supercrit Fluids 46:335–341
6. Sasaki M, Fang Z, Fukushima Y, Adschiri T, Arai K (2000) Dissolution and hydrolysis of cellulose in subcritical and supercritical water. Ind Eng Chem Res 39:2883–2890
7. Sasaki M, Adschiri T, Arai K (2004) Kinetics of cellulose conversion at 25 MPa in sub- and supercritical water. AIChE J 50:192–202

8. Kamio E, Sato H, Takahashi S, Noda H, Fukuhara C, Okamura T (2008) Liquefaction kinetics of cellulose treated by hot compressed water under variable temperature conditions. J Mater Sci 43:2179–2188
9. Mok WSL, Antal MJ (1992) Uncatalyzed solvolysis of whole biomass hemicellulose by hot compressed liquid water. Ind Eng Chem Res 31:1157–1161
10. Sasaki M, Hayakawa T, Arai K, Adichiri T (2003) Measurement of the rate of retro-aldol condensation of D- xylose in subcritical and supercritical water. In: Presented at the proceeding of the 7th international symposium on hydrothermal reactions, pp 169–176
11. Peterson AA, Vogel F, Lachance RP, Fröling M, Antal MJ, Tester JW (2008) Thermochemical biofuel production in hydrothermal media: a review of sub- and supercritical water technologies. Energy Environ Sci 1:32–65
12. Nagamori M, Funazukuri T (2004) Glucose production by hydrolysis of starch under hydrothermal conditions. J Chem Technol Biotechnol 79:229–233
13. Miyazawa T, Funazukuri T (2005) Polysaccharide hydrolysis accelerated by adding carbon dioxide under hydrothermal conditions. Biotechnol Prog 21:1782–1786
14. Miyazawa T, Ohtsu S, Nakagawa Y, Funazukuri T (2006) Solvothermal treatment of starch for the production of glucose and maltooligosaccharides. J Mater Sci 41:1489–1494
15. Liu A, Park YK, Huang Z, Wang B, Ankumah RO, Biswas PK (2006) Product identification and distribution from hydrothermal conversion of walnut shells. Energy Fuel 20:446–454
16. Khuwijitjaru P, Adachi S, Matsuno R (2002) Solubility of saturated fatty acids in water at elevated temperatures. Biosci Biotechnol Biochem 66:1723–1726
17. Holliday RL, King JW, List GR (1997) Hydrolysis of vegetable oils in sub- and supercritical water. Ind Eng Chem Res 36:932–935
18. King JW, Holliday RL, List GR (1999) Hydrolysis of soybean oil in a subcritical water flow reactor. Green Chem 1:261–264
19. Brunner G (2009) Near critical and supercritical water. Part I. Hydrolytic and hydrothermal processes. J Supercrit Fluids 47:373–381
20. Xian Z, Chao Z, Liang Z, Hongbin C (2008) Amino acid production from fish proteins hydrolysis in subcritical Water. Chin J Chem Eng 16:456–460
21. Rogalinski T, Herrmann S, Brunner G (2005) Production of amino acids from bovine serum albumin by continuous sub-critical water hydrolysis. J Supercrit Fluids 36:49–58
22. Saha BC (2004) Lignocellulose biodegradation and applications in biotechnology. American Chemical Society, Washington, pp 2–34
23. Lee S, Shah YT, (eds) (2013) Biofuels and bioenergy processes and technologies. CRC Press, Boca Raton
24. Mosier N, Wyman C, Dale B, Elander R, Lee YY, Holtzapple M et al (2005) Features of promising technologies for pretreatment of lignocellulosic biomass. Bioresour Technol 4:673–686
25. Deguchi S, Tsujii K, Horikoshi K (2006) Cooking cellulose in hot and compressed water. Chem Commun 31:3293–3295. doi:10.1039/B605812D
26. Mochizuki K, Sakoda A, Suzuki M (2000) Measurement of the hydrothermal reaction rate of cellulose using novel liquid-phase thermogravimetry. Thermochim Acta 348(1–2):69–76
27. Olanrewaju KB (2012) Reaction kinetics of cellulose hydrolysis in subcritical and supercritical water. Dissertation, University of Iowa
28. Kruse A, Dinjus E (2007) Hot compressed water as reaction medium and reactant: 2. Degradation reactions. J Supercrit Fluids 41(3):361–379
29. Bobleter O, Pape G (1968) Der hydrothermale abbau von glucose. Monatshefte Fur Chemie 99(4):1560–1567
30. Watanabe M, Aizawa Y, Iida T, Levy C, Aida TM, Inomata H (2005) Glucose reactions within the heating period and the effect of heating rate on the reactions in hot compressed water. Carbohydr Res 340(12):1931–1939
31. Matsumura Y, Yanachi S, Yoshida T (2013) Glucose decomposition kinetics in water at 25 MPa in the temperature range of 448–673 K. Ind Eng Chem Res 45(6):1875–1879

32. Sasaki M, Furukawa M, Minami K, Adschiri T, Arai K (2002) Kinetics and mechanism of cellobiose hydrolysis and retro-aldol condensation in subcritical and supercritical water. Ind Eng Chem Res 41(26):6642–6649
33. Kabyemela BM, Adschiri T, Malaluan RM, Arai K (1997) Kinetics of glucose epimerization and decomposition in subcritical and supercritical water. Ind Eng Chem Res 36(5):1552–1558
34. Yin S, Mehrotra AK, Tan Z (2011) Alkaline hydrothermal conversion of cellulose to bio-oil: influence of alkalinity on reaction pathway change. Bioresour Technol 102(11):6605–6610
35. Yin S, Tan Z (2012) Hydrothermal liquefaction of cellulose to bio-oil under acidic, neutral and alkaline conditions. Appl Energy 92:234–239
36. Chuntanapum A, Matsumura Y (2009) Formation of tarry material from 5-HMF in subcritical and supercritical water. Ind Eng Chem Res 48(22):9837–9846
37. Lü X, Saka S (2012) New insights on monosaccharides' isomerization, dehydration and fragmentation in hot-compressed water. J Supercrit Fluids 61:146–156
38. Abatzoglou N, Chornet E, Belkacemi K, Overend RP (1992) Phenomenological kinetics of complex systems: the development of a generalized severity parameter and its application to lignocellulosics fractionation. Chem Eng Sci 47(5):1109–1122
39. Pińkowska H, Wolak P, Złocińska A (2011) Hydrothermal decomposition of xylan as a model substance for plant biomass waste—hydrothermolysis in subcritical water. Biomass Bioenergy 35(9):3902–3912
40. Aida TM, Shiraishi N, Kubo M, Watanabe M, Smith RL Jr (2010) Reaction kinetics of d-xylose in sub- and supercritical water. J Supercrit Fluids 55(1):208–216
41. Jing Q, Lü X (2007) Kinetics of non-catalyzed decomposition of D-xylose in high temperature liquid water. Chin J Chem Eng 15(5):666–669
42. Oefner PJ, Lanziner AH, Bonn G, Bobleter O (1992) Quantitative studies on furfural and organic acid formation during hydrothermal, acidic and alkaline degradation of D-xylose. Monatsh Chem 123(6–7):547–556
43. Antal MJ Jr, Leesomboon T, Mok WS, Richards GN (1991) Mechanism of formation of 2-furaldehyde from d-xylose. Carbohydr Res 217:71–85
44. Gao Y, Chen HP, Wang J, Shi T, Yang HP, Wang XH (2011) Characterization of products from hydrothermal liquefaction and carbonation of biomass model compounds and real biomass. J Fuel Chem Technol 39(12):893–900
45. Hashaikeh R, Fang Z, Butler IS, Hawari J, Kozinski JA (2007) Hydrothermal dissolution of willow in hot compressed water as a model for biomass conversion. Fuel 86(10–11):1614–1622
46. Miller JE, Evans L, Littlewolf A, Trudell DE (1999) Batch microreactor studies of lignin and lignin model compound depolymerization by bases in alcohol solvents. Fuel 78(11):1363–1366
47. Cheng S, Wilks C, Yuan Z, Leitch M, Xu C (2012) Hydrothermal degradation of alkali lignin to bio-phenolic compounds in sub/supercritical ethanol and water-ethanol co-solvent. Polym Degrad Stab 97:839–848
48. Nielsen RP (2010) The physical chemistry of the CatLiq process. PhD Thesis, Aalborg University, Esbjerg, Denmark
49. Nielsen RP, Olofsson G, Søgaard EG (2012) CatLiq—high pressure and temperature catalytic conversion of biomass: the CatLiq technology in relation to other thermochemical conversion technologies. Biomass Bioenergy 43:2–5
50. Toor SS (2010) Modelling and optimization of Catliq Liquid biofuel process. PhD Thesis, Aalborg University, Aalborg, Denmark
51. Toor SS, Rosendahl L, Rudolf A (2011) Hydrothermal liquefaction of biomass: a review of subcritical water technologies. Energy 36:2328–2342
52. Schmieder H, Abeln J, Boukis N, Dinjus E, Kruse A, Kluth M, Petrich G, Sadri E, Schacht M (2000) Hydrothermal gasification of biomass and organic wastes. J Supercrit Fluids 17:145–153
53. Balat M, Kırtay E, Balat H (2009) Main routes for the thermo-conversion of biomass into fuels and chemicals. Part 2: gasification systems. Energy Convers Manag 50:3158–3168

54. Kruse A, Dinjus E (2007) Hot compressed water as reaction medium and reactant Properties and synthesis reactions. J Supercrit Fluids 39:362–380
55. Kruse A, Dinjus E (2005) Influence of salts during hydrothermal biomass gasification: the role of the catalysed water-gas shift reaction. Z Phys Chem 219:341–366
56. Kruse A, Henningsen T, Sınag A, Pfeiffer J (2003) Biomass gasification in supercritical water: influence of the dry matter content and the formation of phenols. Ind Eng Chem Res 42:3711–3717
57. Gasafi E, Reinecke M-Y, Kruse A, Schebek L (2008) Economic analysis of sewage sludge gasification in supercritical water for hydrogen production. Biomass Bioenergy 32:1085–1096
58. Sinag A, Kruse A, Schwarzkopf V (2003) Formation and degradation pathways of intermediate products formed during the hydropyrolysis of glucose as a model substance for wet biomass in a tubular reactor. Eng Life Sci 3:469–473
59. Boukis N, Galla U, D'Jesus P, Müller H, Dinjus E (2005) gasification of wet biomass in supercritical water: results of pilot plant experiments. In: Proceedings of the 14 European biomass conference. Paris, pp 964–967
60. Iversen SB, Larsen T, Lüthje V, Felsvang K, Nielsen RP, Galla U, Boukis N (2005) CatLiq— a disruptive technology for biomass conversion. In: Proceedings of the 14th European Biomass Conference. Paris, pp 1450–1452
61. Hammerschmidt A, Boukis N, Hauer E, Galla U, Dinjus E, Hitzmann B, Larsen T, Nygaard SD (2011) Catalytic conversion of waste biomass by hydrothermal treatment. Fuel 90:555–562
62. Karagöz S, Bhaskar T, Muto A, Sakata Y (2006) Hydrothermal upgrading of biomass: effect of K_2CO_3 concentration and biomass/water ratio on products distribution. Bioresour Technol 97:90–98
63. Demirbas A (2004) Current technologies for the thermo-conversion of biomass into fuels and chemicals. Energy Sources Part A: Recovery Utilization Environ Eff 26:715–730
64. Balat M (2008) Mechanisms of thermochemical biomass conversion processes. Part 1: reactions of pyrolysis. Energy Sources Part A: Recovery Utilization Environ Eff 30:620–635
65. Balat M (2008) Mechanisms of thermochemical biomass conversion processes. Part 2: reactions of gasification. Energy Sources Part A: Recovery Utilization Environ Eff 30:636–648
66. Balat M (2008) Mechanisms of thermochemical biomass conversion processes. Part 3: reactions of liquefaction. Energy Sources Part A: Recovery Utilization Environ Eff 30:649–659
67. Demirbas A (2004) Conversion of agricultural residues to fuel products via supercritical fluid extraction. Energy Sources Part A: Recovery Utilization Environ Eff 26:1095–1103
68. Demirbas A (2005) Thermochemical conversion of biomass to liquid products in the aqueous medium. Energy Sources Part A: Recovery Utilization Environ Eff 27:1235–1243
69. Demirbas A (2008) Production of biodiesel from algae oils. Energy Sources Part A: Recovery Utilization Environ Eff 31:163–168
70. Kruse A, Gawlik A (2003) Biomass conversion in water at 330–410 & #xB0;C and 30–50 MPa. Identification of key compounds for indicating different chemical reaction pathways. Ind Eng Chem Res 42:267–279
71. Sinag A, Kruse A, Rathert J (2004) Influence of the heating rate and the type of catalyst on the formation of key intermediates and on the generation of gases during hydropyrolysis of glucose in supercritical water in a batch reactor. Ind Eng Chem Res 43:502–508
72. Sinag A, Kruse A, Schwarzkopf V (2003) Key compounds of the hydropyrolysis of glucose in supercritical water in the presence of K_2CO_3. Ind Eng Chem Res 42:3516–3521
73. Kruse A, Meier D, Rimbrecht P, Schacht M (2000) Gasification of pyrocatechol in supercritical water in the presence of potassium hydroxide. Ind Eng Chem Res 39:4842–4848
74. Akiya N, Savage PE (2002) Roles of water for chemical reactions in high-temperature water. Chem Rev 102:2725–2750
75. Savage PE (1999) Organic chemical reactions in supercritical water. Chem Rev 99:63

76. Savage PE, Gopalan S, Mizan TI, Martino CJ, Brock EE (1995) Reactions at supercritical conditions: applications and fundamentals. AIChE J 41:1723–1778
77. Bühler W, Dinjus E, Ederer HJ, Kruse A, Mas C (2002) Ionic reactions and pyrolysis of glycerol as competing reaction pathways in near- and supercritical water. J Supercrit Fluids 22:37–53
78. Yoshida T, Oshima Y, Matsumura Y (2004) Gasification of biomass model compounds and real biomass in supercritical water. Biomass Bioenergy 26:71–78
79. Matsumura Y, Minowa T, Potic B et al (2005) Biomass gasification in near- and super-critical water: status and prospects. Biomass Bioenergy 29:269–292
80. Peterson AA, Vontobel P, Vogel F, Tester JW (2008) In situ visualization of the performance of a supercritical-water salt separator using neutron radiography. J Supercrit Fluids 43:490–499
81. Czernik S, French R, Feik C, Chornet E (2002) Hydrogen by catalytic steam reforming of liquid byproducts from biomass thermoconversion processes. Ind Eng Chem Res 41:4209–4215
82. Trimm DL (1997) Coke formation and minimisation during steam reforming reactions. Catal Today 37:233–238
83. Kruse A, Dinjus E (2007) Hot compressed water as reaction medium and reactant 2. Degradation reactions. J Supercrit Fluids 41:361–379
84. Padban N, Wang W, Ye Z, Bjerle I, Odenbrand I (2000) Tar formation in pressurized fluidized bed air gasification of woody biomass. Energy Fuels 14:603–611
85. Inoue S, Sawayama S, Dote Y, Ogi T (1997) Behaviour of nitrogen during liquefaction of dewatered sewage sludge. Biomass Bioenergy 12:473–475
86. Watanabe M, Inomata H, Osada M, Sato T, Adschiri T, Arai K (2003) Catalytic effects of NaOH and ZrO_2 for partial oxidative gasification of n-hexadecane and lignin in supercritical water. Fuel 82:545–552
87. Watanabe M, Takahashi M, Inomata H (2008) Hydrogen production reaction with a metal oxide catalyst in high pressure high temperature water. J Phys Conf Ser 121:082008
88. Watanabe M, Inomata H, Smith RL, Arai K (2001) Catalytic decarboxylation of acetic acid with zirconia catalyst in supercritical water. Appl Catal A 219:149–156
89. Watanabe M, Inomata H, Arai K (2002) Catalytic hydrogen generation from biomass (glucose and cellulose) with ZrO_2 in supercritical water. Biomass Bioenergy 22:405–410
90. Lee G, Nunoura T, Matsumura Y, Yamamoto K (2002) Comparison of the effects of the addition of NaOH on the decomposition of 2-chlorophenol and phenol in supercritical water and under supercritical water oxidation conditions. J Supercrit Fluids 24:239–250
91. Antal MJ Jr, Xu X (1999) Hydrogen production from high moisture content biomass in supercritical water. In: U.S. DOE hydrogen program review
92. Ahmad MM (2010) Upgrading of bio-oil into high-value hydrocarbons via hydrodeoxygenation. Am J Appl Sci 7:746–755
93. Mercader, FM, Hoogendorn, K (2010) Production of advanced bio-fuels: Co-refning upgraded pyrolysis oil, Berlin
94. Furimsky E (2000) Review: catalytic hydrodeoxygenation. Appl Catal A 199:147–190
95. Furimsky E (2012) Hydroprocessing challenges in biofuels production, catalysis today. Available via online. http://dx.doi.org/10.1016/j.cattod.2012.11.008. Accessed 12 Feb 2013
96. Kersten SRA, van Swaaij WPM, Lefferts L, Seshan K (2007) Options for catalysis in the thermochemical conversion of biomass into fuels. In: Centi G, van Santen RA (eds) Catalysis for renewables: From feedstock to energy production. Chichester, Wiley-VCH
97. Wildschut J (2009) Pyrolysis oil upgrading to transportation fuels by catalytic hydrotreatment. PhD Thesis, Rijkuniversitet Groningen

Chapter 10
Hydrothermal Gasification of Biomass for Hydrogen Production

Jude A. Onwudili

Abstract Over 90 % of the world's current hydrogen production capacity comes from the use of fossil fuels including coal, oil, and natural gas. In recent years, natural gas has become the major source of hydrogen. However, these fossil fuels are finite resources, produce greenhouse gases, and will eventually run out. Biomass presents a proven store of chemical energy, which can be converted into hydrogen by various processes. Hydrogen from biomass will be the world's cleanest fuel, being produced from a renewable carbon-neutral source and its combustion produces no carbon emissions. Hydrothermal gasification has the potential to produce high quality and high yield of hydrogen from, especially, very wet biomass feedstocks and particularly the carbohydrate-rich types. Water is an important reactant for hydrogen production in biomass HTG, thereby avoiding dewatering of wet feedstocks. The chemistry of hydrogen production from biomass HTG largely depends on the reforming of biomass into simple organic molecules that are capable of being gasified into carbon monoxide. Hydrogen is then produced via water-gas shift reaction. The developments of various reactor designs and catalysts for high hydrogen selectivities are ongoing at various research centers around the world. A great deal of knowledge about this process is available, however, there are still challenges regarding the stability and recovery of catalysts, preprocessing of biomass for continuous operations and other process optimization and intensification issues.

J. A. Onwudili (✉)
Energy Research Institute, School of Process, Environmental and Materials Engineering,
The University of Leeds, Leeds LS2 9JT, UK
e-mail: j.a.onwudili@leeds.ac.uk

10.1 Introduction

Biomass represents a large store of the solar and chemical energy. The vast amounts and varieties of biomass in the world indicate that its energy content can be harnessed in different ways to meet the growing current and future energy demands. Terrestrial biomass is found in most parts of the world and has been the source of domestic energy since prehistoric times as wood fuel. The global energy demand is estimated to be about 590 EJ per annum by 2015 and up to 1000 EJ by 2050 [1, 2].

Currently, 81 % of the world primary energy requirement is obtained from fossil fuels; biomass currently provides approximately 50 EJ into the global energy mix [3]. In terms of quantity, biomass is the fourth largest energy resource after oil, coal, and natural gas. In addition, biomass is the single largest renewable resource [2, 4]. The future energy supply potential from biomass in 2050 is currently put at an annual average of 1340 EJ [2], indicating that biomass (both marine and terrestrial) has the capacity to meet the world's entire energy demand. Realistically, it is estimated that biomass resources can provide 10–20 % of the world's primary energy requirement by 2050 [1]. The current technologies used in harnessing biomass energy are shown in Fig. 10.1. Some of these technologies have already been deployed for large-scale commercial production of energy from biomass, while some are different stages of development.

The argument about the cost of energy production from biomass may hold true as long as fossil fuels remain as alternative. However, concerns about sustainability and climate change already makes harvesting energy from biomass a plausible option, driven by technology. The attainment of the biomass energy (bioenergy) goals will depend on factors such as land availability, large-scale production of non-food biomass resources and development of sustainable biomass conversion technologies. In addition, technologies must continue to focus on the different forms in which biomass energy is and/or will be consumed based on demand and technological innovations. Recent developments in marine (algae) biomass cultivation and utilization [5–7] may mitigate the issue of land availability, leaving technological advancement as the last hurdle at the heart of sustainable future biomass energy harvesting. Although the overall impact of large-scale algae production and harvesting on the marine ecosystem is yet to be ascertained.

Hydrogen and methane have been proposed as the main energy carrier and fuel, respectively for the future. Methane is main component of natural gas with a world's current proven reserves estimate of 300 trillion cubic meters. Natural gas currently supplies 21 % of the world energy demand, which is about 103 EJ per year. However, natural gas consumption is set to increase from year to year; for instance an increase of 7.4 % on its consumption was recorded between 2009 and 2010 [3]. Natural gas is nonrenewable and the total proven reserves of natural gas amounts to 9410 EJ of energy and this would last another 91 years at the current consumption rate.

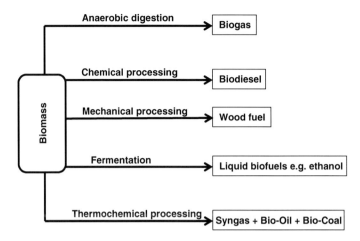

Fig. 10.1 Current biomass-to-energy processing technologies

Current hydrogen production (>90 %) routes include natural gas reforming, coal, and oil gasification, and electrolysis of water. These routes are either dependent on nonrenewable resources or on expensive processes requiring the use of nonrenewable energy. The annual global capacity for hydrogen production is estimated to be 57 billion kg. Given that the energy content of hydrogen is 143 MJ/kg, this gives estimated current global hydrogen energy of 8.15 EJ, which can only account for 1.66 % of the world's energy demand. Coal gasification had been the major source of hydrogen but by 2000, oil accounted for 55 % of global hydrogen production. However, since 2006 natural gas has overtaken oil for hydrogen production, accounting for more than 48 % of global production [8, 9]. The progressive shifting from solid to liquid and finally to gas fuels as well as the associated trends in "decarbonization" implies that eventual transition to hydrogen energy is inevitable [10].

However, hydrogen finds direct and economic uses in the manufacture of ammonia (Haber process), in hydrocracking of heavy fractions of petroleum of low demand into lighter fractions of high economic and industrial values. Ammonia manufacture (60 %), hydrocracking processes (23 %), and methanol production (9 %) are the major uses of the world's current hydrogen production capacity [11]. Hydrogen production for energy purposes would be over and above the current market levels and this calls for concerted efforts at diversifying hydrogen production to include more sustainable sources such as biomass. This will release the amount of hydrocarbons used for hydrogen production and make the petrochemical industry more sustainable itself. One of the major uses of hydrogen as an energy carrier is in fuel cell applications and this is steadily growing market with excellent potentials.

In summary, there are economic, environmental, and political benefits of producing and using hydrogen from biomass compared to fossil-fuel-based processes [12–14]. The economic impacts include job creation, increased capital investment, diversification of energy resources, improving local economy, and international competitiveness. Currently, hydrogen and fuel cells provide nearly 40,000 jobs worldwide and global revenues from hydrogen and fuel cells are estimated at $9.2 billion by 2015 and up to $38.4 billion by 2020 [15]. Political benefits include reduced dependency on foreign energy resources; this engenders enhanced energy security and reduction in international tensions arising from sociopolitical economics associated with fossil fuels. Resources for biomass energy are usually locally produced and hence very reliable. Furthermore, there are many environmental benefits of using biomass energy. These include reduction in greenhouse gases emissions due to its carbon-neutrality and much reduced SOx, NOx, and particulate emissions. In addition, biomass is a renewable resource and is readily available [2, 4, 16].

10.2 Biomass Gasification in Hydrothermal Media for Hydrogen

Hydrothermal gasification (HTG) of biomass for hydrogen is a growing area of research within the scientific/engineering community. Modell and co-workers [17] pioneered the study of biomass gasification in hydrothermal media in the 1980s. Since then, it has been actively researched and currently proposed to be an attractive process based on the use of renewable biomass and can be operated at much lower temperatures compared to conventional fossil-fuel-based gasification, which occurs at temperatures above 1273 K [16]. In the process, water behaves as a reaction medium, a reactant and a catalyst [18–21]. Therefore, the presence of water is desirable indicating that very wet biomass feedstocks can be applied. This presents an incentive in terms of energy savings in the drying of biomass and as well open up a large variety of feedstocks, which otherwise would not be readily considered for thermochemical processing.

Wet biomass and biowastes such as sewage sludge, pulp and paper mill sludge, food wastes and animal droppings, and other agricultural residues have often been confined to biochemical conversion technologies such as anaerobic digestion (AD) and composting. These feedstocks can be directly processed or pretreated prior to conversion to hydrogen using HTG. The advantages of thermochemical conversion route based on HTG of wet waste include fast reaction times, ease of deployment, possibility of applying selective catalysts, and overall versatility in terms of process design for specific products [18].

10.2.1 The Hydrothermal Gasification Process

The special properties of water under hydrothermal conditions such as low dielectric constant, variable ionic product, and low density are very important parameters for hydrothermal biomass conversion. In general, HTG of biomass can

10 Hydrothermal Gasification of Biomass for Hydrogen Production

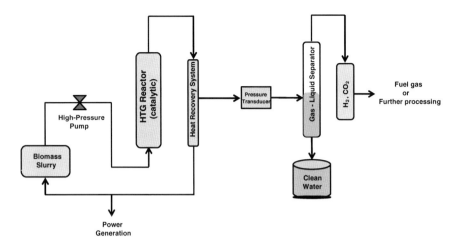

Fig. 10.2 Basic schematic of HTG process for biomass to hydrogen

be carried out within the subcritical and supercritical regions of water. However, research has shown that hydrogen production via HTG occurs much readily under supercritical water conditions [22]. In addition, the ultimate products of any gasification process are all gases and these include hydrogen, methane, C1-C4 hydrocarbons with some carbon dioxide and usually very low yields of carbon monoxide. These are much less complex molecules which can be much more easily handled compared to liquid products (oils) and solid residues (mixtures of char and ash) obtainable from other thermochemical processes, e.g., pyrolysis.

A commercially viable HTG process would be based on a continuous reactor as depicted in Fig. 10.2. For such systems biomass must be delivered in aqueous solutions or slurries using high-pressure pumps. Heat energy recovery is important to improve process economics. The recovered heat can be used to preheat the feed, used to generate electricity or applied for district heating. A pressure transducer is often used to reduce the pressure of the reactor effluents, which are collected in a gas–liquid separator. Clean water is obtained and collected for various uses, while hydrogen gas is recovered from the effluent gases by various means including pressure-swing absorption (PSA) to remove carbon dioxide [23] and membrane separation processes including polymer electrolyte membrane (PEM) to remove methane and other gases [24–26]. A recent study by Fiori et al. [27] demonstrates that HTG of biomass for hydrogen production could be energetically self-sustaining for biomass concentrations of between 15–25 wt% are used. The authors show that the energy balance of conceptual reactor design would provide 150 kWe/1000 g feed of net energy.

Essentially, thermochemical biomass gasification (including HTG) is a thermochemical refining process, which seeks to convert biomass into gas—the ultimate volatile product. Organic nitrogen is converted to ammonia, nitrates, and even nitrogen gas [28, 29], while organic sulfur results in sulfate or other solute sulfur compounds. Catalysts play important roles in the HTG process and these could include product selectivity and reduction in severity of process operating conditions.

10.2.2 Thermochemistry of Hydrogen Production from HTG of Biomass

Hydrothermal gasification of biomass to hydrogen involves a series of complex reaction chemistries. Biomass is a multicomponent feedstock comprising of cellulose, hemicellulose, and lignin. Some biomass can also have high ash contents. The possible interactions of the biomass components and their possible degradation products, significantly increases the complexity of hydrothermal reactions of biomass [16]. However, studies with biomass model compounds indicate that the reactions of carbohydrate-type biomass compounds such as cellulose and glucose can be followed [20]. Ultimately, the overall reaction equation for HTG of biomass can be represented by equation (10.1), which involves the reaction of biomass (represented by glucose).

$$C_6H_{12}O_6 + 6H_2O \rightarrow 12H_2 + 6CO_2 \quad \Delta H_{298K} = +362 \text{ kJ/mol} \quad (10.1)$$

As shown by equation (10.1), the idealized reaction for the production of hydrogen from biomass in the presence of gaseous water under hydrothermal conditions is endothermic. Hence, external heat input is required for the reaction but this is however, not peculiar to the HTG process alone. Hydrogen production from biomass via HTG as represented by the idealized reaction (10.1) underlies the importance of water as a reactant in hydrogen production [18, 19]. The produced hydrogen gas comes from both the biomass and the stoichiometric amount of water used up during the reaction.

Using glucose as biomass feedstock, the maximum yield of hydrogen gas based on reaction (10.1) is 66.7 mol/kg or 133.3 g/kg. This amount of hydrogen is 200 % of the hydrogen content of glucose, showing that the reacting water provides the balance. Evidence of the participation of water in hydrogen production can be found in literature. Table 10.1 compares biomass gasification (including HTG) with other existing and developing processes for hydrogen production [30]. Although, the efficiency and theoretical yield of hydrogen via HTG of biomass are lower compared to conventional processes such as steam reforming of natural gas, it has the advantage of being carbon-neutral process based on renewable resource. In addition, its efficiency can be improved through future research and this will impact positively on the overall cost of the process.

Hydrothermal reactions for hydrogen production must be carried out at reasonably high temperatures. In practice, temperature ranges from 673 K up to 873 K are often used [20, 31–33]. However, equilibrium yields have been calculated at temperatures of up to 1273 K [16]. High temperature favors hydrogen

10 Hydrothermal Gasification of Biomass for Hydrogen Production

Table 10.1 Comparison of various hydrogen production methods [9, 30]

Current hydrogen production methods	Status	Maximum hydrogen yield (mol/kg)	Efficiency (%)	Cost relative to SMR
Steam reforming of methane/natural gas (SMR)	Mature	333	70–80	1
Partial oxidation of heavy oil	Mature	143	70	1.8
Coal gasification	Mature	100	60	1.4–2.6
Partial oxidation of coal	Mature	100	55	–
High-temperature electrolysis of water	R & D	55.5	48	2.2
Biomass gasification (incl. HTG)	R & D	66.7	45–50	2.0–2.4

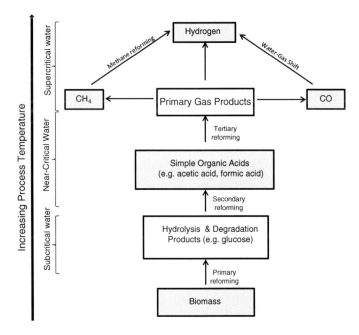

Fig. 10.3 Reaction schemes for hydrogen production from biomass via HTG

production; however a compromise between hydrogen yield and associated costs must be evaluated for sustainable conversion of biomass to hydrogen.

Hydrogen production from HTG of biomass and biomass model compounds is seldom a one-step reaction. There are many elementary component-reactions involved which constitute the idealized equation (10.1) and play significant roles in hydrogen production from biomass. These reactions can often be involved in competing reaction pathways, which can lead to the formation of products other than hydrogen or products whose presence would suppress hydrogen production. Figure 10.3 shows a schematic diagram involving the different reaction processes leading towards hydrogen production. In Fig. 10.3, all other reaction pathways to unwanted products are assumed to be suppressed in order to follow the idealized equation for hydrogen production. Particularly, the major competing reaction pathway involving the formation of 5-hydroxymethylfurfural (5-HMF) inhibits gas formation and hence hydrogen production [34–37].

In this section, each possible reaction capable of producing hydrogen gas or its precursors will be treated as ideal, i.e., competing reactions will be ignored. The main chemical reactions involved include biomass reforming to relevant intermediate liquids and gaseous species, reforming reactions of the intermediates to carbon monoxide and finally the all-important water-gas shift reaction. It has been shown that the important intermediates responsible for gas formation include small acid and aldehyde molecules such as acetic acid, formic acid, lactic acid, levulinic acid, acetic aldehyde, and formic aldehyde [20, 31].

10.2.2.1 Biomass Reforming to Produce Intermediates

$$C_6H_{12}O_6 + 6H_2O \rightarrow 6HCOOH + 6H_2 \quad \Delta H\ 298K = +170\,kJ/mol \quad (10.2)$$

$$C_6H_{12}O_6 + 6H_2O \rightarrow 4HCOOH + 2CO_2 + 8H_2 \quad \Delta H\ 298K = +234\,kJ/mol$$
$$(10.3)$$

$$C_6H_{12}O_6 + 4H_2O \rightarrow 4HCOOH + CH_3COOH + 4H_2 \quad \Delta H 298\ K = +54\,kJ/mol$$
$$(10.4)$$

$$C_6H_{12}O_6 + 2H_2O \rightarrow 2HCOOH + 2CH_3COOH + 2H_2 \quad \Delta H\ 298K = -68\,kJ/mol$$
$$(10.5)$$

$$C_6H_{12}O_6 + 2H_2O \rightarrow 2CH_3COOH + 2CO_2 + 4H_2 \quad \Delta H\ 298K = +2\,kJ/mol$$
$$(10.6)$$

Each of these possible reactions above, except reaction (10.5), has a positive standard enthalpy change (ΔH) values, indicating that external heat input will be required for them to occur. Although, the ΔH values are not the only thermodynamic properties needed to predict the spontaneity of these reactions, they give some indications about the effect of temperature. Indeed, it is expected that the formation of gas products from solid biomass and liquid water, will lead to a dramatic increase in the entropy change (ΔS) of the system. A negative or positive but small ΔH value and a large and positive ΔS value will guarantee a negative Gibb's Free Energy (ΔG), which is a prerequisite for spontaneous reactions. It is well known that the hydrolysis of biomass is spontaneous in subcritical water, which requires heat input. For instance, biomass reforming into simple carboxylic acids such as acetic acid has been well reported in literature during hydrothermal processing of biomass [38–41] under subcritical water conditions. This suggests that hydrothermal reforming reactions of biomass must include acetic acid as a product [20, 28].

The values of ΔH for reactions (10.2) and (10.3), suggest that a large amount of heat energy will be required for biomass to be reformed only into formic acid. On the other hand, the presence of acetic acid in reactions (10.4–10.6) gives either negative ΔH value, or a positive but small ΔH value. Reactions (10.5) and (10.6) are very similar except that in the case of reaction (10.6), the formic acid in reaction (10.5) has been converted to hydrogen gas and carbon dioxide. The difference in ΔH values for reactions (10.6) and (10.5) is due to the extra heat energy released for the conversion of formic acid to H_2 and CO_2 via a subsequent water-gas shift reaction. Experimental evidence [30] suggests that the complete reforming of formic acid, in the form of sodium formate, to hydrogen gas occurs around 673 K. This supports the formation and thermal stability of these simple carboxylic acids within the subcritical water region according to literature [39, 41].

In addition, thermodynamic calculations have shown that the yield of hydrogen gas under subcritical water conditions is usually low [16, 22]. This therefore suggests that reaction (10.5) is more probable than reaction (10.6) as a plausible representation of the idealized reforming of carbohydrate biomass in low temperature hydrothermal media leading to primary products [31].

10.2.2.2 Reforming of Intermediates

$$HCOOH \rightarrow CO + H_2O \quad \Delta H\ 298K = +73\,kJ/mol \quad (10.7)$$

$$CH_3COOH \rightarrow CH_4 + CO_2 \quad \Delta H\ 298K = +15.1\,kJ/mol \quad (10.8)$$

$$CH_4 + H_2O \rightarrow CO + 3H_2 \quad \Delta H\ 298K = +206\,kJ/mol \quad (10.9)$$

The reforming reactions of the intermediate products as represented by the reaction equations (10.7–10.9) are also all endothermic in nature. The formation of methane from hydrothermal biomass gasification is thermodynamically favored in near-critical water conditions, which can be attributed to the decarboxylation of acetic acid under such reaction conditions [22]. The formation of hydrogen under such conditions appears to be due to the direct aqueous reforming of biomass. In addition, the separate reforming of formic acid ($\Delta H298\ K = +32\ kJ/mol$) to hydrogen and carbon dioxide occurs just above the critical point of water [31].

Compared to the combination of reactions (10.8) and (10.9), which represent aqueous reforming of acetic acid ($\Delta H298\ K = +268\ kJ/mol$), reaction (10.8) with a relatively low ΔH value, would be favored. In fact, experimental evidence [30] and thermodynamic equilibrium calculations [42] indicate that the formation of hydrogen via the methane-reforming occurs well into the supercritical water region. The important reactions leading to hydrogen production under hydrothermal conditions are those involving the formation of carbon monoxide. These are equations (10.7) and (10.9) involving the hydrothermal decarbonylation reactions of formic and acetic acids.

10.2.2.3 Hydrogen Gas Production

$$CO + H_2O \rightarrow H_2 + CO_2 \quad \Delta H\ 298K = -41\ kJ/mol \quad (10.10)$$

It has become clear from literature that the production of hydrogen gas can occur at two distinct regions of hydrothermal media. The negative ΔH value of the water-gas shift reaction in equation (10.10) indicates that this reaction can readily be accomplished under any region of the hydrothermal space, once CO has been formed. Catalytic water-gas shift reactions can be carried out at temperatures as low as 433 K [43]. Incidentally, much more energy is required to accomplish the reactions leading to the formation of CO than its reaction with water to form hydrogen gas. On this evidence, water-gas shift reaction is unlikely to be the rate-determining step in hydrothermal hydrogen production. While the decarbonylation of formic acid requires less heat energy, methane reforming reaction with a much larger ΔH value of $+206$ kJ/mol, will require more energy input to be accomplished. In fact, it is almost certain that the formation of CO and its precursors from biomass are the key reactions controlling the eventual formation of hydrogen.

More importantly, much of the hydrogen gas produced during HTG of biomass comes from the complete methane-reforming process (reaction 10.12), which still has a large and positive ΔH value of $+165$ kJ/mol. By comparison, the complete reforming of acetic acid produces four times more hydrogen than that of formic acid.

$$HCOOH \rightarrow CO_2 + H_2 \quad \Delta H\ 298K = +32\ kJ/mol \quad (10.11)$$

$$CH_4 + 2H_2O \rightarrow CO_2 + 4H_2 \quad \Delta H 298K = +165\ kJ/mol \quad (10.12)$$

The above outline of a selection of the possible reaction schemes involved in an idealized HTG of biomass for hydrogen production may give an insight into the development of a sustainable process to maximize the potential of the process. Such optimization and intensification processes can benefit from altering the reaction chemistries to favor hydrogen production. For instance, since hydrogen production from formic acid is less energy-intensive, the process can be designed to convert biomass mainly into this intermediate, while suppressing acetic acid formation. In contrast, the process can be designed to lower the activation energy required for acetic acid reforming, since it appears to be the default daughter product of hydrothermal biomass conversion. Such process designs may require the use of homogeneous or heterogeneous catalysts; a combination of both catalyst types is also possible.

10.3 Factors that Influence Hydrogen Production During HTG

In practice, hydrogen production via HTG of biomass rarely follows the idealized pathways indicated by the series of reactions in the previous section. Generally, organic chemical reactions in hydrothermal media occur through a combination of

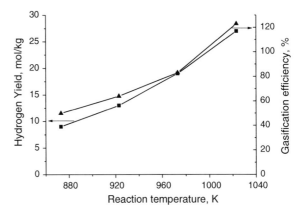

Fig. 10.4 Effect of temperature on hydrogen yield and gasification efficiency, adapted from [16]

ionic and radical mechanisms [28], which may also favor other possible reaction pathways such as char and tar formation. Hydrogen production depends on the gasification of both the hydrogen and carbon atoms in biomass, which means enhanced hydrogen yields would depend on the extent of conversion of the C and H atoms in a large presence of hot-pressurized water. In addition, the presence and concentrations of other elements in biomass could promote or inhibit hydrogen-forming reactions. Research over the years on the subject has shown that the following factors have profound effects on hydrogen production.

10.3.1 Temperature

As with other chemical reactions, temperature has a huge influence on hydrothermal hydrogen production from biomass. This is clear from the preceding section in terms of the positive ΔH values for the listed equations.

Hydrogen production tends to occur in the supercritical water region, except in the presence of some special catalysts. Reactions in the supercritical water region are defined by radical mechanisms due to its physicochemical properties of gas-like diffusivity and liquid-like densities within a single-fluid phase [18, 19, 44]. Such parameters favor biomass gasification reactions, and particularly methane-reforming reactions toward hydrogen production [45, 46]. In essence, increasing reaction temperature leads to increased gasification efficiencies (GE), which in turn leads to increased hydrogen yields as shown in Fig. 10.4.

The influence of temperature on hydrothermal biomass gasification shows increasing trend in the equilibrium yield of hydrogen during sawdust gasification [16]. This shows the importance of reaction temperature on hydrogen yields; however, actual yields tend to be much lower than equilibrium yields. Under the same conditions, methane yields decreases sharply, possibly correlating with its reforming to hydrogen. Indeed, methane yield could usually be higher than

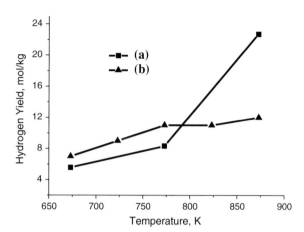

Fig. 10.5 Effect of temperature on hydrogen gas yields from HTG of biomass **a** using a batch reactor [47] **b** using a continuous reactor [48]

hydrogen yield up to a temperature of around 673 K, after which hydrogen production becomes dominant [22].

Guo et al. [49] demonstrated dramatic increase in hydrogen and CO_2 yields with temperatures above 923 K during the HTG of glycerol. The authors attributed it to stronger water-gas shift activity, rather than methane-reforming due to the near-constant yield of methane. However, it is possible that thermodynamic equilibrium yields of gases may depend on actual type of biomass feed. The effect of temperature can also vary depending on the type of reactor employed as demonstrated by Fig. 10.5. Working within similar temperature ranges, it appears the effect of temperature is more profound in a batch reactor [47], compared to a continuous reactor [48].

10.3.2 Biomass Types

Hydrogen production from biomass must be streamlined to functional feedstocks and this has been one of the objectives of scientific research in the last few years. Different researchers have investigated hydrogen production from different types of biomass. The results are overwhelmingly suggestive of the likelihood of using certain biomass types for hydrothermal hydrogen production. For instance, research has found that carbohydrate-type biomass produces more hydrogen than other types such as lipids and proteins [32, 50]. In fact, research by Muangrat et al. [50] and Kruse et al. [51] have demonstrated the ability of proteinous biomass to suppress hydrogen formation possibly due to radical scavenging.

Browning reactions (often called Maillard reaction) between carbonyl groups (e.g., in glucose) and amino-groups (e.g., in amino acids/proteins) are thermodynamically favored within the subcritical water region due to their ionic nature.

10 Hydrothermal Gasification of Biomass for Hydrogen Production 231

Table 10.2 Influence of proteins on hydrogen yield during HTG of glucose at 330 °C, with 1.67 M NaOH, 1.5 wt% H_2O_2

Glutamic acid (wt%)	0	10	50	90	100
Glucose (wt%)	100	90	50	10	0
Hydrogen yield (mol/kg)	13.5	12	11.5	6	4

Adapted from Ref. [50], Copyright 2010, with permission from Elsevier

Such reactions will proceed favorably when a mixture of phytomass and zoomass are reacted together under hydrothermal condition, especially in subcritical water.

Experimental evidence using a batch reactor has confirmed the suppression of hydrogen production from glucose in the presence of amino acids as shown in Table 10.2. This is mostly due to the inhibition of free-radical reaction by the presence of products of Maillard reactions [50, 52]. In addition to nitrogen-groups, the presence of other inorganic elements or groups has been demonstrated to either directly or indirectly affect hydrogen production from HTG of biomass. For instance, in catalytic processes, the presence of sulfur in biomass will lead to the deactivation of some metal catalysts through the formation of strong metal-sulfide bond on the surface of the catalyst causing catalyst poisoning [53].

The ash contents of biomass can vary widely and the quantity as well as composition of the ash has significant effects on the HTG process. The major elements commonly present in biomass ash include Al, Ca, Na, K, Si, Mg in addition to other minor components [54]. The presence of alkaline metals tend to promote hydrogen yield, hence biomass sample with high alkaline ash may produce more hydrogen than those without. The ability of alkaline metals to catalyze hydrothermal biomass gasification may also depend on other factors such as the biochemical composition of the biomass and the presence of other nonmetallic elements like sulfur and nitrogen.

Further research has demonstrated the suitability of carbohydrate-type biomass as candidate feedstock for hydrogen production. Figure 10.6 shows that under similar hydrothermal conditions, different carbohydrate-type biomass can give similar hydrogen yields. For instance, the hydrogen yields from different types of biomass are similar during alkaline HTG [36] and in the absence of any catalyst [52, 55]. Furthermore, in a recent study with different algae strains, Onwudili et al. [54] reported that the carbohydrate-rich macro-alga L. Saccharina, gave the highest yields of hydrogen gas compared to microalgae samples with lower carbohydrate contents.

The high conversion rates of carbohydrate-type biomass may not be unconnected with their high degrees of hydrolyzability under ionic conditions in the subcritical region of water. In essence, the susceptibility of the carbon-carbon bonds to hydrolytic cleavage most likely depends on their "chemical environments." Carbon atoms directly bonded to one or more oxygen atoms are likely to be attacked by nucleophilic species such as OH- ions from water [32]. The lignin fraction of biomass is much more thermally stable than cellulose and hemicellulose. Some researchers believe that the lignin fraction in lignocellulosic biomass

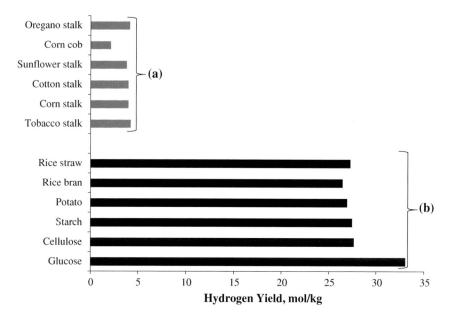

Fig. 10.6 Hydrogen production from different biomass types with or without catalyst **a** 5 wt% biomass feed, 773 K, 24.9–33.6 MPa, no catalyst [52] **b** 6.7 wt% feed, 723 K, 31 MPa, 1.67 M NaOH catalyst [36]

should be better used for the production of renewable and green chemicals such as renewable BTX (benzene, toluene, and xylenes), phenols, and phenolic resins [56].

10.3.3 Feed Concentration

The efficiency of the hydrothermal biomass gasification process is greatly influenced by the feed concentration both in terms of the composition and yields of gas products [36]. Low feed concentrations are invariably favorable for high gasification efficiencies due to the minimization of competing reaction pathways. Feed concentration has a profound effect on the formation of polymerization products leading to tar and char formation during HTG of biomass. Table 10.3 shows the effect of increasing biomass concentration on hydrogen production during the HTG of different biomass samples. The tests were carried out in a batch stainless steel reactor and shows that hydrogen yields decreased for glucose and rice bran when feed concentration exceeded 5 wt%. From literature, it is clear that biomass concentrations of about 5 wt% are suitable for gasification without catalysts [49, 57]. This is arguably one of the biggest challenges for the development of the process in terms of improving thermal efficiency [16]. The formation of tarry materials initiated by the primary formation of 5-HMF [34, 37, 58] will need to be suppressed either by catalysis or by appropriate process design [27]:

Table 10.3 Effect of biomass concentration on gas product composition

Biomass concentration (wt%)	Hydrogen yields (mol/kg)		
	[a]Corn + 2 wt% CMC	[b]Glucose	[b]Rice bran
2.5	–	11	5
3.0	25	–	–
5.0	–	13	11
6.0	20	–	–
10.0	15	10	10.5
12.0	14	–	–
20.0	–	10.6	7.5

[a] Corn cob/CMC at 1023 K [59];
[b] Glucose and rice bran [60]

10.3.4 Pressure

According to basic reaction equilibria (both thermodynamics and kinetics) of the HTG process for biomass, high pressure would not favor gas production, considering all the reaction schemes in Sect. 10.2. High pressure favors the direction of equilibrium reactions with less number of moles of gaseous species. However, some of the physicochemical properties of the hydrothermal reaction media are affected by pressure. For instance, pressures affect the density, dielectric constant, and ionic product of water under hydrothermal condition.

Indeed, the ionic product of water is several orders of magnitude higher in the subcritical region of water than in supercritical and ambient water [18]. This indicates that ionic reactions can be promoted by high pressures for biomass hydrolysis dominates. Therefore, such reactions involving biomass reforming to intermediate products, represented by equation (10.5) can be favored. On the contrary, high pressures suppress radical mechanisms in the supercritical water domain. Demirbas [62] suggested that the increase in hydrogen yields with increased pressure could be due to increased diffusion and mass transfer rates of supercritical water leading to improved biomass gasification efficiency. In essence, the high pressures prevalent in HTG may play a significant role in ensuring that overall HTG is accomplished at much lower temperatures compared with conventional gas-phase gasification. As shown in Fig. 10.7, low pressures favor hydrogen production at high temperatures in a batch reactor [47]. This observation is supported by the work of Alshammari and Hellgardt [63]. On the contrary, in continuous reactors, hydrogen yields increase slightly when pressure is increased [61].

10.3.5 Residence Time and Heating Rate

The amount of time spent by reactants and intermediates within the reaction vessel can greatly influence the yields and composition of gas products. In a batch reactor, heating rates affect the total amount of time the reactants stay in the

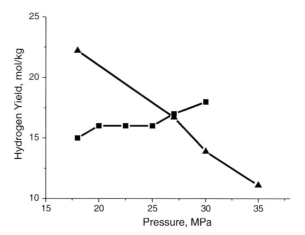

Fig. 10.7 Effects of pressure on hydrogen yields from biomass HTG (*filled triangle*) 0.45 M glucose, 873 K, 1 h, K$_2$CO$_3$ catalyst, batch reactor [47]; (*filled square*) 2 wt% sawdust +2 wt% CMC, 923 K, 27 s, no catalyst, continuous reactor [61]

reactor and this has considerable effect on the formation of key intermediates [64]. Additionally, the thermodynamics of the system and product concentrations could affect hydrogen yield. For instance, hydrogen can be consumed via methanation reactions under favorable reaction conditions including extended reaction times.

The effect of residence time is often greatly magnified in batch processes operated at low temperatures of up to 623 K as shown in Fig. 10.8 [65]. There is also a contributory effect from the biomass type to the influence of residence times. For example, complete HTG of glucose and glycerol can be achieved at very fast reaction times in the order of seconds [61, 66]. Such biomass model compounds are usually the products of biomass reforming via hydrolytic decomposition reactions. In practice, real biomass would first undergo a series of hydrolysis steps to different intermediates. Long residence times under such hydrolytic conditions can initiate polymerization reactions, especially if furfural and 5H-methyl furfural (5-HMF) are formed [37]. The conversion of polymerizable intermediates will most probably lead to slower reaction rates, which could require extended residence times or higher heating rates [61].

Several researchers [47, 61, 65] carried out extensive research on the influence of residence time on hydrogen production using batch reactors. The common observation is that hydrogen production can be promoted by extended reaction times in a batch reactor. In addition, the influence of residence time can be significant if the process is operated at relatively low temperature. However, this can have a huge impact on the cost of the process. The all-important water-gas shift reaction is relatively fast compared to reactions involving biomass reforming and further reforming of intermediate species, especially acetic acid. The presence of certain additives or catalyst may accelerate such reactions and thereby improve reaction rates in such batch systems.

HTG of biomass is seldom affected by residence time at very high reaction temperatures as shown in Table 10.4 [21, 61]. In addition, continuous reactors can achieve high reaction rates for hydrogen production. However, most successful

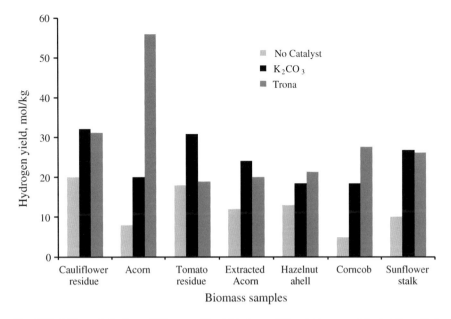

Fig. 10.8 Effect of alkaline additives on H_2 yields from different biomass. Adapted from Ref. [69] Copyright 2011, with permission from International Association of Hydrogen Energy

Table 10.4 Effect of residence time of noncatalytic hydrothermal gasification of biomass samples

Feedstock	Feed conc. (wt%)	Temperature (K)	Reaction time (min)	H_2 yield (mol/kg)	Source
Batch processes					
Glucose	5	603	0	0.5	[65]
Glucose	5	603	60	1.4	[65]
Glucose	5	603	120	3.2	[65]
Continuous processes					
Wood sawdust + CMC	4	923	15	14.2	[61]
Wood sawdust + CMC	4	923	33	16.3	[61]
Wood sawdust + CMC	4	923	40	18	[61]
Glucose	1.8	1013	10	61.1	[21]
Glucose	1.8	1013	30	60	[21]
Glucose	1.8	1013	60	64.1	[21]

continuous systems are operated with dilute (≤5 wt%) biomass concentrations and usually composed of water soluble and/or pumpable feedstocks. Some researchers have attempted to make pumpable biomass slurries by alkaline pretreatment, acid pretreatment, or by mixing finely pulverized biomass with corn starch [66] or carboxymethyl cellulose (CMC) [61] for continuous HTG systems.

Soluble biomass feedstocks, such as glucose and glycerol are usually products obtained from pretreatment of real biomass. Pretreatment of biomass to obtain any water soluble fractions or module compounds for continuous operations is akin to processes occurring during the heat-up stages in batch reactors. High concentrations of biomass could lead to pumping problems and plugging of continuous reactors due to enhanced polymerization and condensation of undesirable reaction products.

10.4 Catalytic HTG for Hydrogen Production

The use of different types of catalyst for hydrogen production during HTG of biomass offers a huge promise for the economic development of the process. In general, catalyst in this process work in several ways to enhance hydrogen production; these include (a) conversion of biomass into intermediates that can be easily gasified (b) lowering the activation energy required to achieve complete gasification (c) selectivity of reaction routes toward hydrogen production.

Catalysts used in HTG of biomass can be broadly but vaguely classified as homogenous and heterogeneous. Essentially, homogeneous catalysts are those that are water soluble, while heterogeneous catalyst are not. However, it can be expected that reactions occurring during HTG could be gas-phase, liquid-phase, or solid-phase. Therefore, a water soluble catalyst will become heterogeneous for gas-phase and solid-phase reactions. In this same way, a heterogeneous catalyst becomes a homogenous catalyst for solid-phase reactions.

10.4.1 Catalyst Supports

Due to the reaction conditions of HTG, it has not been possible to literally deploy largely successful conventional gasification catalysts to the hydrothermal process. Successful catalysts for hydrothermal biomass gasification must adapt to the specific conditions inherent in hydrothermal media. Such conditions include the large presence of water, high pressures, variable ionic product, and dielectric constant. Only a few catalyst supports remain stable under hydrothermal conditions. The use of hydrolyzable catalysts and catalyst supports will need to be avoided if catalysts are to be reused. Elliott et al. [18] found that other chemical transformations of catalyst supports could include phase transition and dissolution. They found that catalyst supports which showed satisfactory stability under hydrothermal conditions include alpha-alumina and zirconia or titania in their monoclinic forms [18].

10.4.2 Alkaline Catalysts/Additives

Nearly in all cases reported in literature, enhanced hydrogen production has been achieved through the use of highly soluble alkaline compounds as homogeneous catalysts. The most commonly used alkaline catalysts include KOH, NaOH, CaO, Trona, $Ca(OH)_2$, K_2CO_3, and Na_2CO_3 [29, 36, 47, 52, 67]. Compared to noncatalytic HTG of biomass, hydrogen yields have been observed to increase up to four times in the presence of some alkaline catalyst [68]. A recent work by Madenoglu et al. [69] is adapted and shown with permission in Fig. 10.8 as a good evidence to the enhanced effect of alkaline additives, including alkaline minerals such as Trona, in the production of hydrogen via HTG.

In general, the mechanistic schemes depicting the catalytic activities of these compounds indicate that they are effective in transforming biomass into simple molecules capable of HTG to product hydrogen. Onwudili and Williams [36] proposed a set of reaction mechanisms for the hydrothermal degradation of glucose into sodium formate and sodium acetate in the presence of sodium hydroxide. Experimental evidence suggests that alkaline additives operate within the subcritical region of water based on the variety of ionic reactions possible within this region. In acting to degrade biomass, these alkaline compounds alter the selectivity of the default reaction chemistry shown in equation (10.5) in favor of the formation of formic acid. This prevents polymerization reactions that could lead to tar formation. Using the gas products obtained from alkaline HTG of biomass, Onwudili and Williams [36] proposed as reaction scheme as follows:

$$C_6H_{12}O_6 + 6NaOH + 3H_2O \rightarrow C_6H_6O_6 \cdot Na_6 \cdot 9H_2O \quad (10.13)$$

$$C_6H_6O_6 \cdot Na_6 \cdot 9H_2O \rightarrow 6NaCOOH + 6H_2 + 3H_2O \quad (10.14)$$

$$6NaCOOH \rightarrow 6CO + 6NaOH \quad (10.15)$$

$$6CO + 6H_2O \rightarrow CO_2 + 6H_2 \quad (10.16)$$

$$6CO_2 + 6NaOH \rightarrow 6NaHCO_3 \quad (10.17)$$

The overall equation,

$$C_6H_{12}O_6 + 6NaOH + 6H_2O \rightarrow 6NaHCO_3 + 12H_2 \quad (10.18)$$

Similar reaction schemes were proposed by Sinag et al. [35] which showed that the formation of hydrogen during K_2CO_3-catalyzed HTG of biomass following the formation of potassium formate as intermediate. The preferential formation of hydrogen at near-critical water conditions indicates that the formate ion becomes the favored product of biomass-reforming reactions under alkaline hydrothermal conditions. However, experimental results indicate that in the presence of NaOH, the gas product consists of about 80–85 % hydrogen and 10–15 % methane [36]. Hence, reaction (10.1) can be rewritten in the presence of sodium hydroxide as follows:

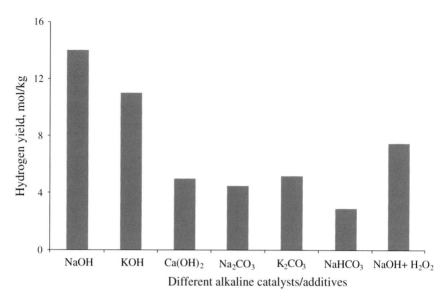

Fig. 10.9 Yields of gas products from glucose gasification in relation to the type of alkaline additive [60, 67]

$$C_6H_{12}O_6 + 4H_2O + 5NaOH \rightarrow 5NaHCO_3 + CH_4 + 8H_2 \quad (10.19)$$

Further research with pure sodium formate and sodium acetates [31] indicates that there are two possible hydrogen formation "windows" in the presence of alkaline additives. The earlier hydrogen formation "window" is completely accomplished at about 673 K, which corresponds to the hydrothermal reactions of sodium formate, shown in reaction (10.20). The latter hydrogen formation "window" occurs near 773 K, which also corresponds to the reactions of sodium acetate, depicted in reaction (10.14). The formation of methane in reaction (10.21) is a prerequisite for its reforming to hydrogen in reaction (10.12).

$$2HCOONa + H_2O \rightarrow 2H_2 + CO_2 + Na_2CO_3 \quad (10.20)$$

$$2CH_3COONa + H_2O \rightarrow 2CH_4 + CO_2 + Na_2CO_3 \quad (10.21)$$

These reactions can all be accomplished fairly quickly within a temperature upper limit of 723 K at the most. In addition, the hydroxides (alkalis, e.g., NaOH and KOH) give much better hydrogen yields than the carbonates (e.g., K_2CO_3, Na_2CO_3) as shown by the work of Muangrat et al. [67] shown in Fig. 10.9. It has been demonstrated that the alkalis are able to capture the carbon dioxide in the gas produced to form soluble carbonates, which are stable under reasonable temperatures ranges. The removal of CO_2 is important in enhancing the efficiency of the water-gas-shift reaction by driving it toward hydrogen production.

For effective hydrogen production, a large quantity of the alkali must be used, usually in a ratio of 1 g biomass/1.2 g alkali [36, 60]. This has cost implications for the HTG process. Also, the high water solubility of alkaline additives is seen as a major drawback for their recovery and reuse. This is because the alkalis are converted to their metal carbonates, which are not as effective as the original alkalis. Research into the recausticization of these carbonates may resolve this challenge.

10.4.3 Metal and Metal Oxide Catalysts

The two most significant metals used for HTG of biomass for hydrogen are nickel and ruthenium on various catalyst-supports [33, 70–80]. The reduced metals are significantly more effective than their corresponding metal oxides for gas production [16, 79]. The metal catalysts often work best in supercritical water medium for gas production, possibly due to a combination of gas-phase and solid-phase radical reactions. High carbon gasification efficiencies of 99.9 % and over have been achieved with ruthenium catalyst of supports such as alpha-alumina, gamma-alumina, carbon and rutile [33, 75].

Hydrogen has been reported as the major gas during catalytic HTG of biomass with metal catalyst in continuous reactors [81, 82]. This also suggests that reactions with high heating rates and short residence times are likely to produce high yields of hydrogen, possibly due to the absence of secondary reactions. However, the use of alkaline compounds as catalyst or co-catalysts has been reported to increase the yield of hydrogen in the product gas [33]. In addition, the use of Raney-nickel or a combination of nickel and sodium metal gives increased hydrogen gas yield [42, 70].

One of the challenges of using metal catalysts in HTG of biomass is that the effective metals such as Ni, Ru and Pd can catalyze both steam reforming reactions of hydrocarbons to hydrogen as well as methanation reaction, which consumes hydrogen [75, 76]. This is similar to the Sabatier reaction. Hence, rather than hydrogen or methane, both gases are obtained during catalysis with ruthenium catalyst as shown in Table 10.5. A cursory look at the table indicates that the selectivities of ruthenium-based catalysts toward hydrogen or methane in a batch reactor could be influenced by reaction temperatures and residence times as well as the catalyst supports. It appears that lower temperatures and longer residence times favor hydrogen production, while methane selectivity increases at higher temperatures [33]. Recently, Zhang et al. [83] showed that hydrogen generation from glucose increased in the presence of ruthenium-modified nickel catalyst. Such catalyst combination is proposed to favor hydrogen production against methanation reaction.

Table 10.5 Effect of ruthenium-based catalysts on hydrogen and methane yields during HTG of biomass in batch reactors [33]

Ruthenium-based catalysts	Feed	Feed loading wt %	Temperature (°C)	Residence time (min)	H_2 yield (mol/kg)
Ru/TiO_2	Sugarcane bagasse	3.3	400	15	3.2
Ru/C	Sugarcane bagasse	3.3	400	15	1.9
Ru/TiO_2	Cellulose	3.3	400	15	2.8
Ru/γ-Al_2O_3	Glucose	2.0	380	15	18
Ru/γ-Al_2O_3	Glucose	2.0	380	60	33
Ru/C	Glucose	2.0	380	15	17
Ru/C	Glucose	2.0	380	60	26
Ru/γ-Al_2O_3	Cellulose	2.0	380	60	34
Ru/C	Cellulose	2.0	380	60	22
Ru/C	P. tricornutum (alga)	2.5	400	60	2.5
Ru/C	P. tricornutum (alga)	5.1	400	67	1.9
Ru/α-Al_2O_3	Glucose	6.7	550	10	10.8
Ru/α-Al_2O_3/NaOH	Glucose	6.7	550	10	21.1
Ru/α-Al_2O_3/CaO	Glucose	6.7	550	10	14.7
Ru/α-Al_2O_3/CaO	Cellulose	6.7	550	10	9.10
Ru/α-Al_2O_3/CaO	Xylan	6.7	550	10	10.7
Ru/α-Al_2O_3/CaO	Sawdust	6.7	550	10	10.4

Adapted from Ref. [33], Copyright 2013, with permission from Elsevier

10.4.4 Carbon Catalyst

Carbon, usually in its activated form has been used for hydrothermal biomass gasification both as a catalyst and as catalyst support [66, 84]. It is clear from literature that the catalytic activity of carbon for HTG is mostly limited to high temperature applications, i.e., within the supercritical region of water with temperatures above 873 K as shown in Fig. 10.10. This in itself poses a challenge with respect to the mechanisms of carbon-catalyzed high temperature HTG, since high temperature alone can give comparable yields of hydrogen [85]. However, carbon shows relatively high stability in supercritical water conditions and can be recovered [86].

In addition, it appears that the catalytic activities of activated carbons depend on the biomass from which they are sourced. As shown in Fig. 10.10, the yields of hydrogen from the supercritical water gasification of 1.2 M glucose solution varied considerably with different types of activated carbons. This could be attributed to the ash contents of the activated carbons; for instance, a high content of alkaline metals may promote hydrogen formation. In addition, it has also been shown that the carbon catalyst itself may be gasified under certain supercritical water conditions to form hydrogen [87].

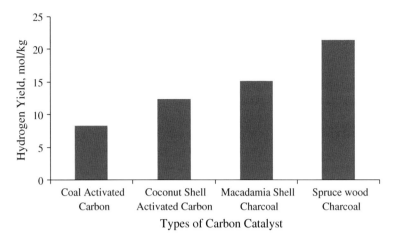

Fig. 10.10 Effect of activated carbons from different sources on hydrogen yields from HTG of 1.2 M glucose solution at 873 K. Adapted from Ref. [85], Copyright 1996, with permission from Elsevier

The activities of carbon catalysts can also depend on the types of biomass and feed concentrations as shown in Fig. 10.11. As mentioned earlier, lower biomass concentrations and carbohydrate-rich biomass types favor hydrogen production. Figure 10.11 indicates that corn starch (10 wt% concentration) could produce hydrogen of up to 31 mol/kg, whereas sewage sludge (3 wt% concentration) produced only 11 mol/kg using the same coconut shell activated carbon as catalyst, within similar supercritical water conditions.

10.5 Considerations for Commercial Hydrogen Production from Biomass HTG

Biomass utilization for energy production is carbon-neutral. Hydrogen production from the HTG of biomass is a potentially cheaper and eco-friendly process based on renewable biomass. For instance, the optimal temperature range required for hydrogen production is much lower compared to conventional gasification. Despite the progress achieved in the past 20 years in this area of research, there are still some issues to be resolved in order to ensure large-scale industrial application of HTG for hydrogen production;

(a) There is a need to streamline the range of feedstocks suitable for HTG of biomass into hydrogen at moderately mild conditions. Research has focused on using all types of biomass; however same research results point toward carbohydrate-type biomass as being suitable for hydrogen production.

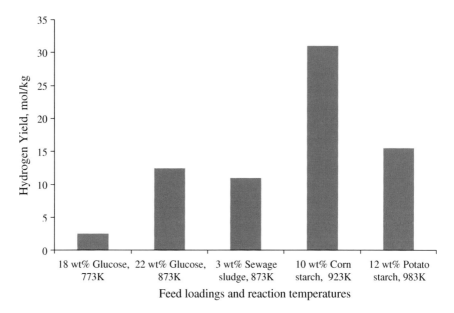

Fig. 10.11 Hydrogen yields from carbon-catalyzed HTG of biomass in continuous reactors [67, 84, 85]

(b) Research shows that biomass needs to be transformed into pumpable slurries for continuous operation of a HTG process. To achieve this goal, there will be a need to develop low-cost preprocessing techniques to convert carbohydrate-type biomass to suitable water soluble hydrogen-forming precursors such as simple carboxylic acids. This will ensure that feedstocks are transformed into appropriate forms for continuous processing into hydrogen. In addition, high concentration of soluble biomass-derived feedstocks can be used to ensure high thermal efficiency of the system.

(c) Commercialization of HTG process for hydrogen production will benefit from a detailed knowledge of the reaction kinetics and how they favor hydrogen production. Current knowledge indicates that conversion of biomass into gasifiable intermediates occur at slow rates, whereas water-gas shift reaction is relatively fast. To improve biomass conversion to hydrogen, there is need to improve the reaction rates involved in the reforming of the original biomass or the intermediate species.

(d) Appropriate catalysts will need to be developed based on the current knowledge and future research. For example, current knowledge indicates that the presence of alkali favor hydrogen production but their recovery is difficult. Metal catalysts such as ruthenium favor simultaneous production of both hydrogen and methane. Development of a synergist catalyst based on alkali and appropriate metals may be a way forward.

References

1. IEA (2008). World Energy Outlook. OECD/IEA. IEA, Head of Communication and Information Office, Paris, France
2. Ladanai S, Vinterback J (2009) Global potential of sustainable biomass for energy. Report ISSN 1654-9406
3. OECD/IEA (2013) Working together to ensure reliable, affordable and clean energy
4. Hall DO, Rosilo-Calle F, de Groot P (1992) Biomass energy: lessons fron case studies in developing countries. Energ Policy 20(1):62–73
5. Stucki S, Vogel F, Ludwig C, Haiduc AG, Brandenberger M (2009) Catalytic gasification of algae in supercritical water for biofuel production and carbon capture. Energy Environ Sci 2:535–541
6. Clarens AF, Resurreccion EP, White MA, Colosi LM (2010) Environmental life cycle comparison of algae to other bioenergy feedstocks. Environ Sci Technol 44:1813–1819
7. National Algae Association (2011) NAA Algae Oil and Biomass White Paper, United States
8. Industrial gases by the chemical economics handbook, SRI (Oct 2007)
9. The hydrogen economy: opportunities, costs, barriers, and R&D needs (2004); National Research Council and National Academy of Engineering. The National Academic Press: Washington, DC
10. Dunn S (2001) Hydrogen futures: toward a sustainable energy system. Worldwatch Institute, Washington, DC
11. The Freedonia Group (2009) Industrial Gases to 2013
12. Silveira S (2005) Bioenergy: realizing the potential. Swedish Energy Agency, Eskilstuna
13. Roberts D (2007) Globalization and Its Implications for the Indian Forest Sector. TIFAC/IIASA Joint Workshop "Economic, Societal and Environmental Benefits Provided by the Indian Forests", New Delhi, India
14. Balat M, Balat M (2009) Political, economic and environmental impacts of biomass-based hydrogen. Int J Hydrogen Energy 35:3589–3603
15. PATH (2011) Annual Report on World Progress in Hydrogen. Partnership for Advancing the Transition to Hydrogen (PATH), Washington, DC
16. Guo L, Cao C, Lu Y (2010) Supercritical water gasification of biomass and organic wastes. In: Momba M, Bux Faizal (eds) Biomass. Sciyo, Croatia. ISBN 978-953-307-113-8, 202
17. Modell M (1985) Gasification and liquefaction of forest products in supercritical water (2005). In: Overend RP, Milne TA, Mudge LK (eds) Fundamentals of thermochemical biomass conversion. Elsevier, London, pp 95–120
18. Elliott DC (2008) Catalytic hydrothermal gasification of biomass. Biofuels Bioprod Biorefin 2:254–265
19. Peterson AA, Vogel F, Lachance RP, Fröling M, Antal MJ Jr, Tester W (2008) Thermochemical biofuel production in hydrothermal media: a review of sub- and supercritical water technologies. Energy Environ Sci 1:32–65
20. Kruse A, Gawlik A (2003) Biomass conversion in water at 330–410 °C and 30–50 MPa. Identification of key compounds for indicating different chemical reaction pathways. Ind Eng Chem Res 42:267–279
21. Susanti RF, Dianningrum LW, Yum Y, Kim Y, Lee BG, Kim J (2012) High-yield hydrogen production from glucose by supercritical water gasification without added catalyst. Int J Hydrogen Energy 37:11677–11690
22. Kruse A (2009) Hydrothermal biomass liquefaction. J Supercrit Fluids 47:391–399
23. Kikkinides ES, Yang RT (1993) Concentration and recovery of CO_2 from flue gas by pressure swing adsorption. Ind Eng Chem Res 23:2714–2720
24. Sedlak JM, Austin JF, La Conti AB (1981) Hydrogen recovery and purification using the solid polymer electrolyte electrolysis cell. Int J Hydrogen Energy 6(1):45–51
25. Lee HK, Choi HY, Choi KH, Park JH, Lee TH (2004) Hydrogen separation using the electrochemical method. J Power Sources 132(1):92–98

26. Ibeh B, Gardner C, Ternan M (2007) Separation of hydrogen from a hydrogen/methane mixture using a PEM fuel cell. Int J Hydrogen Energy 32:908–914
27. Fiori L, Valbusa M, Castello D (2012) Supercritical water gasification of biomass for H_2 production: process design. Bioresour Technol 121:139–147
28. Savage PE (1999) Organic reactions in supercritical water. Chem Rev 99:603–621
29. Yanik J, Ebale S, Kruse A, Saglam M, Yüksel M (2008) Biomass gasification in supercritical water: II. Effect of catalyst. Int J Hydrogen Energy 33(2008):4520–4526
30. T-Raissi A, Block DL (2004) Hydrogen: automotive fuel of the future. IEEE Power Energ Mag 2:40–45
31. Onwudili JA, Williams PT (2010) Hydrothermal reactions of sodium formate and sodium acetate as model intermediate products of the sodium hydroxide-promoted hydrothermal gasification of biomass. Green Chem 12:2214–2222
32. Onwudili JA, Williams PT (2011) Reactions of different carbonaceous materials in alkaline media for hydrogen gas production. Green Chem 13:2837–2843
33. Onwudili JA, Williams PT (2013) Hydrogen and methane selectivity during alkaline supercritical water gasification of biomass with ruthenium-alumina catalyst. Appl Catal B 132–133:70–79
34. Oefner PJ, Lanziner AH, Bonn G, Bobleter O (1992) Quantitative studies on furfural and organic acid formation during hydrothermal, acidic and alkaline degradation of D-xylose. Monatsh Chem 123:547–556
35. Sinag A, Kruse A, Schwarzkopf V (2003) Key compounds of the hydropyrolysis of glucose in supercritical water in the presence of K_2CO_3. Ind Eng Chem Res 42(15):3516–3521
36. Onwudili JA, Williams PT (2009) Role of sodium hydroxide in the production of hydrogen gas from the hydrothermal gasification of biomass. Int J Hydrogen Energy 34:5645–5656
37. Chuntanapum A, Matsumura Y (2009) Formation of tarry material from 5-HMF in subcritical and supercritical water. Ind Eng Chem Res 48:9837–9846
38. Jin F, Zhou Z, Kishita A, Enomoto H, Kishida H, Moriya T (2007) A new hydrothermal process for producing acetic acid from biomass waste. Trans IChemE A: Chem Eng Res Des 85(A2):201–206
39. Jin F, Yun J, Li G, Kishita A, Tohji K, Enomoto H (2008) Hydrothermal conversion of carbohydrate biomass into formic acid at mild temperatures. Green Chem 10:612–615
40. Jin F, Zhou Z, Moriya T, Kishida A, Higashijima H, Enomoto H (2009) Controlling hydrothermal reaction pathways to improve acetic acid production from carbohydrate biomass. Environ Sci Technol 39:1893–1902
41. Hsieh Y, Du Y, Jin F, Zhou Z, Enomoto H (2009) Alkaline pre-treatment of rice hulls for hydrothermal production of acetic acid. Chem Eng Res Des 87:13–18
42. Kruse A, Dinjus E (2003) Hydrogen from methane and supercritical water. Angew Chem Int Ed 42(8):909–911
43. Li Y, Fu Q, Flytzani-Stephanopoulos M (2000) Low-temperature water-gas shift reaction over Cu- and Ni-loaded cerium oxide catalysts. Appl Catal B 27:179–191
44. Broll D, Kaul C, Kramer A, Krammer P, Richter T, Jung M, Vogel H, Zehner P (1999) Chemistry in supercritical water. Angew Chem Int Ed 38(20):2998–3014
45. Sealock LJ, Elliott DC, Baker EG, Butner RS (1993) Chemical-processing in high-pressure aqueous environments. 1. Historical-perspective and continuing developments. Ind Eng Chem Res 32(8):1535–1541
46. Kumar A, Jones DD, Hanna MA (2009) Thermochemical biomass gasification: a review of the current status of the technology. Energies 2:556–581
47. Madenoglu TG, Saglam M, Yuksel M, Ballice L (2013) Simultaneous effect of temperature and pressure on catalytic hydrothermal gasification of glucose. J Supercrit Fluids 73:151–160
48. Cao C, Guo L, Chen Y, Guo S, Lu Y (2011) Hydrogen production from supercritical water gasification of alkaline wheat straw pulping black liquor in continuous flow system. Int J Hydrogen Energy 36:3528–3535

49. Guo LJ, Lu YJ, Zhang XM, Ji CM, Guan Y, Pei AX (2007) Hydrogen production by biomass gasification in supercritical water: a systematic experimental and analytical study. Catal Today 129:275–286
50. Muangrat R, Onwudili JA, Williams PT (2010) Influence of NaOH, Ni/Al$_2$O$_3$ and Ni/SiO$_2$ catalysts on hydrogen production from the subcritical water gasification of model food waste compounds. Appl Catal B 100(1–2):143–156
51. Kruse A, Maniam P, Spieler F (2007) Influence of proteins on the hydrothermal gasification and liquefaction of biomass. 2. Model compounds. Ind Eng Chem Res 46(1):87–96
52. Yanik J, Ebale S, Kruse A, Saglam M, Yuksel M (2007) Biomass gasification in supercritical water. Part 1: Effect of nature of biomass. Fuel 86:2410–2415
53. Osada M, Hiyoshi N, Sato O, Arai K, Shirai M (2007) Effect of sulfur on catalytic gasification of lignin in supercritical water. Energy Fuels 21:1400–1405
54. Onwudili JA, Lea-Langton AR, Ross AB, Williams PT (2012) Catalytic hydrothermal gasification of algae for hydrogen production: composition of reaction products and potential for nutrient recycling. Bioresour Technol 127C:72–80
55. Williams PT, Onwudili J (2006) Subcritical and supercritical water gasification of cellulose, starch, glucose, and biomass waste. Energy Fuels 20:1259–1265
56. Forchheim D, Gasson JR, Hornung U, Kruse A, Barth T (2012) Modeling the lignin degradation kinetics in a ethanol/formic acid solvolysis approach. Part 2: Validation and transfer to variable conditions. Ind Eng Chem Res 51(46):15053–15063
57. Lu Y, Guo L, Zhang X, Ji C (2012) Hydrogen production by supercritical water gasification of biomass: explore the way to maximum hydrogen yield and high carbon gasification efficiency. Int J Hydrogen Energy 37:3177–3185
58. Williams PT, Onwudili JA (2005) Composition of products from the supercritical water gasification of glucose: a model biomass compound. Ind Eng Chem Res 44:8739–8749
59. Lu YJ, Jin H, Guo LJ, Zhang XM, Cao CQ, Guo X (2008) Hydrogen production by biomass gasification in supercritical water with a fluidized bed reactor. Int J Hydrogen Energy 33(21):6066–6075
60. Muangrat R, Onwudili JA, Williams PT (2010) Reaction products from the subcritical water gasification of food wastes and glucose with NaOH and H$_2$O$_2$. Bioresour Technol 101:6812–6821
61. Lu YJ, Guo LJ, Ji CM, Zhang XM, Hao XH, Yan QH (2006) Hydrogen production by biomass gasification in supercritical water: a parametric study. Int J Hydrogen Energy 31:822–831
62. Demirbas A (2004) Hydrogen-rich gas from fruit shells via supercritical water extraction. Int J Hydrogen Energy 29(12):237–1243
63. Alshammari YM, Hellgardt K (2012) Thermodynamic analysis of hydrogen production via hydrothermal gasification of hexadecane. Int J Hydrogen Energy 37:5656–5664
64. Sinag A, Kruse A, Rathert J (2004) Influence of the heating rate and the type of catalyst on the formation of key intermediates and on the generation of gases during hydropyrolysis of glucose in supercritical water in a batch reactor. Ind Eng Chem Res 43(2):502–508
65. Muangrat R, Onwudili JA, Williams PT (2011) Alkaline subcritical water gasification of dairy industry waste (Whey). Bioresour Technol 102:6331–6335
66. Antal MJ, Allen SG, Schulman D, Xu X (2000) Biomass gasification in supercritical water. Ind Eng Chem Res 39:4040–4053
67. Muangrat R, Onwudili JA, Williams PT (2010) Influence of alkali catalysts on the production of hydrogen-rich gas from the hydrothermal gasification of food processing waste. Appl Catal B 100(3–4):440–449
68. Watanabe M, Inomata H, Osada M, Sato T, Adschiri T, Arai K (2003) Catalytic effects of NaOH and ZrO$_2$ for partial oxidative gasification of n-hexadecane and lignin in supercritical water. Fuel 82(5):545–552
69. Madenoglu TG, Boukis N, Saglam M, Yuksel M (2011) Supercritical water gasification of real biomass feedstocks in continuous flow system. Int J Hydrogen Energy 36:14408–14415

70. Elliott DC, Sealock LJ Jr, Baker EG (1993) Chemical processing in high-pressure aqueous environments. 2. Development of catalysts for gasification. Ind Eng Chem Res 32:1542–1548
71. Osada M, Sato T, Watanabe M, Shirai M, Arai K (2005) Catalytic gasification of wood biomass in subcritical and supercritical water. Combust Sci Technol 177:1–16
72. Osada M, Sato O, Arai K, Shirai M (2006) Stability of supported ruthenium catalysts for lignin gasification in supercritical water. Energy Fuels 20:2337–2343
73. Lee IG, Ihm SK (2009) Catalytic gasification of glucose over Ni/Activated charcoal in supercritical water. Ind Eng Chem Res 48:1435–1440
74. Yamaguchi A, Hiyoshi N, Sato O, Bando KK, Osada M, Shirai M (2009) Hydrogen production from woody biomass over supported metal catalysts in supercritical water. Catal Today 146:192–195
75. Osada M, Yamaguchi A, Hiyoshi N, Sato O, Shirai M (2012) Stability of catalysts for lignin gasification in supercritical water. Energy Fuels 26:3179–3186
76. Azadi P, Khan S, Strobel F, Azadi F, Farnood R (2012) Hydrogen production from cellulose, lignin, bark and model carbohydrates in supercritical water using nickel and ruthenium catalysts. Appl Catal B 117–118:330–338
77. Azadi P, Khodadadi AA, Mortazavi Y, Farnood R (2009) Hydrothermal gasification of glucose using Raney nickel and homogeneous organometallic catalysts. Fuel Process Technol 90:145–151
78. Zhang L, Champagne P, Xu C (2011) Supercritical water gasification of a aqueous by-product from biomass hydrothermal liquefaction with novel Ru modified Ni catalysts. Biorefin Technol 102:8279–8287
79. Hao X, Guo L, Zhang X, Guan Y (2005) Hydrogen production from catalytic gasification of cellulose in supercritical water. Chem Eng J 110:57–65
80. May A, Salvado J, Torras C, Montane D (2010) Catalytic gasification of glycerol in supercritical water. Chem Eng J 160:751–759
81. Byrd AJ, Pant KK, Gupta RB (2008) Hydrogen production from glycerol by reforming in supercritical water over Ru/Al$_2$O$_3$ catalyst. Fuel 87:2956–2960
82. Takuya Y, Yoshito O (2004) Partial oxidative and catalytic biomass gasification in supercritical water: a promising flow reactor system. Ind Eng Chem Res 43:4097–4104
83. Zhang L, Xu C, Champagne P (2012) Activity and stability of a novel Ru modified Ni catalyst for hydrogen generation by supercritical water gasification of glucose. Fuel 96:541–545
84. Xu X, Antal MJ (1998) Gasification of sewage sludge and other biomass for hydrogen production in supercritical water. Environ Prog 17:215–220
85. Xu X, Matsumura Y, Stenberg J, Antal MJ Jr (1996) Carbon-catalyzed gasification of organic feedstocks in supercritical water. Ind Eng Chem Res 35:2522–2530
86. Yanagida T, Minowa T, Shimizu Y, Matsumura Y, Noda Y (2009) Recovery of activated carbon catalyst, calcium, nitrogen and phosphate from effluent following supercritical water gasification of poultry manure. Bioresour Technol 100:4884–4886
87. Azadi P, Farnood R (2011) Review of heterogeneous catalysts for sub- and supercritical water gasification of biomass and wastes. Int J Hydrogen Energy 36(16):9529–9541

Part IV
Hydrothermal Conversion of Biomass into Other Useful Products

Chapter 11
Review of Biomass Conversion in High Pressure High Temperature Water (HHW) Including Recent Experimental Results (Isomerization and Carbonization)

Masaru Watanabe, Taku M. Aida and Richard Lee Smith

Abstract In this chapter, we briefly explain unique properties of high pressure high temperature water (HHW). In high pressure media, concentration of reactant can be controlled by changing temperature and pressure, and the reaction rate (also product distribution) can be controlled. In addition, in the presence of solvent (water is concerned here), the properties of the solvent can also be adjusted by pressure and temperature, and the control of solvent properties can help to improve the reaction rate and selectivity. Some of important reactions occurring in the high pressure high temperature water (HHW) media are summarized and the relationship between the reactions and the products is roughly categorized into gasification, liquefaction, and carbonization. Briefly, over 400 °C, radical reaction is dominant and thus gasification (small fragment formation) occurs. Between 200 and 400 °C, both ionic and radial reactions competitively occur and biomass conversion can be controlled widely by changing temperature and pressures. Therefore, production of chemical block for industries is performed in the temperature range. Below 200 °C, namely low temperature and high density of water (liquid phase of water), hydrolysis and dehydration are favored because ionic reactions are predominant. Through dehydration between molecules (high concentration condition is preferred), carbonization is also developed. Concerning each product category, our research topics are briefly overviewed. Finally, our recent experimental results for isomerization of glucose and carbonization of biomass are roughly introduced.

M. Watanabe (✉) · R. L. Smith
Research Center of Supercritical Fluid Technology, Tohoku University, Sendai, Japan
e-mail: meijin@scf.che.tohoku.ac.jp

M. Watanabe · T. M. Aida · R. L. Smith
Department of Environmental Study, Tohoku University, Sendai, Japan

11.1 Introduction

Biomass is the single sustainable organic resource on the earth. From the ancient time, human being has utilized biomass as fuel after drying. Wood has been used to build house and furniture. Crops and beans have been important as food. Leaves have provided medicine and drinks. Someone invented pulping process and human can also use biomass as material effectively and widely. Biomass has supported human life at various aspects.

Disadvantage of biomass is climate sensitive and widely spread. It is difficult for human to get biomass at a constant yield for all four seasons and all years. It takes cost for collection of biomass at a place. In addition, if the amount of harvest takes over that of cultivating, biomass at an area should be consumed and the area becomes desert, that is one of the reasons why some ancient civilization disappeared.

To overcome the issue related to biomass (unstable harvesting) pointed out above, human being found fossil fuels and succeeded to raise industrial revolution. Fossil fuels were cheap and seemed not to have limitation. Petroleum is, in particular, liquid and easy to use as transport fuel and chemical resources. But, as everybody knows, it has been revealed that fossil fuels are consumable and it can be gained at the limited area. These imperfections cause some new economical and political issues. Furthermore, fossil fuels are accumulation of ancient CO_2 under the soil and thus consumption of them provides excess amount of CO_2 in the air. From these economical, ecological, and political reasons, usage of fossil fuels has to be limited such as premium fuel for long-range transportation.

Plant biomass consists of organics (carbohydrate, lignin, lipid, protein, their hybrid, and their small constituents), mineral, and a large quantity of water. Before the industrial revolution, people did not have modern equipment. At that time, biomass was simply burned or converted after drying, which requires a lot of energy because of high latent heat of water vaporization. To decrease input energy for biomass upgrading, water in biomass had better be heated up to an appropriate temperature and pressure by keeping liquid phase, not vaporization, and thus one must manage high pressure high temperature water (HHW) to use biomass effectively.

Nowadays, huge number of industries utilize high pressure high temperature water (HHW) including supercritical state; for example, steam turbine at fuel-fired power plant is operated at supercritical state, artificial quartz is synthesized in supercritical water operated at two zones of temperature, woody chip at pulping is digested at hydrothermal condition in the presence of alkali. Then, one can access high pressure high temperature water (HHW) operation in these days, if they want, but there are still many problems to use it for biomass application from economical (still expensive) and mechanical (corrosion, plugging, solid treatment, and so on) point of view. To overcome the drawbacks, various kinds of application of hydrothermal biomass upgrading have to be reported and cost reasonable process at business stage has to be demonstrated.

In this chapter, some of characteristics of high pressure high temperature water (HHW) are roughly explained, basic reaction for biomass conversion in water environment is then described and our few recent results are finally overviewed.

11.2 High Pressure High Temperature Water (HHW)

11.2.1 Advantages of High Pressure Media

Some advantages of high pressure technology are to reduce energy loss of expansion. Figure 11.1 shows an example. As mentioned above already, water vaporization at atmospheric pressure loses a huge quantity of energy as kinetic energy (when 1 kg of water at 100 °C is vaporized completely, 2.6 MJ of energy is required). On the other hand, heat-up of 1 kg of liquid water up to 300 °C without phase change requires only 1.2 MJ.

The other advantage is to keep or control density of fluid by controlling temperatures. Density of fluid is basic property and governs physical properties such as dielectric constant, viscosity, and so on. Therefore, flexible control of fluid density achieves wide control of solubility and reactivity.

The controllable of solubility as well as density of reactants brings high flexibility for reaction equilibrium known as Le Chatelier's principal. In addition, the adjustability of concentration by changing temperature and pressure (namely density) gives high operability in reaction dynamics (kinetics) as well as reaction equilibrium. Concentration (the amount of reactant in reaction environment: density and phase behavior are closely related) is the major factor for the reaction rate and product distribution. One of the typical examples is hydrocarbon cracking. The cracking is radical chain reaction. Simply, it consists of 4 types of reactions: initiation, radical decomposition, H abstraction, and termination (Eqs. 11.1–11.4, respectively).

$$S \xrightarrow{k_i} 2\beta \tag{11.1}$$

$$\alpha \xrightarrow{k_\beta} \beta + \gamma \tag{11.2}$$

$$\beta + S \xrightarrow{k_H} \beta H + \alpha \tag{11.3}$$

$$2 \text{radicals}(\alpha \text{ or } \beta) \xrightarrow{k_T} \text{products} \tag{11.4}$$

Here, S is reactant, α and β are radicals (α is relatively large radical which is produced by H abstraction, β is smaller radicals produced from initiation and radical decomposition), and γ is small product formed from radical decomposition. The rate constants of each reaction are k_i (initiation), k_β (radical decomposition), k_H (H abstraction), and k_T (termination). When the concentration of S is high

Fig. 11.1 Enthalpy of water at saturation condition

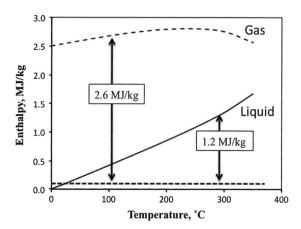

enough, hydrogenated product (βH) is main product, while γ is main at low S concentration. To consider the concentration dependence, the overall rate constant is developed with steady-state approximation for the concentration of radicals (α and β). As a result, Eq. 11.5 is obtained.

$$k_s = \frac{k_\beta k_H}{k_\beta + k_H[S]} \sqrt{\frac{k_i[S]}{k_T}} \qquad (11.5)$$

Here, k_s is overall rate constant of the cracking of S. Equation 11.5 tells us that k_s depends on the concentration of S: k_s increases at low concentration region with an order of 0.5 to S concentration, k_s is not sensitive to S concentration at mild concentration region, and k_s decreases at high S concentration with -0.5 order to S concentration. For hydrocarbon cracking, supercritical water (up to 500 °C) is inert as reactant but phase behavior of hydrocarbon is affected by the existence of supercritical water [1]. When supercritical water dissolve some amount in dense hydrocarbon phase, the concentration of hydrocarbon in the dense phase is reduced. The reduction of concentration enhances the rate of hydrocarbon cracking and thus supercritical water works as cracking promoter. One has to notice that the promotion of the reaction is not the change of reaction but only the change of concentration of reactant.

At high pressure media, pressure itself sometimes promotes reaction rate. When a reaction pass through an activated complex, the reaction rate of the reaction is promoted with increasing pressure if volume of the activated complex (activation volume) is smaller than that of the reactant.

11.2.2 Special Features of Water at High Pressure High Temperature (HHW)

Water is ecological friendly and essential solvent on the earth. It has wider application if high pressure environment is available. When reaction field is in a solvent, solubility of a solute in the solvent is the critical factor of the reaction rate; for example, the rates of oxidation and hydrogenation are limited by the solubility of oxygen and hydrogen, respectively. One of famous examples for usefulness of supercritical water as reaction field is supercritical water oxidation. In supercritical state of water, organic compounds such as hydrocarbon and oxygen (of which solubility is restricted at ambient condition) are miscible and thus complete oxidation of organics can be achieved at short reaction time [2]. There have also been several reports that show effectiveness of supercritical water as reaction environment for hydrogenation [3, 4].

Furthermore, some features of water are unique and these result in attractive media for biomass conversions. The rate of some reactions of polar molecules is affected by dielectric constant as shown in Eq. 11.6 (Kirkwood equation) [5]. Suppose $A + B \Rightarrow AB \Rightarrow C$. A and B are reactants, AB is activated complex and C is product. Sometimes AB has higher polarity than reactants.

$$\ln k_{\text{sol}} = \ln k_0 - \frac{N_A}{4\pi\varepsilon_0 RT}\left(\frac{\varepsilon-1}{2\varepsilon+1}\right)\left(\frac{\mu_{AB}^2}{r_{AB}^3} - \frac{\mu_A^2}{r_A^3} - \frac{\mu_B^2}{r_B^3}\right) \tag{11.6}$$

Here, k_{sol} is rate constant in a solution, k_0 is rate constant in vacuum, and N_A is Avogadro number. Dipole moment of the reactants (A and B) and the active complex are denoted as μ_i and radius of molecules is denoted as r_i ($i = A, B,$ and AB: the reactants and the activated complex). According to this equation, when the dipole moment of activated complex is high enough compared with that of the reactants, the rate of the reaction is accelerated at the solvent having high dielectric constant. Dielectric constant of water can be adjusted by density and temperature, as shown in Eq. 11.7 [6].

$$\varepsilon = 1 + \left(\frac{A_1}{T}\right)\rho + \left(\frac{A_2}{T} + A_3 + A_4 T\right)\rho^2 + \left(\frac{A_5}{T} + A_6 T + A_7 T^2\right)\rho^3$$
$$+ \left(\frac{A_8}{T^2} + \frac{A_9}{T} + A_{10}\right)\rho^4 \tag{11.7}$$

Here A_i ($i = 1$–10) is fitting parameter. Indeed, there are some examples that can be explained well based on Kirkwood analysis [7, 8].

Additionally, the other feature of water is high dissociation into proton and hydroxyl ion. The dissociation of water is deeply related to control of many kinds of reactions such as acid–base reactions. Typical biomass molecules such as cellulose, hemicellulose, lignin, protein, and so on, have reactive chemical bonds such as ether, which can be attacked by water molecules. The reaction of the bond with water can be enhanced by proton or hydroxyl ion and thus, if these ions could provide effectively by controlling temperature and pressure, the reaction occurs without catalyst [9]. Ion product (Kw), which is index of water dissociation, can be controlled by temperature and water and thus is also governed by density and temperature as shown in Eq. 11.8.

$$\log K_W = A + \frac{B}{T} + \frac{C}{T^2} + \frac{D}{T^3} + \left(E + \frac{F}{T} + \frac{G}{T^2}\right)\log \rho \qquad (11.8)$$

Here A–G are fitting parameters [10]. Some organic reactions such as hydrolysis, dehydration, decarboxylation, and so on, are controlled by acid and base catalyst as well as temperature. For high pressure high temperature water (HHW) media, density could be controlled by temperature and pressure. Therefore the acid–base reactions could be controlled by these parameters, in particular at supercritical state because density can be changed continuously by changing the operation parameters (temperature and pressure) there.

Ion product of water affects surface acidity or basicity of solid catalysts. Of course, it is the best way of performing reactions in high pressure high temperature water (HHW) that the rate of the reaction can be controlled by only temperature and pressure. However, the concentrations of proton and hydroxyl ion in neutral water are always the same and adjustability of acidity and basicity of water must be limited. To expand the controllability of acid–base property of water, heterogeneous catalyst would be better than homogeneous catalyst because of the existence of both acid and base sites on the surface and wide controllability of acid–base properties (of course, heterogeneous catalysts are favored from the feasibility point of view, such as recyclability, reusability, and environmentally friendliness). Indeed, the research activities of using the heterogeneous catalyst for controlling biomass reactions in high pressure high temperature water (HHW) are increasing year by year. According to the detailed studies of hydration of olefin in the presence of solid acid catalyst (acidic metal oxide such as MoO_3) in sub and supercritical water, the surface acidity increased with increasing ion product with an order of 0.45 [11]. It was considered that proton concentration of bulk water contributed the amount of surface proton on the acid catalyst. Thus, to control acid–base reactions either in the presence or absence of catalyst (both homogeneous and heterogeneous), density of water is essential key parameter.

11.3 Biomass Conversion in High Pressure High Temperature Water (HHW)

In water, both ionic and radical reactions are developed and these are important for biomass conversion. By controlling these reactions, biomass component can be transformed into intended products selectively. Here basic matters for biomass reactions (such as kind of ionic reactions that have to be headed) are roughly mentioned and expected biomass conversion process is briefly reviewed.

11.3.1 Ionic Reactions

Typically, acid–base reactions are ionic reactions. Unlike in the case of petroleum (hydrocarbons), biomass has various function groups such as hydroxyl, carboxyl, carbonyl, ether, ester, and so on. In oil refinery for petroleum, attachment of functional group in hydrocarbons is the key reaction, on the other hand, detachment of functional group is selectively controlled in biomass refinery because of richness of function groups.

For the detachment, acid and base reactions are quite important. For example, it is well known that detachment of hydroxyl group, namely hydration, is promoted by proton, that is, acid reaction. Some kinds of carboxyl groups are detached by alkali catalyst (depending on substituted group). Ether and ester groups are hydrolyzed and promoted by both acid and base catalysts. Carbohydrates have ether and hydroxyl groups. Lipids have ester and carboxyl groups. Proteins have peptide groups, which are decomposed by acid and base catalysts. Lignin has ether and hydroxyl groups as well as carbohydrates, but the reactivity of these function groups are different because substituted group around the function group is different.

It is also well known that some types of reaction can be controlled by acid–base catalysts; for example, isomerization of glucose, Cannizzaro reaction, retro-aldol condensation, water–gas shift reaction are promoted by base catalyst, while alkylation such as combination between alcohol and phenols promoted by acid catalyst.

To enhance controllability of the ionic reactions, addition of catalyst is quite useful with control of temperature and pressure. Typical homogenous catalyst (mineral acids and alkali catalyst, in particular, H_2SO_4 as acid catalyst, and NaOH and KOH as alkali catalysts are often used) can also work in high pressure high temperature water (HHW). As mentioned just before, the number of the reports for the acid–base reactions in the presence of heterogeneous catalysts has increased recently. As well as typical acid catalyst such as zeolite (silica–alumina), ion exchange resin, sulfated zirconia, and so on, it was reported that MoO_3 [12], TiO_2 [13, 14], and Nb_2O_5 [15] worked as Bronsted and Lewis acid–base catalyst in high pressure high temperature water (HHW). Still useful catalytic system for the high pressure water media has been being looked for.

11.3.2 Radical Reactions

Thermal decomposition (pyrolysis) of organic molecules including biomass compounds is radical chain reaction. The contribution of pyrolysis is favored over 300 °C because thermal energy for homolytic decomposition of chemical bond in biomass can be gained at that temperature. If pyrolysis of biomass consists of the same elementary steps as that of hydrocarbons described above, the overall rate constant depends on the concentration. In this case, water acts as promoter or inhibitor by adjusting the concentration of biomass (so it is quite important to know phase behavior of biomass compound with water, but still lack of the knowledge about it). At higher temperature, main reaction pathway of biomass gasification is pyrolysis and the concentration of biomass in water is one of the key factors of the gasification. To enhance the radical decomposition, metal catalysts such as nickel, ruthenium, and so on, have been employed.

Both radical and ionic reactions for the same reactants are developed in parallel in high pressure high temperature water (HHW). One of the examples is glycerol conversion. As Bühler et al. [16] pointed out, glycerol conversion into acrolein was favored at high pressure (namely high water density) compared with that at low pressure (lower water density). To consider the reaction behavior, it was suggested that the rate of ionic reaction (acrolein formation) was enhanced with increasing water density (proton concentration), while the rate of radical reactions such as allyl alcohol formation was insensitive to water density. As shown in this example, in high pressure high temperature water (HHW), radical reaction can be also controlled by changing the contribution of ionic reaction with water density and temperature.

The other important radical reaction on biomass conversion is oxidation. For biomass conversion into useful compounds, partial oxidation is sometimes useful. There are several reports about hydrogen formation from biomass through partial oxidation [17, 18]. The oxidants are oxygen and water. To improve the selectivity of hydrogen formation, some catalysts were selected, for example, ruthenium oxide, zinc oxide, zirconium oxide, alkali hydroxide, and so on.

11.3.3 Small Review of Biomass Conversion Process

Based on the huge number of basic researches, biomass conversion in high pressure high temperature water (HHW) has been tried to industrialize. Here it is roughly categorized into several process, both from raw material-oriented point of view and from product-oriented point of view.

As repeatedly mentioned, biomass (plant biomass) typically consists of protein, carbohydrate (sugar, cellulose, and hemicellulose), lipid, polyphenol (including lignin), and mineral. The solubility and reactivity of the component are organized by temperature. Figure 11.2 shows the schematic diagram of biomass modification in high pressure high temperature water (HHW). Protein is hydrolyzed into amino acid and solubilized into water up to 200 °C. Sugar is dissolved in water and starts to decompose over 100 °C; isomerization and dehydration of sugar proceed up to 200 °C, retro-aldol condensation is favored over 200 °C, and gasification of sugar meaningfully occurs over 400 °C. Hemicellulose is hydrolyzed into sugars at around 200 °C, cellulose is dissolved in water at around 350 °C, and cellulose is significantly hydrolyzed over 350 °C. By using these two-step reactions (hemicellulose hydrolysis at 200–300 °C and cellulose hydrolysis at 350 °C or higher),

11 Review of Biomass Conversion in High Pressure High Temperature 257

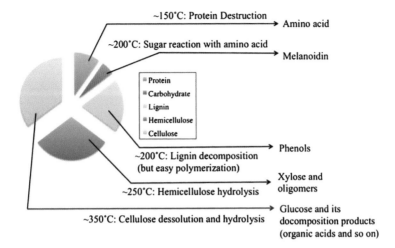

Fig. 11.2 Biomass modification in high pressure high temperature water (HHW)

a venture company starts business of sugars production from plant biomass [19]. Hydrolysis of lipid happens and lignin decomposition begins at 300 °C. Almost all water soluble minerals (Na, K, Ca, and so on) are dissolved in water up to 300 °C, while dissolution of water insoluble minerals such as Si requires high temperatures (over 400 °C at high water density).

Target material and compound from biomass are various but the appropriate condition for the target can be roughly categorized. Hot water has been used for extraction of ingredients of biomass and the reactive separation method (such as hydrolysis-sugar separation, as mentioned just above) by use of high pressure high temperature water (HHW) has been checked out [20]. Below 200 °C, separation of protein via hydrolysis would also be achieved.

Figure 11.3 shows three demonstrated large-scale processes for biomass conversion in high pressure high temperature water (HHW). Dehydration of carbohydrate is favored around 200 °C, cellulose is not hydrolysed at this temperature range, and thus carbonization of biomass is suited at around 200 °C. Recently, hydrothermal carbonization (HTC), which was invented by Dr. Bergius, the winner of Novel Prize in Chemistry, is rediscovered. Carbonaceous materials are widely applied for artificial coal, solid adsorbents, catalysts, templates of hollow metal oxides, battery materials, medical application, and so on. HTC is an energy efficient method for producing hydrochar from biomass because the process consists in the physicochemical conversion of biomass confined with water typically below critical point of water, 374 °C, under the self-generated pressure [21, 22]. Literature outlines a series of effect of raw materials on the characteristics of the hydrochars mainly. There have been many studies concerning carbonization of monosaccharides [23–27], cyclodextrins [28], cellulose [29], and some real biomass mainly consisted of carbohydrates [30–34]. The produced hydrochar in hydrothermal conditions can be used as a fuel [35], a catalyst after functionalization

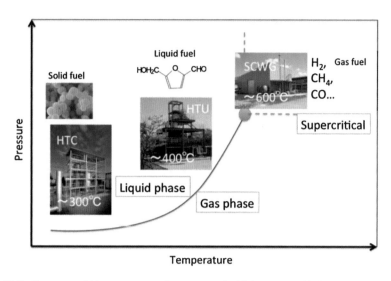

Fig. 11.3 Category of biomass conversion process in high pressure high temperature water (HHW)

[36–38], and template (or support) of hybrid materials [39–44]. In the methods of preparing these materials, hydrothermal carbonization is gaining increasing interest since the drying step can be eliminated and the process is energetically favorable compared with burning biomass. There have been huge numbers of basic researches and several venture companies that have begun business of the process [45–47].

In biomass refinery, chemical block such as furan, aldehyde, alcohol, carboxylic acid, phenols, and so on, which are useful for various industries, has to be provided from biomass. Sugar, which is a center compound in biomass refinery, can be converted into these useful chemicals in high pressure high temperature water (HHW) from 200 to 400 °C. Many researchers have studied to produce furan compound, mainly 5-hydroxylmethylfurfural (HMF), which is also considered to be precursor of hydrothermal carbonization, from sugars (typically glucose and fructose). It is well known that acid catalyst is effective to produce HMF from fructose [13, 14, 48]. It is found that microwave irradiation is useful for rapid production of HMF [14]. From glucose, HMF yield is not high because of low efficient in isomerization of glucose into fructose. In our laboratory, to overcome this issue, some base catalysts have been tested for glucose isomerization and some results will be introduced in the next section. Production of carboxylic acid (organic acid) from biomass has to be investigated widely. In the case of acid production, the starting material is not only sugars but also glycerol, which is byproduct of biodiesel production. Lactic acid can be produced from glucose with high yield in alkaline hydrothermal condition [49]. Lactic acid production from glycerol is much higher than that from glucose even at the similar reaction condition [50, 51]. Formic acid can be selectively formed from glucose via partial oxidation in the presence of base catalyst [52]. For organic acid production, we have also studied conversion of alginic acid, which is main component of sea algae and is carbohydrate having carboxylic group in constituent sugar unit (uronic acid) [53, 54]. Aldehyde such as acrolein is important chemical block for producing polymer and could be selectively obtained from glycerol [55]. Phenols can be obtained from lignin in principle because it consists of phenyl–propane units, however, due to polymerization of phenol unit, the yield of single-ring phenol is quite low. To improve the yield of phenols, high yield of double-ring phenol can be obtained by adding single-ring phenol compounds (such as cresol) as radical capping agency [56–58].

Liquid fuels can be produced from biomass through hydrothermal reaction in the presence of alkali at around 350 °C, which is known as hydrothermal upgrading (HTU), as shown in Fig. 11.3. Therefore, liquefaction of biomass (that is, production of water soluble compounds or liquid fuels) can be operated at around 200–400 °C, depending on the target compounds. Homogeneous and heterogeneous catalysts are employed to improve selectivity and the selection of suitable catalyst has been studied by many researchers [59].

To produce fuel gases from biomass, higher temperature (over 400 °C) is chosen and it is called supercritical water gasification (SCWG), as shown in Fig. 11.3. The appropriate temperature depends on the kind of biomass and the selected catalyst. Carbohydrate-rich biomass can be gasified at around 400 °C in the presence of noble metal catalysts and alkali catalyst [59, 60]. High content of lignin, which is resist compound for gasification, requires severe condition (over 300 °C) and active catalysts such as Ru metal. By employing partial oxidation, the condition for gasification can be moderated and glucose can be converted into hydrogen gas via formic acid formation by partial oxidation in the presence of base catalyst [61]. Still, there are quite a lot of basic researches for gasification and also many bench/pilot plants of biomass gasification in supercritical water have been tested around the world.

11.4 Recent Topics of Biomass Conversion in Our Laboratory

11.4.1 Experimental Procedures

Before explaining the experimental results, the experimental procedures we have typically employed are shown. We also conducted experiments of biomass conversion by a flow apparatus, but the results explained here were gained through the experiments conducted by batch reactors. Thus, we briefly introduce our batch experimental apparatus and procedures.

Below 200 °C, experiments by microwave heating were performed with the apparatus as shown in elsewhere [62]. Figure 11.4 shows the schematic diagram of

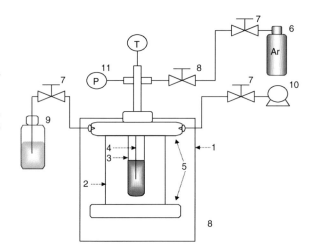

Fig. 11.4 Apparatus for carbonizing materials. *1* Microwave oven, *2* Polycarbonate outer tube, *3* Thick-walled glass reactor, *4* Thermocouple, *5* PEEK cap, *6* Ar gas, *7* Valve, *8* Valve for gas replacement, *9* Cooling water, *10* Vaccum pump, *11* Pressure indicator

the apparatus. The microwave is multimode (Shikoku Keisoku, model: μ-Reactor) and the reactor (inner volumes are 10–96 mL) was composed of an inner thick-wall Pyrex glass tube (Taiatsu Techne. Corporation, model: HPG series). The glass tube was covered with an outer polycarbonate tube and two PEEK screw caps with special seal joint were used to fix glass tube and a polycarbonate (PC) tube. The thermocouple was inserted into the reactor through a stainless steel sleeve. When the microwave irradiation was started, the reaction mixture could be heated up to a targeted temperature within 1 min. After the given reaction time, the reactor was cooled down (by introducing room temperature water to the space between outer and inner tube) to below 100 °C within 40 s.

Over 200 °C of reaction temperature, a stainless steel reactor was also used. The details of the reactor are as follows: a 1/2 inch tube (inner diameter was 8.5 mm) was constructed by two connectors at both ends; one of the ends was capped with a 1/2 inch cap, another was connected by 1/2-1/16 inch connector, and the end of 1/16 inch was capped with a 1/16 inch plug. 316-stainless steal mainly consists of Fe, Cr, Ni, and Mo. These metals have a catalytic effect on some reactions even in sub and supercritical water. To reduce catalytic effect of these metals, inner wall of reactor was oxidized to be covered with its oxide layer, that is, 3 g of 10 wt% of hydrogen peroxide aqueous solution was loaded in a reactor and the reactor was heated up to 400 °C for 30 min. After loading sample, water, and additive, Ar gas was used to purge air in the reactor. The reactor was submerged in a molten salt (50 % $NaNO_3$–50 % KNO_3) bath at controlled temperature. Heat-up time was around 90 s. After passing the expected reaction time, the reactors were taken out of the bath and rapidly cooled down to room temperature in a cold water bath within 1 min.

For both the cases, the reactor was opened after sampling of gaseous products and washed with water to recover the liquid products and solid (including heterogeneous additive). The recovered aqueous solution including solid material was filtered with a membrane to separate the solid from the liquid products.

Liquid products were analyzed by HPLC and TOC and yields were defined on a carbon basis. The shape of the carbonaceous materials was observed by SEM using a Hitachi S-4800 microscope. XPS analysis was carried out with a Shimazu AXIS-ULTRA spectrometer using monochromatic Al Kα radiation. Ultimate analysis is analyzed by CHNS corder.

11.4.2 Overview of Our Recent Researches

In our laboratory, gasification, liquefaction (including separation and chemical block formation), and hydrothermal carbonization have been investigated.

For gasification, we recently reviewed our experimental researches [61]. To achieve low economical and energetic cost for gasification process of wet biomass, we consider that effective route of biomass has to be designed. Partial oxidation with an appropriate catalyst at low temperature is our target. For several kind of biomass (glucose, glycerol, sugarcane bagasse, and microalgae), effectiveness of partial oxidation method was investigated and confirmed appropriateness of this method. As confirmed the product in partial oxidation of glucose, formic acid is selectively produced. We proposed two-step reaction of hydrogen formation from biomass; biomass is partially oxidized into formic acid at 200 °C and the produced formic acid is decomposed into hydrogen at 300 °C. To improve the process, it was confirmed that alkali catalyst for partial oxidation and ZnO for formic acid decomposition are effective.

For liquefaction (particularly chemical production), we have recently focused on multistep process of HMF formation from glucose in high pressure, high temperature. The first-step of HMF production from glucose is isomerization of glucose into fructose. Alkaline-earth metal carbonates are effective for isomerizing lactose [63]. We used alkaline-earth metal carbonate for glucose isomerization and studied catalyst and reaction conditions. Catalysts selected were $CaCO_3$, $MgCO_3$, and dolomite ($CaMg(CO_3)_2$). For comparison, ZrO_2 and TiO_2 acid–base solid catalysts were used. Solution and catalyst were loaded into glass reactors and heated in a microwave oven. Figure 11.5 shows fructose selectivity against glucose conversion at 160 °C for the catalysts studied and those from the literature [64]. In all cases, fructose selectivity decreased with increasing glucose conversion. For glucose conversions greater than 50 %, fructose selectivity was the highest with $CaCO_3$ and dolomite. At 50 % conversion, fructose selectivity was in the order of $MgCO_3$ > dolomite > $CaCO_3$ > acid–base hybrid catalyst [64]. Probably, this can be explained by isomerization being inhibited by acid sites and accelerated by base sites. Therefore, one of the most important factors for isomerization seems to be the ratio of base to acid sites. The $MgCO_3$, dolomite and $CaCO_3$ have about the same or higher selectivity compared with Souza et al. who studied NaOH or basic hybrid catalysts [64]. The $MgCO_3$, dolomite and $CaCO_3$

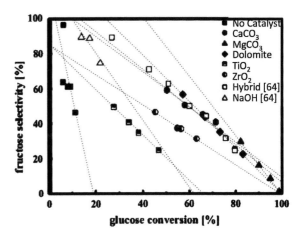

Fig. 11.5 Fructose selectivity for isomerization of glucose with solid base catalysts in water. (5 g aq. sol. 2 wt% glucose, cat. 0.1 g, 160 °C, 3–30 min rxn time)

solid base catalysts are readily available, have good processing characteristics and seem to be effective for glucose isomerization. Still we continue to study the isomerization (glucose into fructose) for the selection of catalyst and reaction condition in addition to optimization of the total process (fructose into HMF and total effectiveness).

In next sections, the recent results of carbonization of glucose and alkali lignin are roughly introduced as basic study on hydrothermal carbonization (HTC) process shown in Fig. 11.3.

11.4.3 Hydrothermal Carbonization of Glucose in the Presence of Sulfuric Acid

There have been a great number of reports regarding hydrothermal carbonization. While glucose carbonization proceeds via isomerization to fructose, and dehydration to HMF, there are few details on the yields that can be obtained from carbonaceous materials from hydrothermal carbonization. It is expected that acid addition increases carbonaceous material yield, because acid has the potential to promote glucose dehydration. In addition, studying surface functional groups is useful to apply those carbonaceous materials as catalysts and templates. In this work, we studied the carbonization of glucose in hydrothermal conditions and examined the effect of sulfuric acid on product yields and surface functional groups of the carbonized materials.

The yields and rate of reacted carbon of each product (glucose: Glc or glucose with sulfuric acid: Glc + H_2SO_4) are shown in Fig. 11.6. SEM images are shown in Fig. 11.7. As shown in Fig. 11.6, 10 % product yield was obtained for sample in the presence of sulfuric acid (Glc + H_2SO_4), which was 1.6 times higher than the sample without sulfuric acid (Glc), which had a yield of 6 %. The 60 % rate of reacted carbon in sample Glc + H_2SO_4 was 1.3 times higher than that obtained for

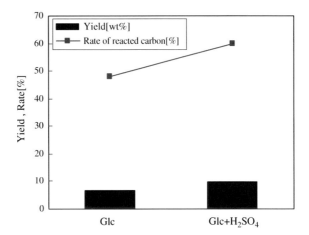

Fig. 11.6 Products yields and rate of reacted carbon (180 °C, 3 h, glucose: Glc or glucose with sulfuric acid: Glc + H_2SO_4)

sample Glc, which had a rate of 47 % (Fig. 11.6). Here, the rate of reacted carbon is the ratio of carbon in the produced hydrothermal char to the carbon in the reacted glucose.

Images of the products are shown in Fig. 11.7, where it can be seen that both products had spherical shapes. The particle size of sample Glc + H_2SO_4 was five times larger than that of sample Glc. It is possible that sulfuric acid might promote nucleation and growth according to the increased rate of reacted carbon and the larger particle sizes.

The surface functional groups were confirmed by XPS analysis (Fig. 11.8). Two different oxygen groups in the O 1 s region were observed: C–OH at 532.0 eV and –COOR at 533.5 eV. Sample Glc + H_2SO_4 exhibited a relatively lower C–OH peak intensity than that of sample Glc. In the C 1 s region, three different carbon groups were observed: C–C/C=C at 284.6 eV, C–O at 285.7 eV, and –COOR at 287.2 eV. Sample Glc + H_2SO_4 had relatively lower intensities for the C–C/C=C peaks than those of sample Glc. The O/C ratio from surface atomic ratio decreased from 30 to 24 % by adding sulfuric acid. From this result, it seems that sulfuric acid promoted dehydration and carbonization of glucose thus resulting in an increase of yield and particle size.

Research is still going to know the effects of conditions and additives on the yield and shape of carbonization.

11.4.4 Alkali Lignin Carbonization

Lignin structure is complex and constituted by different substituted phenyl–propane units bonded mainly by C–C or C–O–C. Remarkable chemical stability of lignin plays a major role in the biomass resistivity. In general, hydrothermal conversion of lignin produces solid, organic and/or aqueous liquid, and gas phases, the yields

Fig. 11.7 SEM images of carbonized glucose (180 °C, 3 h, glucose: Glc or glucose with sulfuric acid: Glc + H_2SO_4)

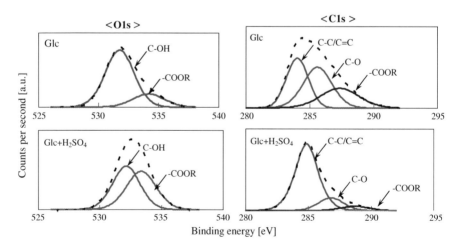

Fig. 11.8 XPS spectra of O 1 and C 1 s of carbonized materials (180 °C, 3 h, glucose: Glc or glucose with sulfuric acid: Glc + H_2SO_4)

depending on the operating parameters. In supercritical water, lignin is partially hydrolyzed and degraded producing energetic gases (low hydrocarbons, hydrogen, CH4, and so on), variable quantities of solids and a liquid phase containing different organic species (phenolic compounds, cresols, cathechol, and so on) [56–58]. The previous works suggested that carbonization of lignin in supercritical water progressed via a reaction between phenolic structure and active small molecule like formaldehyde [56, 57].

Here, alkali lignin carbonization in sub (300 and 350 °C) and supercritical water (400 °C) was performed by batch type reactors. To clarify the effect of water, carbonization of alkali lignin in the presence of water was compared with that in the absence of water (pyrolysis). The effects of reaction time and water loading on the carbonization were also investigated. Addition of formalin (consists of formaldehyde, methanol, and water) was also attempted to promote the carbonization.

The carbonization was analyzed by ultimate analyses (also van Krevelen analysis) and SEM observation.

Table 11.1 lists the experimental conditions with some experimental results. Alkali lignin (0.3 g of alkali lignin in the reactor) was carbonized with water (pure water or formalin aqueous solution) or without (pyrolysis), as indicated in Table 11.1. The experimental results were divided into three series, as listed in Table 11.1: series A is the effect of reaction time at 300 and 400 °C (in the presence of 1.0 g of water), series B is the effect of water loading at 300, 350 and 400 °C (for 30 min of reaction time), and series C is the effect of formalin at 300, 350, and 400 °C (for 30 min of reaction time). The run numbers of each experiment are indicated in Table 11.1.

Water loading was varied from 0 g (namely without water), 0.5, 1.0, 2.0 to 3.0 g (Run No. 1–20). The loaded density of water in the reactor, defined as mass of water/inner volume of the reactor (6 cm^3) were 83.33 kg/m^3 for 0.5 g, 166.67 kg/m^3 for 1.0 g, 333.33 kg/m^3 for 2.0 g and 500 kg/m^3 for 3.0 g of water loading. Reaction temperatures were 300, 350, and 400 °C. At subcritical water (300 and 350 °C in the presence of water), phase separation of water occurred. At 300 °C, saturated vapor was 47 kg/m^3 of density and liquid was 712 kg/m^3. Thus, some part of liquid water phase was formed at 300 °C (water pressure was 8.6 MPa). Also, at 350 °C, saturated water vapor density was 11 kg/m^3 and that liquid density was 580 kg/m^3. Therefore, when 1.0 g of water was loaded in the reactor (loaded density of water was 167 kg/m^3) the gas–liquid phase separation occurred and the system pressure was 16.5 MPa. At supercritical water condition (400 °C), water always forms single phase and pressure in the reactor changed corresponding to water density (18 MPa at 83 kg/m^3, 25 MPa at 167 kg/m^3, 30 MPa at 330 kg/m^3, and 37 MPa at 500 kg/m^3).

In the presence of formalin (Run No. 21–23), 1.0 g of formalin aqueous solution was loaded and then the amount of water was considered to be 0.6 g based on the concentration of formaldehyde (about 35 wt%) and methanol (about 5 wt%).

Reaction time was basically 30 min (Run No. 3, 8, and 11–23), but in the presence of 1.0 g of water (Run No. 1–10), reaction time ranged from 10 min to 50 min, as listed in Table 11.1.

Table 11.1 Experimental condition and results

T, °C	Run No.	Reaction time, min	Solid, wt%	C, wt%	H, wt%	N, wt%	S, wt%	O, wt%	O/C[a]	H/C[a]
(A) Effect of reaction time (water loading: 1.0 g)										
300	1	10	43.1	70.9	5.85	0.26	1.08	22.0	0.23	0.99
	2	20	41.5	71.8	5.72	0.25	0.89	21.3	0.22	0.96
	3	30	39.6	72.8	5.71	0.26	0.94	20.3	0.21	0.94
	4	40	38.4	73.4	5.75	0.25	0.74	19.9	0.20	0.94
	5	50	39.5	74.0	5.64	0.28	0.91	19.2	0.19	0.91
400	6	10	30.1	79.6	4.92	0.23	0.7	14.6	0.14	0.744
	7	20	31.3	83.2	4.90	0.23	0.65	11.0	0.10	0.71
	8	30	38.2	84.4	4.82	0.21	0.68	9.88	0.09	0.69
	9	40	30.0	83.0	4.83	0.19	0.68	11.4	0.10	0.70
	10	50	30.1	83.7	4.79	0.2	0.58	10.7	0.10	0.69
(B) Effect of water loading (reaction time: 30 min)										
300	11	0	55.8	62.7	4.53	0.13	0.45	32.2	0.39	0.87
	12	0.5	43.9	72.2	5.33	0.28	0.74	21.5	0.22	0.89
	13	2.0	30.9	75.6	6.08	0.27	0.88	17.2	0.17	0.96
	14	3.0	36.9	74.2	5.89	0.26	0.91	18.7	0.19	0.95
350	15	0	53.7	66.0	3.91	0.14	0.42	29.5	0.34	0.71
	16	1.0	37.1	76.6	4.80	0.25	0.52	17.8	0.17	0.75
400	17	0	30.1	74.7	4.55	0.16	0.56	20.1	0.20	0.73
	18	0.5	31.3	82.7	4.59	0.23	0.74	11.7	0.11	0.67
	19	2.0	30.0	83.3	5.09	0.24	0.76	10.7	0.10	0.73
	20	3.0	30.1	85.2	5.86	0.22	0.72	8.0	0.07	0.83
(C) Effect of Formalin (Reaction time: 30 min, Formalin: 1.0 g)										
300	21	0.6	72.8	73.5	6.51	0.13	0.76	19.1	0.19	1.06
350	22		70.2	74.5	5.15	0.12	0.56	19.7	0.20	0.83
400	23		58.1	75.9	1.25	0.09	0.49	22.2	0.22	0.20

[a] Atomic ratio

The effect of reaction time at 300 and 400 °C in the presence of 1.0 g of water (namely series A, Run No. 1–10, in Table 11.1) is firstly mentioned. As seen in Table 11.1 (at series A), the solid yields at 400 °C was 5–10 % (Run No. 6–10) lower than those at 300 °C (Run No. 1–5). With increasing reaction temperature, the amount of generated hydrochar was reduced and the amounts of H and O decreased. The amount of N after the reaction was around 0.2 wt% regardless of the reaction condition and it was higher than that before the reaction (ultimate analysis of alkali lignin is as follows: C: H: N: S: O [wt%] = 51.4 : 4.87 : 0.096 : 2.43 : 41.18). On the other hand, the amount of S was sensitive to the reaction conditions (reaction time and temperatures) and it decreased with increasing the severity (reaction time and temperature).

The experiments of series B in Table 11.1 (Run No. 11–20) were conducted to investigate the effect of the amount of water on the hydrochar formation at the temperatures (300, 350, and 400 °C) for 30 min. The experiments of Run No. 3 and 8 in series A were also noticed because these experiments were also conducted for 30 min in the presence of 1.0 g of water. Compared with the experiments in the absence of water (namely pyrolysis) at different reaction temperatures (Run No. 11, 15, and 17), the hydrochar yields were drastically reduced by adding water. The hydrochar yield was not sensitive to water loading in supercritical water (Run No. 8, 18–20). The amounts of H, C, and O in the absence of water were higher than those in the presence of water at all the reaction temperatures. With increasing water loading, O content increased, C increase, and H slightly increased. The amount of N decreased after the reaction, while S decreased, at all the reaction conditions.

In the presence of formalin (series C in Table 11.1: Run No. 21–23), the hydrochar yields at all the reaction time were remarkably higher compared with the other reaction conditions (series A and B). One of the reasons of the highest hydrochar yield is probably because of the existence of formalin (formaldehyde and methanol). Some part of formaldehyde in the formalin would react with lignin fragment to form the hydrochar or its precursor, as proposed in the literature [56–58]. This probably resulted in the decrease of N and S contents in the solid (see in Table 11.1) because formaldehyde has no N and S.

As seen in all the experiments, N in the hydrochar was higher than that in the raw material but S was lower. This was probably due to its chemical stability. N in lignin was strongly bonded in the lignin structure and remained in the hydrochar. On the other hand, S was physically absorbed or weekly bonded as sulfo group and was easily detached from lignin structure into the solution.

Figures 11.9 and 11.10 show van Krevelen diagram of the hydrochar obtained at all the experiments. In these figures, directions of dehydration, decarboxylation, and demethanisation processes were also indicated as straight lines. The arrows of dehydration, decarboxylation, and demethanisation were also shown in Fig. 11.9 and they indicate the dominant direction of these reactions. In general, van Krevelen diagram allows considering qualitatively reaction tendencies.

In Fig. 11.9, the temperature dependence of the produced hydrochar at each reaction condition for 30 min was plotted. The raw material was also plotted.

Fig. 11.9 Effect of additive on van Krevelen diagram for 30 min of reaction

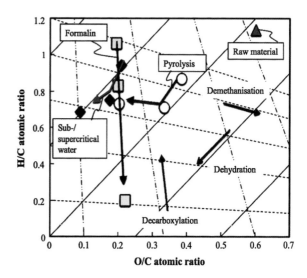

The three different reaction atmospheres are drawn: pyrolysis (namely carbonization in the absence of water at 300, 350, and 400 °C: Run No. 11, 15, and 17, respectively), sub/supercritical water (carbonization in the presence of 1.0 g of water at 300, 350, and 400 °C: Run No. 3, 16, and 8, respectively), and formalin (in the presence of formalin at 300, 350, and 400 °C: Run No. 21, 22, and 23, respectively).

In Fig. 11.9, the bold arrows with the plots show the increase in temperature at each reaction atmosphere. From Fig. 11.9, dehydration was dominant for pyrolysis and sub/supercritical water cases, while reverse reaction of demethanisation (namely C–C bond formation) seemed to be main reaction pathway for formalin case. The pyrolysis process was probably controlled by dehydration up to 350 °C and, at 400 °C, decarboxylation appears to be predominant at some degree. For sub/supercritical water in Fig. 11.9, at 300 °C, significant reduction of the O/C occurred at constant H/C. The similar results were reported in a previous study [65]. Lignin (H/C atomic ratio was 1.09 and O/C was 0.30) was hydrothermally carbonized at temperature ranges from 180 to 250 °C (lower temperature than in this study) for 4 or 17 h (longer reaction time compared with this study) [65]. The H/C of the produced hydrochar were not so large different (0.97–1.05) but the O/C was largely lower (0.14–0.21), respectively. In this case, it was considered that deoxygenation (for example, by decarboxylation, decarbonylation, or C–C bond formation) was developed to form the hydrochar. In our study, at higher temperature (300 °C), deoxygenation was dominant but, at the higher temperature (350 and 400 °C), dehydration became main pathway to form hydrochar by condensation of polyaromatic structure between the aromatic rings. In the case of formalin, with increasing temperature, the H/C value drastically decreased while O/C was almost constant. It was possibly due to disappearance of methyl group in the hydrochar or its precursor.

Fig. 11.10 Effect of reaction time and water loading on van Krevelen diagram

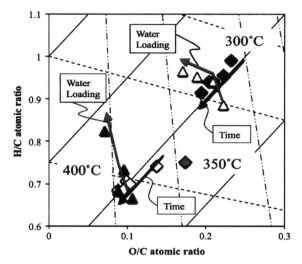

Figure 11.10 was enlarged plot at 0–0.3 of O/C atomic ratio (X axis) and 0.6–1.1 of H/C atomic ratio (Y axis). In this figure, the reaction time dependence (Run No. 1–10) and the water loading dependence (Run No. 12, 13, 16, and 18–20) were plotted. For the reaction time dependence, the arrow indicates the prolongation of reaction time. Concerning the effect of water loading, the arrow shows the climb of water loading. Figure 11.10 tells us the reaction time dependency at 300 and 400 °C. The major reaction pathway seems to be the dehydration of lignin leading to the hydrochar with a high composition in C. Dehydration gradually evolved with the reaction time at 300 °C while the effect of reaction time was not remarkable at 400 °C.

Figure 11.10 also shows the effect of water loading on reaction pathway. With the increase of water loading at 300 and 400 °C, the O/C ratio decreased compared with the H/C ratio. The influence of the water loading was higher in supercritical condition (at 400 °C) that at 300 °C. The sensitive reaction pathway to increase water loading was probably detachment of some carboxyl or carbonyl groups would be promoted at higher water loading.

The shape of the hydrochar was quite sensitive to reaction atmosphere (namely pyrolysis, sub/supercritical water, or formalin) but not to reaction temperature or time. Here we only show typical SEM images at each reaction atmosphere. Figure 11.11 shows the SEM images of the hydrochars obtained at pyrolysis at 300 °C, subcritical water (1.0 g of water) at 300 °C, supercritical water (3.0 g of water) at 400 °C, and formalin at 400 °C. For all the cases, reaction time was 30 min. The hydrochar of pyrolysis was large shred-type pieces and sponge-like rubbles, which were a mass of irregular particles of different sizes, more or less coalescent. As well as pyrolysis, the SEM image of the hydrochar obtained in the presence of formalin shows shred-type pieces and sponge-like rubbles. The structure of the hydrochar obtained in subcritical water (300 °C) condition was

Fig. 11.11 Typical SEM photographs at different conditions for 30 min of reaction

agglomerates of microspheres of different sizes. In supercritical water at higher water loading, the hydrochar was microspheres with a smaller size distribution and locally coalescent.

Based on these experimental results, the carbonization mechanism of alkali lignin is discussed from chemical and physical point of view. From chemical point of view (from Table 11.1, Figs. 11.9 and 11.10), water promotes C–C bond formation with the formation of polyaromatic ring structure in the hydrochar and higher water loading assists further to remove oxygen atom via carbonyl or carboxyl group elimination to increase aromaticity further. Formaldehyde in formalin helps to form C–C bond such as methyl or methylene group in the hydrochar and the alkyl groups decompose at higher temperature. From physical point of view (from Fig. 11.11), phase behavior of the precursor of the solid products influences the shape of the produced solid. The structures of carbonaceous materials from soluble biomass were typically spherical microparticles shown in this chapter already and the literature [66]. From the SEM photograph, in water, spherical solids were formed and smaller particles increased with increasing water loading. On the other hand, shred-type pieces were formed at pyrolysis and sponge-shape solid was produced at formalin experiment. Rahmani et al. [67] reported that a shape of coke produced from heavy crude oil was droplet (like sphere particle) or coalescence structure according to phase behavior of the precursor. When the coke precursor forms droplet in a dispersion media, the produced solid (coke) has spherical shape. This mechanism would be adopted the hydrochar formation in sub and supercritical water. Alkali lignin formed small droplets (as see in Fig. 11.11) and dehydration and polymerization progressed to form hydrochar and thus the formed solid was droplet-like sphere. On the other hand, the hydrochar precursor itself is a dispersion media (so-called phase inversion), for example, dense liquid phase containing small molecule as solute, and the formed hydrochar has coalescence or sponge-like structure because the solute in the hydrochar precursor could be escaped after hydrochar formation. Without water or in the presence of formalin, alkali lignin thermally converted into polycondensated material and forms liquid phase by itself at the bottom of reactor with some small molecules (by-products, methanol in formalin, water, and so on) as solutes. The solute in the solid went away to the rinsing with solvent when the hydrochar was recovered from the reactor and thus the final carbonaceous material had the sponge-like shape.

11.5 Conclusion

In this chapter, we explained some unique features of high pressure media and high pressure high temperature water (HHW). Due to favorable for ionic reaction at liquid phase (or high density of water), carbonization, and liquefaction has been studied at this temperature and pressure region (subcritical water below 400 °C). On the other hand, over 400 °C, radical reactions are predominant and thus gasification (small fragment formation) has been investigated. In addition to physical properties of water, catalyst and phase behavior are key points for developing highly controlled biomass process and our recent experimental results (isomerization and carbonization) show the point clearly. To understand and control these reactions in more detail, we have to know the reactivity (concentration of reactant, physical properties of water, and catalyst) and phase behavior.

References

1. Watanabe M, Kato S, Ishizeki S, Inomata H, Smith RL Jr (2010) Heavy oil upgrading in the presence of high density water: Basic study. J Supercrit Fluids 53:48–52
2. Savage PE (1999) Organic chemical reactions in supercritical water. Chem Rev 99:603–622
3. Adschiri T, Shibata R, Sato T, Watanabe M, Arai K (1998) Catalytichydrodesulphurization of dibenzothiophene through partial oxidation and a water-gas shift reaction in supercritical water. Ind Eng Chem Res 37:2634–2638
4. Arai K, Adschiri T, Watanabe M (2000) Hydrogenation of Hydrocarbons through Partial Oxidation in Supercritical Water. Ind Eng Chem Res 39:4697–4701
5. Akiya N, Savage PE (2002) Roles of water for chemical reactions in high-temperature water. Chem Rev 102:2725–2750
6. Uematsu M, Franck EU (1980) J Phys Chem Ref Data 9:1291–1306
7. Marrone PA, Arias TA, Peters WA, Tester JW (1998) Solvation effects on kinetics of methylene chloride reactions in sub- and supercritical water: theory, experiment, and Ab initio calculations. Phys Chem. A 102:7013–7028

8. Salvatierra D, Taylor JD, Marrone PA, Tester JW (1999) Kinetic study of hydrolysis of methylene chloride from 100 to 500 °C. Ind Eng Chem Res 38:4169–4174
9. Sasaki M, Kabyemela B, Malaluan R, Hirose S, Takeda N, Adschiri T, Arai K (1998) Cellulose hydrolysis in subcritical and supercritical water. J. Supercrit. Fluid 13:261–268
10. Marshall WL, Franck EU (1981) J Phys Chem Ref Data 10:295–304
11. Tomita K, Oshima Y (2004) Enhancement of the catalytic activity by an ion product of sub and supercritical water in the catalytic hydration of propylene with metal oxide. Ind Eng Chem Res 43:2345–2348
12. Tomita K, Koda S, Oshima Y (2002) Catalytic hydration of propylene with MoO_3/Al_2O_3 in supercritical water. Ind Eng Chem Res 41:3341–3344
13. Watanabe M, Aizawa Y, Iida T, Aida TM, Levy C, Sue K, Inomata H (2005) Glucose reactions with acid and base catalysts in hot compressed water at 473 K. Carbohydr Res 340:1925–1930
14. Qi X, Watanabe M, Aida TM, Smith RL Jr (2008) Catalytical conversion of fructose and glucose into 5-hydroxymethylfurfural in hot compressed water by microwave heating. Catal Commun 9:2244–2249
15. Nakajima K, Baba Y, Noma R, Kitano M, Kondo JN, Hayashi S, Hara M (2011) $Nb_2O_5 \cdot nH_2O$ as a heterogeneous catalyst with water-tolerant lewis acid sites. J Am Chem Soc 133:4224–4227
16. Bühler W, Dinjus E, Ederer EJ, Kruse A, Mas C (2002) Ionic reactions and pyrolysis of glycerol as competing reaction pathways in near- and supercritical water. J Supercrit Fluids 22:37–53
17. Watanabe M, Inomata H, Osada M, Sato T, Adschiri T, Arai K (2003) Catalytic effects of NaOH and ZrO_2 for partial oxidative gasification of n-hexadecane and lignin in supercritical water. Fuel 82:545–552
18. Yoshida T, Oshima Y (2004) Partial oxidative and catalytic biomass gasification in supercritical water: a promising flow reactor system. Ind Eng Chem Res 43:4097–4104
19. http://renmatix.com/
20. Ohira H, Torii N, Aida TM, Watanabe M, Smith RL Jr (2009) Rapid separation of shikimic acid from Chinese star anise (Illicium verum Hook. f.) with hot water extraction. Sep Purif Technol 69:102–108
21. Titirici MM, Thomas A, Antonietti M (2007) Back in the black: hydrothermal carbonization of plant material as an efficient chemical process to treat the CO_2 problem? New J Chem 31:787–789
22. Hu B, Wang K, Wu L, Yu SH, Antonietti M, Titirici MM (2010) Engineering carbon materials from the hydrothermal carbonization process of biomass. Adv Mater 22:813–828
23. Wang Q, Li H, Chen L, Huang X (2001) Monodispersed hard spherules with uniform nanopores. Carbon 39:2211–2214
24. Yao C, Shin Y, Wang LQ, Windisch W Jr, Samuels WD, Arey BW, Wang C, Risen WM Jr, Exarhos GJ (2007) Hydrothermal dehydration of aqueous fructose solutions in a closed system. J Phys Chem C 111:15141–15145
25. Mi Y, Hu W, Dan Y, Liu Y (2008) Synthesis of carbon micro-spheres by a glucose hydrothermal method. Mater Lett 62:1194–1196
26. Baccile N, Laurent G, Babonneau F, Fayon F, Titirici MM, Antonietti M (2009) Structural characterization of hydrothermal carbon spheres by advanced solid-state MAS 13C NMR investigation. J Phys Chem C 113:9644–9654
27. Sevilla M, Fuertes AB (2009) Chemical and structural properties of carbonaceous products obtained by hydrothermal carbonization of saccharides. Chem Eur J 15:4195–4203
28. Shin Y, Wang LQ, Bae IT, Arey BW, Exarhos GJ (2008) Hydrothermal syntheses of colloidal carbon spheres from cyclodextrins. J Phys Chem C 112:14236–14240
29. Sevilla M, Fuertes AB (2009) The production of carbon materials by hydrothermal carbonization of cellulose. Carbon 47:2281–2289

30. Titirici MM, Thomas A, Yu SH, Mueller JO, Antonietti M (2007) A direct synthesis of mesoporous carbons with bicontinuous pore morphology from crude plant material by hydrothermal carbonization. Chem Mater 19:4205–4212
31. Heilmann SM, Jader LR, Sadowsky MJ, Schendel FJ, von Keitz MG, Valentas KJ (2011) Hydrothermal carbonization of distiller's grains. Biomass Bioenerg 35:2526–2533
32. Falco C, Caballero FP, Babonneau F, Gervais C, Laurent G, Titirici MM, Baccile N (2011) Hydrothermal carbon from biomass: structural differences between hydrothermal and pyrolyzed carbon via 13C solid state NMR. Langmuir 27:14460–14471
33. Heilmann SM, Davis HT, Jader LR, Lefebvre PA, Sadowsky MJ, Schendel FJ, von Keitz MG, Valentas KJ (2010) Hydrothermal carbonization of microalgae. Biomass Bioenerg 34:875–882
34. Sevilla M, Macia-Agullo JA, Fuertes AB (2011) Hydrothermal carbonization of biomass as a route for sequenstration of CO2: chemical and structural properties of the carbonized products. Biomass Bioenerg 35:3152–3159
35. Paraknowitsch JP, Thomas A, Antonietti M (2009) Carbon colloids prepared by hydrothermal carbonization as efficient fuels for indirect carbon fuel cells. Chem Mater 21:1170–1172
36. Demir-Cakan R, Baccile N, Antonietti M, Titirici MM (2009) Carboxylate-rich carbonaceous materials via one-step hydrothermal carbonization of glucose in the presence of acrylic acid. Chem Mater 21:484–490
37. Demir-Cakan R, Makowski P, Antonietti M, Goettmann F, Titirici MM (2010) Hydrothermal synthesis of imidazole functionalized carbon spheres and their application in catalyst. Catal Today 150:115–118
38. Xiao H, Guo Y, Liang X, Qi C (2010) One-step synthesis of novel biacidic carbon via hydrothermal carbonization. J Solid State Chem 183:1721–1725
39. Sun X, Li Y (2004) Colloidal spheres and their core/shell structures with noble-metal nanoparticles. Angew Chem Int Ed 43:597–601
40. Yu SH, Cui X, Li L, Li K, Yu B, Antonietti M, Coelfen H (2004) From starch to metal/carbon hybrid nanostructures: hydrothermal metal-catalyzed carbonization. Adv Mater 16:1636–1640
41. Qian HS, Yu SH, Luo LB, Gong JY, Fei LF, Liu XM (2006) Synthesis of uniform Te@carbon–rich composite nanocables with photoluminescence properties and carbonaceous nanofibers by the hydrothermal carbonization of glucose. Chem Mater 18:2102–2108
42. Cui X, Antonietti M, Yu SH (2006) Structural effects of iron oxide nanoparticles and iron ions on the hydrothermal carbonization of starch and rice carbohydrates. Small 2:756–759
43. Titirici MM, Antonietti M, Thomas A (2006) A generalized synthesis of metal oxide hollow spheres using hydrothermal approach. Chem Mater 18:3808–3812
44. Yu J, Yu X (2008) Hydrothermal synthesis and photocatalytic activity of zinc oxide hollow spheres. Environ Sci Technol 42:4902–4907
45. http://www.ava-co2.com/web/pages/en/home.php?lang=EN
46. http://www.suncoal.de/en
47. http://www.terranova-energy.com/en/home.php
48. Qi X, Watanabe M, Aida TM, Smith RL Jr (2008) Catalytic dehydration of fructose into 5-hydroxymethylfurfural byion-exchange resin in mixed-aqueous system by microwave heating. Green Chem 10:799–805
49. Yan X, Jin F, Tohji K, Moriya T, Enomoto H (2007) Production of lactic acid from glucose by alkaline hydrothermal reaction. J Mater Sci 42:9995–9999
50. Kishida H, Jin F, Zhou Z, Moriya T, Enomoto H (2005) Conversion of glycerin into lactic acid by alkaline hydrothermal reaction. Chem Lett 34:1560–1561
51. Shen Z, Jin F, Zhang Y, Wu B, Kishita A, Tohji K, Kishida H (2009) Effect of alkaline catalysts on hydrothermal conversion of glycerin into lactic acid. Ind Eng Chem Res 48:8920–8925
52. Jin F, Yun J, Li GM, Kishita A, Tohji K, Enomoto H (2008) Hydrothermal conversion of carbohydrate biomass into formic acid at mild temperatures. Green Chem 10:612–615

53. Aida TM, Yamagata T, Watanabe M, Smith RL Jr (2010) Depolymerization of sodium alginate under hydrothermal conditions. Carbohydr Polym 80:296–302
54. Aida TM, Yamagata T, Abe C, Kawanami H, Watanabe M, Smith RL Jr (2012) Production of organic acids from alginate in high temperature water. J Supercrit Fluids 65:39–44
55. Watanabe M, Iida T, Aizawa Y, Aida TM, Inomata H (2007) Acrolein synthesis from glycerol in hot-compressed water. Bioresour Technol 98:1285–1290
56. Saisu M, Sato T, Watanabe M, Adschiri T, Arai K (2003) Conversion of lignin with supercritical water-phenol mixtures. Energy Fuels 17:922–928
57. Okuda K, Man X, Umetsu M, Takami S, Adschiri T (2004) Efficient conversion of lignin into single chemical species by solvothermal reaction in water-p-cresol solvent. J Phys: Condens Matter 16:S1325–S1330
58. Okuda K, Umetsu M, Takami S, Adschiri T (2004) Disassembly of lignin and chemical recovery—rapid depolymerization of lignin without char formation in water–phenol mixtures. Fuel Process Technol 85:803–813
59. Peterson AP, Vogel F, Lachance RP, Froling M, Antal MJ Jr, Tester JW (2008) Thermochemical biofuel production in hydrothermal media: A review of sub- and supercritical water technologies. Energy Environ Sci 1:32–65
60. Osada M, Sato T, Watanabe M, Shirai M, Arai K (2006) Catalytic gasification of wood biomass in subcritical and supercritical water. Combust Sci Tech 178:537–552
61. Watanabe M, Aida TM, Smith RL Jr, Inomata H (2012) Hydrogen Formation from Biomass Model Compounds and Real Biomass by Partial Oxidation in High Temperature High Pressure Water. J Jpn Petrol Inst 55:219–228
62. Watanabe M, Qi X, Aida TM, Smith RL Jr (2012) Microwave apparatus for kinetic studies and in situ observations in hydrothermalor high-pressure ionic liquid system. In: The Development and application of microwave heating, InTech, 2012. http://dx.doi.org/10.5772/45625
63. Lecomte J, Finiels A, Moreau C (2002) Kinetic study of the isomerization of glucose into fructose in the presence of anion-modified hydrotalcites. Starch 54 (2002):75–79
64. Souza ROL, Fabiano DP, Feche C, Rataboul F, Cardoso D, Essayem N (2012) Glucose–fructose isomerisation promoted by basic hybrid catalysts. Catal Today 195:114–119
65. Dinjus E, Kruse A, Troger N (2011) Hydrothermal carbonization 1. Influence of lignin in lignocelluloses. Chem Eng Technol 34:2037–2043
66. Titirici MM, White RJ, Falco C, Sevilla M (2012) Black perspectives for a green future: hydrothermal carbons for environment protection and energy storage. Energy Environ Sci 5:6796–6822
67. Rahmani S, McCaffrey W, Elliott JA, Gray MR (2003) Liquid-phase behavior during the cracking of asphaltenes. Ind Eng Chem Res 42:4101–4108

Chapter 12
Hydrothermal Carbonization of Lignocellulosic Biomass

Charles J. Coronella, Joan G. Lynam, M. Toufiq Reza and M. Helal Uddin

Abstract Hydrothermal carbonization (HTC) of lignocellulosic biomass is a pretreatment process to homogenize and densify diverse biomass feedstocks. The solid product is hydrophobic and friable with ultimate analysis similar to that of lignite, and is easily made into durable, dense pellets. Byproducts include aqueous sugars, acids, carbon dioxide, and water. The process consists of treatment in hot (180–280 °C) compressed water for short contact times, and has been demonstrated on woody biomass, agricultural residues, and grasses. HTC reactions include hydrolysis, dehydration, decarboxylation, condensation, polymerization, and aromatization. Nearly all hemicellulose is removed and converted to simple sugars and furfural. Cellulose begins to react at 200 °C, and produces oligosaccharides, glucose, 5-HMF, and organic acids. Lignin is relatively inert. HTC reactions are relatively fast, with reaction times measured in minutes. Both hemicellulose and cellulose degrade by apparent first-order reaction kinetics, where hemicellulose exhibits an activation energy of 30 kJ mol^{-1}, and that of cellulose is 73 kJ mol^{-1}. There has been a flurry of research on HTC published recently, but little commercial activity. Innovative design is required for commercialization, and costs may be high, due to high pressure operation. However, as demand for biomass increases, HTC will surely play a role in enhancing supply chain logistics.

12.1 Introduction

Progressing toward sustainable, renewable energy production is essential for our world in the next century. Fossil fuels, with their irreversible generation of greenhouse gases, cannot provide us with energy indefinitely. Biomass, which

C. J. Coronella (✉) · J. G. Lynam · M. T. Reza · M. H. Uddin
Chemical and Materials Engineering Department, University of Nevada, MS 170, Reno, NV 89557, USA
e-mail: coronella@unr.edu

takes in CO_2 as it grows and uses sunlight as its power source, can play a pivotal role in weaning us from fossil fuels.

Biomass encompasses many types of plant materials. To avoid the food-versus-fuel conflict, only nonfood lignocellulosic biomass is considered appropriate for sustainable energy production. Such biomass includes woody biomass, grassy biomass, and agricultural residues from food production. Lignocellulosic biomass is a many-layered package including lignin, cellulose, hemicellulose, extractives, and inorganics (ash). Converting lignocellulosic biomass to useful forms of fuel presents challenges in handling, storage, and processing. As harvested (often on a seasonal basis), hydrophilic biomass has a high moisture content often rendering it perishable, as well as heavy and thus expensive to transport. Drying can improve storage properties, but is energy intensive. A process to make biomass more hydrophobic would be advantageous. In addition to having high moisture content, biomass is low in energy density compared to fossil fuels. Lignocellulosic biomass has a higher heating value (HHV) of 14–21 MJ/kg, while coal's HHV ranges from 23–28 MJ/kg, and crude oil has an HHV of roughly 45 MJ/kg [1, 2]. Increasing the energy density of biomass would make its use more practical, and would facilitate supply chain logistics. Liberating the cellulose monomer glucose from the lignocellulosic bundle would allow its use in fermentation to ethanol, presently widely added to gasoline.

Applying HTC to biomass overcomes the challenges posed by biomass conversion. HTC makes biomass hydrophobic, enhancing its drying and storage properties, increases its energy density, and releases glucose and other sugar monomers into an aqueous phase, permitting their subsequent upgrading to chemicals and fuels [3–8]. Immersing biomass in liquid water, a treatment especially amenable to moist biomass feedstocks, at temperatures of 180–280 °C, at pressures of 1–5 MPa for as little as 5 min can dramatically improve the fuel properties of the biomass, as well as convert hemicellulose and cellulose to available sugar monomers. The HTC process has the potential to give biomass a significant role in reducing the use of fossil fuels.

12.2 Background

HTC is also known as hydrothermal carbonification, coalification, hydrothermal coalification, hydrothermal treatment, hot-compressed-water hydrolysis, liquid hot water pretreatment, aqueous pyrolysis, aqueous phase carbonization, hydrothermolysis, and wet torrefaction. Its solid product can be called hydrochar, char, HTC char, biochar, bio-coal, or biocarbon. This multiplicity of names for the same process and its solid product can make finding useful data in the literature challenging. HTC produces a liquid byproduct which contains sugar monomers and their decomposition products, and organic acids, as well as CO_2 gas.

Friedrich Bergius [9, 10] first investigated what he called hydrothermal carbonification of biomass or its components in 1913. By heating cellulose in water under pressure, his team obtained a black, coal-like substance. Further research and dispute about whether cellulose or lignin contributed more to "coalification" continued until the 1980s, with the van Krevelen diagram for biomass and coal being the most useful concept produced [11–17]. Dirk Willem van Krevelen evaluated biomass and coal by plotting hydrogen:carbon atomic ratios as a function of the oxygen:carbon atomic ratios, with points closer to the graph origin representing higher fuel value. Thus, the effectiveness of HTC could be determined in terms of fuel enhancement [18–20]. Niesner et al. [21] also pioneered an effective method using HPLC to identify the hydrothermal degradation products of hemicellulose and cellulose from wheat straw.

In the 1980s, the National Renewable Energy Laboratory (NREL) sponsored a Lawrence Berkeley Laboratory project that used conditions similar to HTC, but since liquefaction was the object of the study, the project was discontinued [22, 23]. In 1982, Fuel published Herman Ruyter's work, which emphasized "coalification's" removal of CO_2 and H_2O to improve fuel value and increase hydrophobicity. Ruyter reported results for various biomass from autoclave work, where temperature was varied from 175 to 350 °C, with pressure being the vapor pressure of water at these temperatures. He called attention to the complexity of the mechanisms involved, but nevertheless attempted to quantify the "enrichability" of a biomass as a function of O/C ratio, HHV, and mass yield, the ratio of biochar produced to raw feedstock [24]. In 1983, Bonn et al. [25] described their use of a 2-stage hydrothermolytic process used on poplar wood and wheat straw. The first stage at 180 °C removed the easily hydrolysable polysaccharides (likely hemicelluloses), while the second stage at 260 °C hydrolyzed cellulose to glucose. The overall reaction time was less than 1 h. Schwalde et al. [26] reported their use of HPLC and gel permeation chromatography in determining more accurately the carbohydrates produced from HTC of cotton waste materials. To simplify the complicated reaction scheme of HTC, researchers in the late 1980s and early 1990s performed high temperature/high pressure studies on pure components of biomass, such as cellulose or lignin [27–29]. However, the matrix nature of lignocellulosic biomass prevented easy answers to the complexities of HTC.

In recent years, HTC has become a process more widely recognized to enhance biomass' energy density, friability for homogenized handling characteristics of varying biomass, and hydrophobity for improved storage and transportation properties. In addition, sugar monomer production (at HTC temperatures less than 230 °C) provides a valuable byproduct for possible conversion to liquid biofuels. Practices of HTC vary, but all such processes use subcritical water at temperatures above 150 °C, below which reaction does not substantially proceed. HTC temperatures of 180–280 °C appear to be most practical. Pressures above the water vapor pressure at a given reaction temperature must be maintained, so that the water remains in its liquid state. The biomass feedstock must be submerged in the liquid water, rather than floating above it, with subcritical liquid water serving as solvent, reactant, and catalyst. Reaction times vary from as short as 1 min to as long as 72 h. Water to biomass (w/w) ratios of 5:1–20:1 may be used. Batch processing and continuous processing have been used, with the latter considered economically feasible at large scale. Two-stage processes, which use a lower temperature to obtain sugar monomers, and then a higher one to encourage biochar formation have been proposed. For the HTC solid product, biochar, the main process outputs of concern are mass yield, the ratio of biochar produced to raw feedstock, and energy densification, the ratio of the HHV of the biochar to the HHV of the raw feedstock. Optimization of process conditions for HTC, particularly for specific biomass, remains an active area of research. The knowledge thus gained will be a helpful guide to commercialization for this valuable step in the renewable biomass to energy conversion process.

12.3 HTC of Biomass Components

12.3.1 Cellulose

Cellulose is the most abundant renewable organic material on earth, with an annual production of over 50 billion tons [30]. Plants typically consist of 30–50 % cellulose on a dry basis. Cellulose is a straight chain polysaccharide consisting of glucose residues linked with $\beta-1\rightarrow4$ glycosidic bonds, with hundreds to thousands of glucose repeating units. These glycosidic bonds can be broken by protonation of the oxygen between glucose units or the cyclic oxygen [31]. Hydroxyl groups on the glucose units hydrogen bond with oxygen units on the same or adjacent polymer chains, so that a group of chains form microfibrils of high rigidity and crystallinity. However, microfibrils are formed by microcrystals connected and surrounded by amorphous cellulose chains of varying lengths [30, 32, 33]. Amorphous portions of cellulose polymers often are in random coil conformations [34].

The amorphous parts of cellulose can be hydrolyzed to oligomers of 4–13 degrees of polymerization (DP) using HTC at temperatures as low as 100 °C, but temperatures higher than 150 °C are necessary to break amorphous cellulose oligomers to their monomer components. Crystalline cellulose portions are much less reactive under HTC conditions, due to their strong intra- and intermolecular hydrogen bonds, so that temperatures above 180 °C are required to begin decomposition to glucose monomer [30]. With increased reaction temperature and short reaction time, DP of oligomer varies from 1 (monomer) to a maximal of 23 at 230 °C, 25 at 250 °C, and 28 at 270 °C, suggesting that the glycosidic bonds are attacked randomly on cellulose surfaces [35]. At temperatures above 300 °C, glucose monomer completely decomposes within 5 s, so HTC is impractical if glucose production is desired [36].

At temperatures above 200 °C and times over 5 min, cellulose oligomers progress to glucose monomers that undergo isomerization to fructose, which produces 5-hydroxymethyl-2-furfural (5-HMF), which can be converted to jet fuel, and other products [37]. The main products of cellulose hydrolysis with HTC are glucose, fructose, erythrose, dihydroxyacetone, glyceraldehyde, pyruvaldehyde, and glucose oligomers. Longer reaction times permit increased decomposition products of glucose to form [38]. Cellulose oligomers can also crosslink to form a solid product, particularly at HTC temperatures of 270 °C, and above [35]. For HTC with a reaction time of 240 min at temperature below 200 °C, cellulose can form aggregates of microspheres of about 2.0 μm in diameter [39, 40].

Cellulose particle size and DP can be decreased with ball milling, causing a substantial increase in reactivity with HTC. Hydrogen bond disruption in the crystalline structure may be responsible for this phenomena [41].

12.3.2 Hemicelluloses

Hemicelluloses are the second most common natural polysaccharides, after cellulose, making up 10–35 % of biomass [42]. The polymer chains of hemicelluloses have short branches, are amorphous, and consist of several possible monosaccharide units, including pentoses (xylose and arabinose) and hexoses (glucose, mannose, and galactose). It is commonly acetylated and has side chains of uronic acid and the 4-O-methyl ester [43, 44]. Its low crystallinity due to its branched nature permits it to be soluble in water and generally more reactive than cellulose. Hemicellulose heterocyclic ether bonds are vulnerable to hydrolysis reactions catalyzed by hydronium ions, which are produced under HTC conditions [45]. Hemicelluloses' degree of polymerization varies from 50–200. Xylan is often used as a model hemicellulose compound.

HTC has been performed on beech wood xylan while varying temperature between 180 and 300 °C and reaction time between 0 and 30 min [46]. When the xylan-water mixture was heated to the reaction temperature, a procedure taking ~15 min, and then immediately quenched, xylose and xylose oligomer yield increased from 14 % of xylan reactant at 180 °C to a maximum of 40 % at 220 °C. A second maximum of 42 % occurred at 235 °C, with a rapid decline in yield at higher temperatures. Looking at the effect of reaction time at 220 °C, <5 min reaction time was found to give optimal yield. When HTC was performed on xylan at 235 °C, a very short reaction time (heating to temperature and immediately quenching) gave nearly twice the xylose and xylose oligomer yield compared to a 5 min reaction time [46]. Apparently, temperatures of less than 230 °C and relatively short reaction times are required to hydrolyze hemicellulose using HTC [4, 7, 8, 45]. When HTC was performed on hemicellulose-rich bagasse at 200 °C, only 10 min was needed to maximize xylan to xylose monomer and oligomer yields, with oligomers formed first subsequently decomposing into monomers [38, 46]. Increasing reaction times past 10 min showed reduced xylose yield due to dehydration to 2-furaldehyde (furfural) and other degradation products [38]. Decreasing pH was also found with increasing reaction time for both beech wood xylan and hemicellulose-rich sugarcane bagasse [43, 46]. This phenomena is likely due to liberation of acetyl groups or other organic acid-generating side chains [47, 49]. At lower HTC temperatures (150–160 °C), hemicellulose-rich wheat bran, brewery spent grain, corn cobs, and Eucalyptus wood exhibit xylo-oligomers of structures characteristic of the original biomass, and furfural production is minimal [49].

12.3.3 Lignin

Lignin, the other main component in lignocellulosic biomass, is a three-dimensional, intricately branched structure of high molecular weight (typically > 10,000 Da). It has many hydroxyl and methoxy substituted phenylpropane units. The three most common of these monomers are p-coumaryl, coniferyl, and sinapyl based with ether bonds predominating between these units [38]. Different types of biomass have different ratios of these three monomers, with grasses and softwoods having predominately coniferyl, and hardwoods having a mix of coniferyl and sinapyl. Covalent bonds link lignin with hemicelluloses, enhancing cellulose microfibril protection [50].

Lignin appears to be relatively unreactive under HTC conditions. HTC of lignin (dealkali) at 225, 245, and 265 °C reaction temperatures for 20 h showed lignin to be the least easily decomposed of the main lignocellulosic components, with increasing temperature causing slight decreases in mass yield of solid products [51]. This high yield could be due to lignin's stable phenolic structure. Alternatively, part of the lignin may be decomposed and then undergo self-condensation reactions to reform as insoluble cross-linked polymers [51–54].

When hydrothermal pretreatment is performed on lignin, reaction temperatures higher than typical HTC conditions (180–300 °C) are frequently used. In one such study of alkali lignin, temperatures of 280, 370, 380, and 390 were used with reaction times of 0, 30, 60, 90, 120, and 240 min [55]. At the subcritical reaction temperatures (<374 °C) and the shortest reaction time (immediate quenching,) the main products identified in the liquid phase were guaiacol and vanillin. The solid-phase mass yield was found to be 48.9 % for 280 °C and 25.3 % at 370 °C. Mass yield decreased for these temperatures with increasing reaction time [55]. In another study, a treatment temperature of 304 °C produced 36 % dissolution after 10 min, and 365 °C gave a 90 % dissolution [29].

12.4 Reaction Chemistry

A useful way to depict the effects of both HTC time and temperature is by means of a van Krevelen diagram. This diagram is shown as Fig. 12.1, which plots atomic H/C ratio versus atomic O/C ratio, as commonly used to evaluate the energy quality of solid fuels [20]. Loblolly pine pretreated for 5 min by HTC at 200, 230, and 260 °C are designated as HTC 200, HTC 230, and HTC 260 [56].

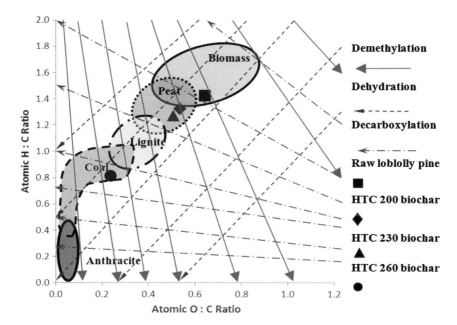

Fig. 12.1 A van-Krevelen diagram of HTC biochars for 5 min reaction time with major reaction lines. Reprinted from [56] with permission

Raw loblolly pine can be found in the biomass region, whereas HTC 200 and HTC 230 are in the peat area and HTC 260 is in the lignite region according to the van Krevelen diagram. It can also be noted that dehydration is the predominant reaction during HTC according to the van Krevelen diagram, although decarboxylation also has some effect. HTC 200, HTC 230, and HTC 260 are in a straight line corresponding to the dehydration reactions [39, 57].

12.4.1 Reaction Mechanisms

12.4.1.1 Hydrolysis

Hydrolytic reactions are the major solid surface reactions, where water reacts with cellulose or hemicellulose and breaks ester and ether bonds (mainly $\beta-(1-4)$ glycosidic bonds), resulting in a wide range of products including soluble oligomers like (oligo-) saccharides from cellulose, and hemicellulose (Fig. 12.2). With increased reaction time, these oligomers further hydrolyze into simple mono- or disaccharides (e.g., glucose, fructose, xylose). On the other hand, 5-HMF and other intermediates might further hydrolyze into simple acids like levulinic, acetic, and/or formic acid [9].

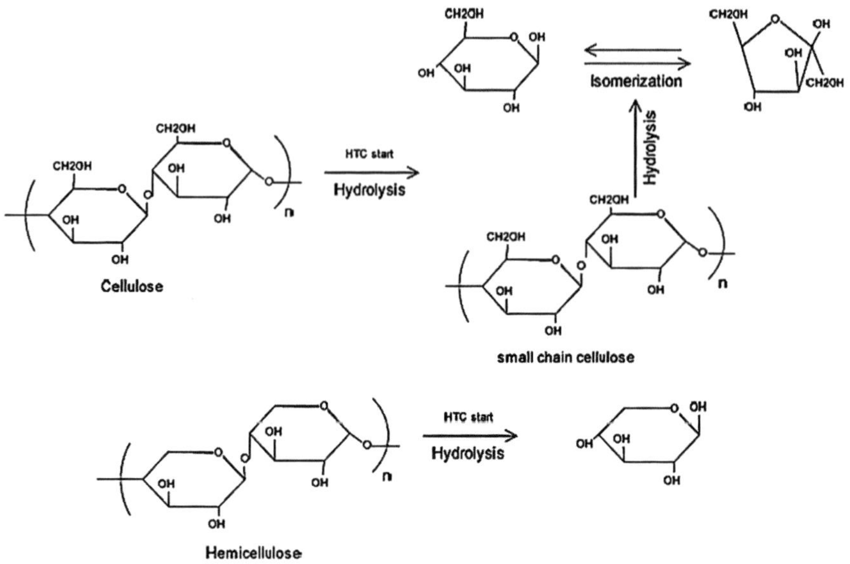

Fig. 12.2 Pathways of cellulose and hemicellulose hydrolysis for HTC. Reprinted from [56] with permission

Hemicellulose starts hydrolyzing at HTC temperatures above 180 °C, but cellulose hydrolysis starts above 230 °C. Liquid water at 200 °C hydrolyzes the β-(1-4) glycosidic bonds of hemicellulose [56, 58], which degrades into sugar monomers, then further degrade into furfurals and other compounds, including 2-furaldehyde [45]. Cellulose can degrade into oligomers, a portion of which hydrolyzes into glucose and fructose due to hydrolysis. Other biomass components like extractives, which are monomeric sugars (mainly glucose and fructose) along with various alditols, aliphatic acids, oligomeric sugars, and phenolic glycosides, are very reactive in hydrothermal media [59]. Lignin and inorganic components are very stable and probably remain unchanged by HTC at 200–260 °C, as discussed above [60].

12.4.1.2 Dehydration and Decarboxylation

Dehydration during HTC can be result from both chemical and physical processes. The physical process is well-known as dewatering, where the residual water is ejected from the biomass during HTC due to the increased hydrophobicity of HTC biochar [3]. Chemical dehydration happens due to elimination of hydroxyl groups [39]. The main reason for this significant decrease in oxygen is the reduction of carboxyl groups, mainly from extractives, hemicellulose, and cellulose. Dehydration and decarboxylation happen simultaneously (Fig. 12.3). One possible path for decarboxylation is the degradation of extractives, hemicellulose, and cellulose.

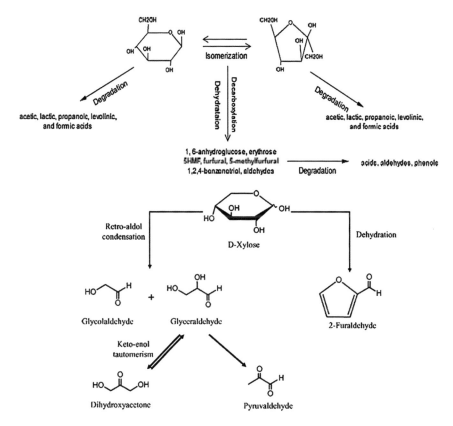

Fig. 12.3 Pathways of degradation, dehydration, and decarboxylation of cellulose, and hemicellulose monomers. Reprinted from [56] with permission

Under hydrothermal conditions, they can degrade into monomers like acetic acid, formic acid, or furfurals, which can further degrade into CO_2 and H_2O [39, 45]. For example, one mole of glucose in an appropriate environment can be converted into 6 moles of CO_2 and 6 moles of H_2O. Thus, the ratio (r) of mol CO_2 to mol H_2O is one for glucose. Similarly, r is defined by other researchers as 0.2–1.0 for cellulose depending on the HTC conditions [9]. Another possible reaction causing dehydration is the degradation of hydrolyzed products from biomass into furfurals like 5-HMF, erythrose, and aldehydes. For instance, each mole of 5-HMF production from glucose yields two moles of water. Moreover, the polymerization of hydrolyzed intermediates can yield water, too. For example, the retro-condensation of 5-HMF into aldol condensation or keto-enol condensation of n monomers yields n moles of water [39]. At the same time, aromatization or polymerization takes place, which also produces significant amounts of water [9].

12.4.1.3 Condensation, Polymerization, and Aromatization

Some of the intermediate compounds (5-HMF, anhydroglucose, furfural, erythrose, 5-methyl furfural) produced from dehydration and decarboxylation reactions of monomers are themselves highly reactive. These intermediates undergo condensation, polymerization, and aromatization as shown in Fig. 12.4. Thus, a linear polymer like cellulose can be converted to a cross-linked polymer similar to lignin. Condensation reactions of monosaccharides are slower, since cross-linked polymerization competes with recondensation to oligosaccharides [9]. Condensation polymerization is most likely governed by step-growth polymerization, which is enhanced by higher temperatures and reaction times [38, 39]. It thus is likely that the formation of HTC-biochar during HTC is mainly characterized by condensation polymerization and aromatization, specifically aldol condensation [38]. Moreover, condensing fragments within the biomass matrix are able to 'block' remaining biomacromolecules, thus preventing water access and subsequent hydrolysis, a phenomenon that makes the remaining HTC biochar hydrophobic [45].

12.5 Water Production in the HTC Process

Higher hydrophobicity may contribute to water production with HTC. Lignocellulosic components, being diverse polymers, can be degraded into many components in the HTC process. This results in a reaction mixture with many chemical reactions between different components, making it difficult to model the degradation kinetics. Although literature studies show that hydrolysis, dehydration, decarboxylation, condensation/polymerization, and polymerization reactions occur in parallel in the HTC process, primarily dehydration reactions are found after the initial hydrolysis. To simplify the reaction model and understand the main reactions, Uddin [61] developed an experimental and theoretical approach for the quantification of dehydration reaction of loblolly pine. Figure 12.5 shows that the water production or dehydration increases almost linearly with increasing reaction temperature over the temperature range considered for both 5 and 30 min reaction times. However, the average water production for 5 and 30 min reaction time at 200 °C is negative. Hydrolysis of hemicellulose and cellulose requires one mole of water to produce one mole of dissolved monomeric sugar. On the other hand, two moles of those monomers (e.g. glucose) in the liquid can form one mole of dimer (e.g., sucrose) and one mole of water, or one mole of those monomers can further degrade into lower carbon component and water. If there is no other means of water production in the process then only dehydration or polymerization or degradation to lower carbon component of those dissolved sugars can balance the water used for hydrolyzation. Thus, the possible explanation for a negative water production at 200 °C is that the dehydration reactions are not significant in this case, and further degradation/polymerization of monomers is slow. Cellulose

Fig. 12.4 Pathways of condensation, polymerization, and aromatization of active intermediates. Reprinted from [56] with permission

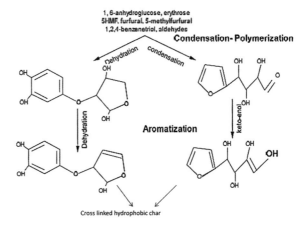

Fig. 12.5 Water production variations of loblolly pine with temperature and time. Reprinted from [61] with permission

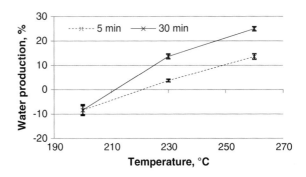

decomposition would not start until the temperature is at/over 230 °C. Because the decomposition of hemicellulose starts at a lower temperature (180 °C) than that of cellulose (230 °C) and the degradation of glucose rapidly increases at temperatures >230 °C, the degradation of glucose would be unavoidable if hemicellulose and cellulose are hydrolyzed together [36, 62–64]. For instance, the water production increases from −8 to ~4 % from 200 to 230 °C for 5 min reaction time. This might be the result of simultaneous hydrolysis of hemicellulose and cellulose polymers and decomposition of sugar monomers. A further increase of temperature from 230 to 260 °C increases the dehydration rate, and there is about two times more water production from ~4 to 13.6 % at 260 °C for 5 min residence time.

Based on the solid mass yield, the study also reported that with a long reaction time, the conversion of biomass from low carbon to high carbon content may slow down within 5–10 min, with side reactions occurring in the liquid phase during the remaining reaction time.

12.6 Process Water Recycling

To perform HTC in the most environmentally friendly manner would require recycling of process water, both to reduce the use of fresh water resources and the need for waste-water treatment. Typical HTC practice is to add about four times more water than biomass by weight [6, 19, 63–67]. In an intensive bioenergy production scenario, the supply of process water will be one of the key factors in industrial practice. For example, to produce one metric ton of dry bio-coal from loblolly pine at 260 °C assuming a 60 % mass yield in a batch process would require 8.34 metric tons of water per batch [6]. This large demand for process water can be minimized by recycling it after filtration for a batch or continuous processing system. Another advantage of recycling is that process heat could be recovered, significantly reducing external heating costs. In addition, wastewater treatment costs could be reduced since less wastewater would be generated per kg of feedstock. For both batch and continuous processes, optimization of process water recycling is very important because biomass pretreatment could be affected by the presence of sugars and inorganic materials remaining in the liquid solution. It would be expected that an increase of sugar concentration in the liquid would increase mass transfer resistance for sugar monomers produced from hemicelluloses and cellulose degradation. This hindering of sugar monomer movement to the liquid phase could impede the energy densification aspect of the HTC process.

Stemann et al. [68] first addressed the prospect of water reuse by investigating the recyclability of process water for poplar wood HTC. They found about a 3 % increase of mass yield and 5±1 % increase of HHV for poplar wood at a reaction temperature of 220 °C with a 4 hr reaction time. These researchers expected that the total organic carbon content (TOC) in the liquid phase would increase with recycle number but the results indicated a change in liquid and solid reaction chemistry for later recycles. Stemann et al. [68] described this as an unidentified polymerization of substances.

A similar study was done by Uddin [61] for loblolly pine at three different temperatures (200, 230, and 260 °C) for a reaction time of 5 min. The mass yields data obtained for the three different temperatures with recycling increased about 5 % after the first recycle. The mass yields remained the nearly same for the rest of the liquid recycle experiments compared to the first recycle run. However, HHV values appeared to be approximately constant with increasing recycle numbers for all temperature studied. The water soluble carbonaceous substances are concentrated with successive recycles, but not as much as would be expected to increase the mass transfer resistance, thus increasing the mass yield of the char. The equilibrium moisture content (EMC) results indicated unchanged hydrophobicity of the biochars with recycling process water.

Thus, the recycling of HTC process water appears to be feasible. Recycling will increase mass yield, but not irreparably. If implemented, HTC water recycling would reduce both environmental damage and costs.

Table 12.1 EMC of the HTC biochar pellets at different relative humidities. Reproduced from Ref. [3] by permission of Elsevier

	Treatment temperature (°C)	EMC(%) at $H_R = 11.3\ \%$		EMC(%) at $H_R = 83.6\ \%$		EMC(%) at $H_R = 100\ \%$
		Pellets	Biomass	Pellets	Biomass	Pellets
Raw	–	2.6	3.5 ± 0.5	17.6	15.6 ± 0.9	29.8
HTC	200	1.5	1.8 ± 0.5	12.4	12.8 ± 0.7	27.3
	230	1.0	0.9 ± 0.3	8.6	8.2 ± 0.7	12.6
	260	0.7	0.4 ± 0.3	4.7	5.3 ± 0.03	7.1

12.7 Hydrophobicity

One measure of hydrophobicity is EMC. EMC is defined as the moisture content in the biomass that is in thermodynamic equilibrium with the moisture in the surrounding atmosphere at a given relative humidity, temperature, and pressure. Acharjee et al. [3] performed HTC on loblolly pine at 200, 230, and 260 °C for 5 min. Table 12.1 shows EMC data for these HTC biochar. Their EMC values are lower than those for raw biomass at all relative humidities (HR) tested. This indicates that pretreated biomass, is more hydrophobic than raw biomass. Moreover, it becomes more hydrophobic with increasing process temperature. The explanation offered for this behavior is a combination of removal of aqueous extractives and hemicellulose, which are more hydrophilic than other fractions of biomass, and hydrolysis of -OH groups, thereby leaving behind fewer sites for hydrogen bonding.

12.8 Chemical Characteristics of HTC Biochar from Attenuated Total Reflectance: Fourier Transform Infrared Spectroscopy

HTC biochar has different chemical characteristics than raw biomass. FTIR is an important tool for studying chemical bonds. From Fig. 12.6, raw loblolly pine shows strong bonds at 910, 1050, 1180, 1260, 1390, 1510, 1613, and 1735 cm^{-1} in the fingerprint region. These correspond to aromatic carbon hydrogen bonds of hemicellulose (C–H), alcohol groups of glucose (C–OH), glycosidic bonds of cellulose (C–O–C), aryl-alkyl ethers of lignin (O–CH$_3$), aromatic acids of hemicelluloses (C–H), ketone groups of hemicelluloses (C=O), aromatic carbon skeletons (C=C), and carboxylic acid groups of hemicellulose (C=O), respectively. A broad aliphatic hydrocarbon or wax (–CH) and aliphatic hydroxyl bond for bound water (–OH) can be observed at 2840–2920, and 3100–3500 cm^{-1} respectively [65]. Both the peaks are found flattened with increasing HTC temperature. In the spectrum of raw loblolly pine, the band at 1725 cm^{-1} due to

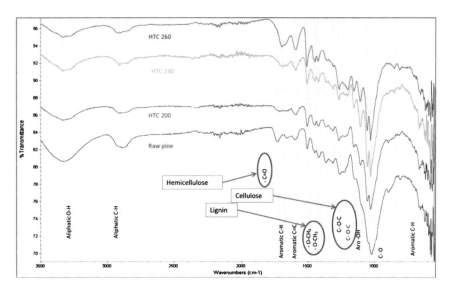

Fig. 12.6 ATR-FTIR spectroscopy of raw loblolly pine and loblolly pine pretreated at 200, 230, and 260 °C. Spectrum closest to the x axis corresponds to raw pine and those farther away are HTC 200, HTC 230, and HTC 260 spectra, respectively. Reprinted from [56] with permission

hemicelluloses [69], cannot be found in the HTC biochar spectrums, possibly due to the degradation of hemicellulose. The band at 1640 cm^{-1} could be assigned to the C=C stretching of aromatic rings or the bending mode of adsorbed water. Only the raw biomass spectrum shows that absorbance, meaning that the raw biomass is more hydrophilic than the pretreated. The peak at 1495 cm^{-1} (–O–CH$_3$) is the characteristic peak for lignin, and the peak is sharpens with increasing HTC temperature, indicating lignin enrichment. The peak at 1374 cm^{-1} weakened with HTC reaction temperature. The spectral peak on the shoulder at 1264 cm^{-1} (C–O stretching of the ether linkage) is the signature of lignin and it sharpens with increasing temperature [57]. Thus, FTIR-ATR of loblolly pine and its biochars from different reaction temperatures give some idea of what kinds of chemical changes occur in biomass with HTC. Such knowledge is useful in modeling the kinetics of the process.

12.9 Kinetics

Chemical reaction kinetics deals with the rates of chemical processes where the potential barrier constitutes the activation energy of the process, and determines the rate at which it will occur at a given temperature. When the barrier is low, the thermal energy of the reactant molecules is high enough to overcome the barrier so that they convert to reaction products rapidly. However, when the barrier is high, only a few reactant molecules will have sufficient energy, so that reaction is slower. Average reactant molecules' thermal energy increases with temperature, so that reaction rate increases with temperature.

HTC of lignocellulosic biomass follows a very complex reaction mechanism, resulting in difficulty in predicting accurate activation energies for the various reactions. For this reason, simplified models of the reactions of cellulose, hemicellulose, and lignin are used when investigating the kinetics of the hydrolysis of lignocellulosic materials.

12.9.1 Kinetic Models of Hemicellulose

The models proposed in the literature use pseudo homogeneous irreversible first-order reactions. Based on the first degradation model proposed by Saeman [70] for cellulose, researchers come up with a two-step first-order reaction model describing hemicellulose hydrolysis kinetics in hydrothermal condition. However, after observing the hydrolysis reaction rate, which decreased significantly after about 70 % conversion, Kobayashi and Sakai [71] introduced a model that included two types of hemicellulose, one fast hydrolyzing and one slow, each with its own kinetic constant. According to this model, the hydrolysis of the hemicellulosic fraction follows the following path [45, 72, 73]:

$$\begin{matrix} \text{Fast hydrolyse} \xrightarrow{\text{yields}} \\ \text{Slow hydrolyse} \xrightarrow{\text{yields}} \end{matrix} \text{Oligomers} \xrightarrow{k1} \text{Monomers} \xrightarrow{k2} \text{Monomer products} \quad (12.1)$$

Sarkar et al. [74] measured the degradation of xylose, a hemicellulose monomers, in rice straw in a very dilute acid solution at 100 °C and reported the rate constants (k1 = 0.094 and k2 = 0.006, min^{-1}). A similar study conducted by Rodriguez et al. [75] for wheat straw at 130 °C showed the rate constants (k1 = 0.1068 and k2 = 0.00026, min^{-1}).

12.9.2 Kinetic Models of Cellulose

The dilute acid hydrolysis of cellulose was described by Saeman [71] to understand the kinetics of lignocellulosic materials. The two-step degradation proposed by Saeman was two consecutive first-order reactions:

$$\text{Cellulose polymers} \xrightarrow{k1} \text{Glucose monomers} \xrightarrow{k2} \text{Monomer degradation products}$$
$$(12.2)$$

There are two types of hydrogen bonds in cellulose molecules: those that form between the C3 OH group and the oxygen in the pyranose ring within the same molecule and those that form between the C6 OH group of one molecule and the oxygen of the glucosidic bond of another molecule. Ordinarily, the beta-1,4 glycosidic bonds themselves are not too difficult to break. However, because of these hydrogen bonds, cellulose can form very tightly packed crystallites. These crystals are sometimes so tight that water cannot penetrate. This fact was missing in Saeman [70] hydrolysis model of cellulose. Because the amorphous cellulose hydrolyzes almost instantaneously to glucose, an initial glucose concentration must be assumed. This factor is easily incorporated into the original model, and all subsequent models have done so. To incorporate both crystalline and amorphous cellulose, Conner et al. [76] proposed an improved model of glucose degradation in hydrothermal condition. Mok and Antal [77] confirmed that a part of cellulose was nonhydrolyzable under hydrothermal reaction. However, this study concluded that this is due to an acid-catalyzed parasitic pathway that competes with the acid-catalyzed hydrolysis pathway and that there is no significant chemical change in the cellulose itself. Formation of intermediates in hydrothermal pretreatment of cellulose and hemicellulose has been well examined in several studies [76–80].

12.9.3 Kinetic Models for Lignin

Lignin being a large cross-linked polymer can be degraded into tens of components resulting in a large number of chemical reactions between different components in the reaction mixture, which make it difficult to model the degradation kinetics. Bobleter and Concin [78] first proposed two phase degradation kinetics of poplar lignin. The first phase reactions are very rapid, even less than 1 min, where lignin is degraded into soluble fragments, mainly consisting of low molecular components (e.g., monomers) and higher molecular components (oligomers). In the second phase (slower), the condensation reactions between soluble components and/or insoluble polymers produce insoluble polymers and char. In both cases, they assumed first-order degradation and condensation.

$$\text{Lignin} \xrightarrow{k1} \text{Monomers} + \text{Oligomers} \xrightarrow{k2} \text{Insoluble polymers} + \text{char} \quad (12.3)$$

According to Bobleter and Concin, the insoluble polymer exists during the first phase of quick reaction as a homogeneous suspension, which also reacts with the reactive soluble part.

Based on this model, Zhang et al. [81] proposed similar kinetic model for the hydrothermal degradation of Kraft pine lignin, where they assumed that a part of the soluble fragments reacts with all soluble components in the reaction mixture to form the insoluble polymers and char in the second phase. The estimated activation energy for lignin degradation is 37 kJ/mol for a temperature range of 300–380 °C. However, it should be noted that this model could not be directly used in hydrothermal treatment of the other biomass. Modification of the model is required to ensure the accuracy of model prediction when applied to the other biomass.

12.9.4 Model Proposed by Reza et al. [107] for Loblolly Pine

Under specific HTC conditions the lignin, cellulose, and hemicellulose fractions in lignocellulosic biomass follow distinct paths [82]. For pure components, reaction paths and their corresponding activation energies are available in the literature. However, when packaged in the lignocellulosic biomass matrix, these components may express different activation energies. One reaction can affect the other, resulting in a change of activation energy for the individual components. Reza et al. [60] proposed a simple kinetic model, consisting of two parallel first-order reactions, to calculate the activation energies of cellulose and hemicellulose of loblolly pine assuming the water extractive reactions are instantaneous.

$$\text{Extractives} \rightarrow \text{Soluble} + \text{Gas} \tag{12.4}$$

$$\text{Hemicellulose} \rightarrow \text{Soluble sugars} + \text{Gas} \tag{12.5}$$

$$\text{Cellulose} \rightarrow \beta \, \text{Bc} + (1 - \beta)(\text{Soluble sugar} + \text{Gas}) \tag{12.6}$$

where Bc, represents solid products from cellulose decomposition. Cellulose does not decompose completely and the mass yield of biochar from cellulose is denoted by the parameter β in the third reaction. The rates of decomposition of both hemicellulose (H(t)) and cellulose (C(t)) are both described by the first-order reaction kinetics, where k_1 and k_2 are the rate constants for the two reactions.

$$\frac{dH(t)}{dt} = k_1 H(t) \tag{12.7}$$

$$\frac{dC(t)}{dt} = -k_2 C(t) \tag{12.8}$$

The solution of these equations is shown below, with the initial component mass at time zero indicated by a zero subscript. The expression for the functions of H(t), C(t), and Bc(t) are function of reaction time but the function of L(t) is expressed as a constant as lignin is considered an inert component.

$$H(t) = H_0 e^{-k_1 t} \tag{12.9}$$

$$C(t) = C_0 e^{-k_2 t} \tag{12.10}$$

$$B_c(t) = \beta C_0(1 - e^{-k_2 t}) \qquad (12.11)$$

$$L(t) = L_0 \qquad (12.12)$$

where H_0, C_0, and L_0 represent the initial mass of hemicellulose, cellulose, and lignin in loblolly pine, respectively. If M(t) represents the mass of unreacted biomass plus solid-phase reaction products at time t, M(t) can be written by

$$M(t) = H(t) + C(t) + B_c(t) + L(t) \qquad (12.13)$$

To express the mass yield of biomass Y(t), Eq. (12.13) can be rewritten as Eq.12.14

$$Y(t) = \frac{M(t)}{M_0} = Y_{H0}e^{-k_1 t} + Y_{C0}e^{-k_2 t} + \beta Y_{C0}(1 - e^{-k_2 t}) + Y_{L0} \qquad (12.14)$$

To determine the mass yield of biomass by equation (12.14), the value of β is needed at any time t. If t tends to infinity then equation (12.14) will be written as Equation (12.13), without considering aqueous extractives:

$$\lim_{t \to \infty} Y(t) = \beta Y_{C0} + Y_{L0} \qquad (12.15)$$

Equation (12.15) is independent of HTC temperature if β is a temperature-independent parameter. To evaluate β and determine its dependency on HTC temperature, HTC reactions for extended periods (10, 15, and 30 min) were performed at 200, 230, and 260 °C and the value calculated was 0.54.

To calculate the rate constants k_1 and k_2, the mass yields of loblolly pine were measured experimentally for various reaction times at a specific HTC temperature.

The best fit for rate constants can be obtained by minimizing the objective function $F(k_1, k_2)$.

$$F(k_1, k_2) \equiv \sum_{i=1}^{n} \left(Y_i^{\text{experimental}} - Y_i^{\text{model}}(k_1, k_2) \right)^2 \qquad (12.16)$$

where Yi experimental represents the experimental mass yield and Yi model represents the mass yield as calculated in Eq. 12.14.

An Arrhenius plot (Fig. 12.7) of these rate constants generated the degradation activation energies of hemicellulose and cellulose of loblolly pine in the HTC process. The activation energies were reported to be 29 kJ·mol^{-1} for hemicellulose (rate constant k_1), and 77 kJ·mol^{-1} for cellulose (rate constant k_2). The higher activation energy for cellulose compared to that of hemicellulose is consistent with data reported elsewhere [45, 82], indicating the greater recalcitrance of cellulose. The numerical values for activation energies are smaller than those for pure hemicellulose and cellulose (129–215 kJ/mol) in this temperature range [82, 83].

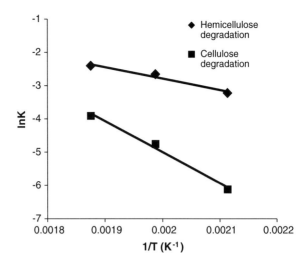

Fig. 12.7 Arrhenius plot for HTC reaction kinetics. Reprinted from Ref. [60], Copyright 2013, with permission from Elsevier

Instantaneous degradation of aqueous extractives likely reduces the pH of the solution and thus enhances degradation of hemicelluloses and cellulose [4], while hemicellulose degradation results in even greater acid production, further catalyzing cellulose reactions. As a result, the individual activation energies for both hemicellulose and cellulose are lower than those for the pure forms not encased in a lignocellulosic biomass package. Lignin, in this simple kinetic model, is assumed inert, but since the activation energies of hemicellulose and cellulose are lower in lignocellulosic biomass than in their pure form, so lignin activation energy in biomass may also be reduced. Torrefaction is another practice for thermally carbonizing lignocellulosic biomass. Prins et al. [84] reported a kinetic study of torrefaction of lignocellulosic biomass assuming similar first-order reactions for hemicelluloses and cellulose, and reported activation energies of 76 and 152 kJ·mol^{-1}, respectively. Compared to the (dry) torrefaction results (Yan et al. [85]), HTC kinetics in this study are more rapid, likely due to the reactivity of high temperature water.

12.10 HTC of Lignocellulosic Biomass Types

Lignocellulosic biomass comprises many different types. Different biomass have varying proportions of cellulose, hemicelluloses, and lignin, with the latter two categories containing various chemical components and side groups. HTC process variables may need to be optimized to obtain a desired result for a given lignocellulosic biomass. This section describes literature data on HTC of "woody" biomass, "grassy" biomass, and biomass obtained from agricultural residues.

12.10.1 HTC of Woody Biomass

"Woody" biomass is composed of 38–55 % cellulose, 15–30 % hemicelluloses, 15–30 % lignin, 2–5 % extractives, and less than 5 % inorganics (ash) [86]. Acetylated galactoglucomannans are the main softwood hemicelluloses [87], while hardwood hemicelluloses are predominately xylans [88]. Lignin type differentiates hardwoods from softwoods, with softwoods possessing guaiacyl phenylpropanoid almost entirely, and hardwoods a mixture of syringyl and guaiacyl phenylpropanoids. Hardwoods also have large pores, or vessels, absent in softwoods.

12.10.1.1 HTC of Hardwoods

Softwoods are generally less expensive than hardwoods, which are valued for construction as well as fuel. Nevertheless, HTC has been investigated for the hardwoods chinquapin, aspen and forest mangrove. With a flow-through reactor with temperature increased by stages to 285 °C, Ando et al. [89] found for the hardwood chinquapin that hemicellulose was solubilized by 180 °C, and cellulose by 230 °C, with a reduction in solvent pH to 4. The end pH value for the aqueous phase indicates production of organic acids from the biomass, reflecting the type of hemicellulose it contains. The 30 % lignin in the raw biomass was mostly solubilized, as well. These authors claimed that hardwood lignin was easier to decompose than softwood lignin. For aspen (Populus tremuloides) and a flow-through reactor with two temperature stages, Bonn et al. [25] reported that hemicellulose was hydrolyzed at 180 °C after 30 min, but cellulose required a reaction temperature of 265 °C for 30 additional minutes. A final mass yield of only 5 % was obtained, indicating that most lignin must have been degraded. For a 300 °C for 30 min batch hydrothermal pretreatment of forest mangrove (acacia magium), Nonaka et al. [90] found the solid product to be more hydrophobic, a 91 % mass yield, but an energy densification of 140 %. Thus, polymerization of solubilized components is likely to occur at this high HTC temperature to achieve such an enhanced energy densification with a high mass yield.

12.10.1.2 HTC of Softwoods

The effects of HTC have been investigated on softwoods, including Japanese cedar (Cryptomeria Japonica). Phaiboonsilpa et al. [91] used a two-step semi-flow hot-compressed water first at 230 °C for 15 min and then at 280 °C for 30 min on Japanese cedar. At 230 °C, nearly all the hemicelluloses were hydrolyzed, as well as about half the lignin, with cellulose only 10 % hydrolyzed. The 280 °C for 30 min step removed the remaining cellulose from the solid residue, leaving only 12 % as a lignin residue. In a flow reactor, HTC was performed by Ando et al. [89] on Japanese cedar at 180 °C for 20 min, then 285 °C for 7 min. Again, hemicelluloses were hydrolyzed at 180 °C and about half of the lignin, with a lignin residue of 12 % remaining, nearly twice as much as for the same process done with hardwood. The pH value for this softwood was about 4.5, slightly higher than for the hardwood chinquapin described in the previous paragraph. Nonaka et al. [90] investigated hydrothermal pretreatment of Japanese cedar using both flow and batch methods at 300 °C for a 30 min reaction time. Both reactor types gave the same mass yield of 89 %, and similar energy densifications of 140 %, as well as increased hydrophobicity. Thus, scaling up to a more economical continuous HTC process may be quite feasible. Overall, Japanese cedar softwood gives results similar to hardwood, except that its lignin is more resistant to hydrolysis.

Using a batch reactor at 200, 230, or 260 °C for 5 min, Yan et al. [8] performed HTC on loblolly pine, a softwood of less than 1 % ash from the southeastern United States. They reported that EMC, a measure of hydrophobicity, decreased with increasing HTC reaction temperature. EMC was reduced to less than a third of the raw loblolly pine's value. Mass yield also decreased, from 90 to 57 %, with increasing reaction temperature. Energy densification increased from 110 to 140 % with increased reaction temperature. As with hardwood and Japanese cedar, all hemicellulose is solubilized by HTC even at the lowest temperature of 200 °C. Cellulose is hydrolyzed more (from 22 to 64 %) with increasing HTC reaction temperature. For loblolly pine, lignin does not decompose at 200 °C, and only undergoes 15 % conversion at 230 °C and 23 % at 260 °C. However, Yu and Wu [30] have suggested that repolymerization of cellulose, producing a pseudo-lignin compound detected by fiber analysis, may account for the apparent inert behavior of lignin in HTC. Loblolly pine has an unusually high amount of cellulose (55 %), so if solubilized fragments polymerize into a pseudo-lignin, the effect could mask lignin decomposition.

Lynam et al. [92] reported similar results for mass yield and energy densification for HTC of loblolly pine for the same reaction conditions, as well as remarking on the friability of the solid product. Scanning electron microscope (SEM) images of loblolly pine raw and pretreated at 200, 230, and 260 °C indicate the stripping of hemicellulose from the lignocellulosic matrix at 200 °C, and the stripping of cellulose at 260 °C. In addition, Lynam et al. (2013) determined that reaction temperature was the most significant variable for HTC, compared to reaction time, water to biomass ratio, and loblolly pine particle size. The work of Lynam et al. [4] on loblolly pine suggested that increasing temperature caused more dehydration of glucose to 5-HMF, a product of greater HHV, which would deposit in the pores of the now friable, higher surface area biochar to increase its HHV. This work indicated that loblolly pine HTC at 230 °C for 5 min reduced pH to 3 in the solvent.

Kang et al. [51] investigated pine wood meal HTC at 225, 245, and 265 °C using a reaction time of 20 h. Mass yield for all three reaction temperatures was ~55 %, but energy densification increased from 148 to 162 % with increased temperature. Since Reza et al. [60] suggest that reactions in the solid product are complete within 5 min and that mass yield for loblolly pine reaches a minimum of 55 %, 20 h may not be an economical option for a reaction time.

Using a mix of Jeffrey Pine and White Fir (called Tahoe Mix) obtained from the Tahoe Forest in California, USA, Hoekman et al. [19] investigated reaction temperatures of 215, 235, 255, 275, and 295 °C for 30 min reaction times, as well as 5, 10, 30, and 60 min reaction times at a reaction temperature of 255 °C, using a batch reactor with stirring. Increasing temperature decreased mass yield from 69 to 50 %, while increasing time for a 255 °C HTC temperature only decreased mass yield from 58 to 52 %. Again, the significance of reaction temperature compared to time is illustrated. With increasing temperature for a 30 min reaction time, energy densification increased from 111 to 145 %. With increasing reaction time at 255 °C, energy densification increased from 123 to 143 %. The end pH for all reaction conditions was about 3. Lynam et al. [92] also investigated Tahoe Mix, using 200, 230, and 260 °C and a 5 min reaction time in an unstirred batch reactor. Mass yield at similar temperatures, although showing a decrease with increasing temperatures, was ~15 % higher. This discrepancy in mass yield is likely due to the shorter reaction time or to the lack of stirring in the reactor.

Overall, softwoods give results for HTC similar to hardwoods, except that softwood lignin is more resistant to hydrolysis. HTC hydrolyzes the hemicellulose of both types of woody biomass at temperatures near 180 °C and cellulose at higher temperatures. To remove substantial amounts of lignin requires even higher temperatures. The solvent end pH value decreases to 3–4 due to organic acid formation.

12.10.2 HTC of Biomass from Grasses

Although grasses may vary in cellulose, hemicellulose, and lignin proportions, nearly all of them have more ash content than woody biomass, which can cause difficulties when used as an energy resource. In recent years, interest in switchgrass, which is ~7 % ash, as an energy crop has intensified. Pretreating it using HTC was studied by Lynam et al. [92] at 200, 230, and 260 °C with a 5 min reaction time. Likely due to the percentage of hemicellulose (31 %) being much higher than that of a typical biomass, HTC of switchgrass gave a mass yield at 230 and 260 °C much lower than that of the softwoods loblolly pine and Tahoe mix. For a 260 °C reaction temperature, a mass yield of only 30 % was achieved. However, for reaction temperatures of 230 and 260 °C, energy densifications were reported to be 112 and 133 %, respectively.

HTC of miscanthus, also known as elephant grass, was investigated by Chiaramonti et al. [93] at temperatures in the range of 150–230 °C. These authors describe HTC reaction conditions in terms of "severity", a mathematical combination of reaction time and temperature used. Since no significant interactions between time and temperature with HTC have been reported, such a severity approach has its difficulties [92]. Miscanthus also has high hemicellulose (27 %) and an ash of 3 %. With HTC, mass yields of 70–90 % were obtained with a rapid reduction in hemicellulose with increasing HTC temperature and time, while

12 Hydrothermal Carbonization of Lignocellulosic Biomass

cellulose and lignin remained intact. To summarize, grassy biomass, despite having higher ash than woody biomass, appears to react in similar ways, except that lower mass yield was found with HTC, particularly for switchgrass.

12.10.3 HTC of Agricultural Residues

Agricultural residues vary widely in type and composition, from straws, corn stover, and sugar cane bagasse to rice hulls, sunflower stems, and coconut fiber. HTC can make solid fuel out of these byproducts that are required to grow food, so that land resources can be used to produce both food and fuel simultaneously.

HTC of wheat straw, of 6.5 % ash, has been investigated by Petersen et al. [58] using a 50 kg/hour continuous flow system at temperatures of 185, 195, and 205 °C for 6 or 12 min. As might be expected for HTC temperatures near 200 °C, cellulose and lignin remain in the solid product, while hemicellulose is hydrolyzed. The amount of hemicellulose remaining in the solid product varies from 70 to 30 % as temperature and time is increased from185 to 205 °C and from 6 to 12 min. This sensitivity of hemicellulose conversion to slight changes in temperature and time points to their importance in HTC processes for straw-type biomass. Using rye straw (6 % ash), Rogalinski et al. performed HTC using both a batch reactor and a continuous reactor [94, 95]. In the batch reactor HTC at 190 °C for 120 min gave a mass yield of 75 %. With the continuous reactor, temperature ranged between 120 and 310 °C and residence time between 1.0 and 14.5 min, while pressure was held at 100 bar. Increasing temperature and time decreased mass yield from 95 to 50 %. These authors reported that rye straw particle size and water to biomass ratio seemed to have no influence on mass yield.

Lynam et al. [92] investigated the effects of HTC on the agricultural residues of corn stover and rice hulls) at 200, 230, and 260 °C with a 5 min reaction time. For both biomass, mass yield at 200 and 230 °C was reported to be about 90 and 75 %, respectively. Energy densification for the two were near 100, 105, and 125 % at 200, 230, and 260 °C, respectively. These similarities may relate to the high ash content of corn stover, 7 %, and rice hulls, 21 %. However, at 260 °C, the 43 % mass yield of corn stover was much lower than for rice hulls or the woody biomass loblolly pine and Tahoe mix. At this higher HTC reaction temperature, corn stover behaved much more similarly to switchgrass. A primary agricultural residue, corn stalks and leaves would seem less closely related to the secondary agricultural residue rice hulls, the siliceous husks of rice grains and more closely related to grassy biomass. Mosier et al. [96] investigated corn stover using a continuous HTC process at reaction temperatures of 170, 180, 190, and 200 °C. Slight decreases in mass yield were seen with reaction time increase from 5 to 20 min, but mass yield decreased from 90 % at 170 °C to 60 % at 200 °C. This kind of sensitivity of mass yield to temperature may be a characteristic of grassy agricultural residues.

Sugarcane bagasse has been widely studied as an energy resource. Chen et al. [97] performed HTC on bagasse at 180 °C for 5, 15 and 30 min using a batch reactor. Increasing reaction time decreased mass yield from 70 to 61 %, but had no effect on energy densification, which was 110 %. Hemicellulose was solubilized at this temperature. Increasing temperature from 200 to 280 °C and using a semi-batch reactor, Sasaki et al. [98] demonstrated that extractives and hemicelluloses were extracted from bagasse between 200 and 230 °C, while cellulose was solubilized between 230 and 280 °C.

HTC of more exotic agricultural residues have been investigated. Lui et al. [99] performed HTC at 150–350 °C in a batch reactor for 30 min on coconut fiber (~5 % ash) and dead eucalyptus leaves (~7 % ash) due to their high production potential in Singapore. With increasing reaction temperature, mass yield declined for both biomass from 90 % at 150 °C to 30 % at 350 °C, with little effect on mass yield seen above 300 °C. Energy densification increased from 134 % at 220 °C to 160 % at 300 °C, but declined with further temperature increase for both biomass. The researchers claimed higher hydrophobicity with increasing HTC reaction temperature. Roman et al. [100] studied the effects of HTC on sunflower stems and walnut shells at 190 and 230 °C for reaction times of 20 or 45 h. Results showed that increasing time from 20 to 45 h had no effect, as one might expect for HTC. For sunflower stems, increasing temperature decreased mass yield from 40 to 29 % and increased energy densification from 149 to 176 %, exhibiting a remarkable increase in HHV. For walnut shells (~6 % ash), increasing temperature decreased mass yield from 50 to 35 %, while energy densification from 118 to 150 %. Considering the low temperatures, HTC seems quite effective in enhancing the solid products of these less studied biomass.

In summary, the diversity of agricultural residues makes them hard to generalize. However, agricultural residues perform more similarly to grassy biomass than woody biomass when undergoing HTC. Their sensitivity to small reaction temperature changes is noteworthy. Mass yields for these biomass tend to be lower, particularly at higher reaction temperatures. Agricultural biomass are also similar to grassy biomass in that both have higher ash (inorganic) content than woody biomass.

12.11 Inorganic Analysis of Lignocellulosic Biomass and Its Biochars

Plants acquire inorganics, which are necessary for their metabolic pathways, from the soil in which they are grown. Inorganics are in the form of inorganic salts, bound to the organic structure by ionic bonds, or possibly covalent bonds, in a cross-linked matrix [101]. Woody biomass contains less inorganic content than grasses or agricultural residues [102, 103]. Due to their high melting points, the inorganics usually remain in the ash, which may have a negative impact on biomass firing or even co-firing with coal [104]. During combustion, ash must be removed from the boiler, as it increases the complexity in co-firing as well as lowers the efficiency of the boiler. Moreover, sodium, potassium, calcium, and other metals can cause slagging and fouling, resulting in lower power plant efficiency [101, 105]. Chloride, which is very corrosive for stainless steel, can also react with alkalis and silicates to form an undesirable stable slag [103]. Heavy metals like mercury, lead, arsenic, chromium, copper, zinc, and selenium are scarce in biomass. Slagging is the formation of molten or partially fused deposits on furnace walls or convection surfaces exposed to radiant heat. Fouling is defined as the formation of deposits on convection surfaces such as superheaters and reheaters [101]. The viscosity of the coal ash slag determines the diffusivity of ions within the slag that affects its corrosivity [106].

12.11.1 Inorganic Analysis for Grassy Biomass and Agricultural Residues

Reza et al. performed a detailed inorganic analysis for four different biomass (i.e., corn stover, miscanthus, switch grass, and rice hulls), which have relatively high inorganic content. HTC treatment resulted in a reduction of slagging tendency for every biomass except corn stover (Table 12.2). An intermediate tendency of slagging for raw feedstocks, HTC 230, and HTC 260 are found for corn stover, while, HTC 200 has the highest slagging tendency. Fouling index data indicated that the fouling tendency improved in HTC 260 for all biochars compared to the starting biomass feedstocks. With respect to ratio slag viscosity (IV), only raw corn stover showed a medium slagging tendency, which was low for the other raw biomass and every HTC biochar. Cl was found to be high in raw and HTC 200 switch grass, but was low for every other HTC biochar. In terms of alkali index (IA) only raw corn stover, miscanthus, switch grass, and HTC 200 corn stover showed a probable tendency for slagging. But with increased HTC temperature, the alkali index was found to be low for every biochar.

12.11.2 Heavy Metal Analysis of HTC Biochar

Heavy metals like Hg, Pb, Cd, Cr, Cu, Zn, As, Ni, Ag, and Se exist in trace amounts in biomass. Nevertheless, in using HTC biochar and handling its ash, data about heavy metal content is essential due to environmental issues. Table 12.3 shows the heavy metal content in HTC biochar for corn stover, miscanthus, switch grass, and rice hulls. Hg and Se concentrations are below the detection limit of Inductively Coupled Plasma Atomic Emission Spectroscopy (ICP-AES). Heavy metals are found in this range: 1–20 mg Ni, 5–17 mg Ag, 27–34 mg Pb, 6–45 mg Zn, 3–14 mg Cu, 1–44 mg As, 9–52 mg Cd, and 2–14 mg Cr per kg of raw biomass. Only Cd for all raw biomass and As for raw rice hull exceeds the soil protection act limit. These heavy metals have high melting points and as a result they are concentrated in the ash after combustion, sometimes by an order of magnitude. There could be compliance issues in disposal of raw biomass ash. Low

Table 12.2 Slagging, fouling, alkali, and ratio-slag indices, Cl content, definition, and their limits [103]

Slagging/fouling index	Expression	Limit
Slagging index	$I_s = (B/A) * S^d$ $S^d = \%$ of S in dry fuel	$I_s < 0.6$ low slagging inclination $I_s = 0.6$–2.0 medium $I_s = 2.0$–2.6 high $I_s > 2.6$ extremely high
Fouling index	$I_F = (B/A)*(Na_2O + K_2O)$	$I_F \leq 0.6$ low fouling inclination $0.6 < I_F < 40$ medium $I_F \geq 40$ high
Alkali index	$I_A = (Na_2O + K_2O)$ in kg/GJ	$0.17 < I_A < 0.34$ slagging/fouling probable $I_A \geq 0.34$ slagging/fouling is certain
Slag viscosity index	$I_V = (SiO_2*100)/$ $(SiO_2 + MgO + CaO + Fe_2O_3)$	$I_V > 72$ low slagging inclination $65 \leq I_V \leq 72$ moderate $I_V < 65$ high
Chlorine content	Cl as received (%)	$Cl < 0.2$–0.3 low slagging inclination $0.2 < Cl < 0.3$ medium $0.3 < Cl < 0.5$ High $Cl > 0.5$ extremely high

Group A: Fe_2O_3, CaO, MgO, Na_2O, or K_2O. Group B: SiO_2, Al_2O_3, & TiO_2. Reprinted from Ref. [107], Copyright 2013, with permission from Elsevier

Table 12.3 Slagging and fouling indices for HTC biochar. Reprinted from Ref. [107], Copyright 2013, with permission from Elsevier

Biomass	Condition	I_S	I_F	I_V	Cl	I_A
Corn stover	Raw	medium	high	medium	low	probable
	HTC 200	high	medium	low	low	probable
	HTC 230	medium	medium	low	low	low
	HTC 260	medium	medium	low	low	low
Miscanthus	Raw	low	medium	low	low	probable
	HTC 200	Medium	medium	low	low	low
	HTC 230	low	medium	low	low	low
	HTC 260	low	low	low	low	low
Switch grass	Raw	medium	medium	low	high	probable
	HTC 200	medium	medium	low	high	low
	HTC 230	low	medium	low	low	low
	HTC 260	low	medium	low	low	low
Rice hull	Raw	low	medium	low	low	low
	HTC 200	low	low	low	low	low
	HTC 230	low	low	low	low	low
	HTC 260	low	low	low	low	low

temperature HTC treatment is effective for heavy metal reduction. All the heavy metals except Pb and As are found concentrations less than 15 mg per kg of HTC 200 for every biomass [107]. Pb and As are relatively inert to HTC reactions.

In summary, HTC can be effective in reducing slagging and fouling, as well as hazardous heavy metals concentrations for biomass. If biomass is to be used for co-firing, using HTC on the higher ash content grassy and agricultural residues would be beneficial, both economically and environmentally.

12.12 Energy Balance: Heat of Reaction

Design of an industrial-scale reactor requires knowledge of the heat of reaction. For continuous design, other considerations will impact the amount of heat added or removed, especially including heat integration, water recycle, and amount and temperature of makeup water. Here, we consider only the heat of reaction, heat released (or consumed) by the chemical reactions of HTC of lignocellulosic biomass.

There has been little literature published in this area. Yan et al. [7] attempted to calculate the heat of reaction from the heats of formation of all reactants and products. The reaction products were simplified in the analysis. All sugars were considered to consist of either glucose (at 200 °C) or 5-HMF at higher temperatures. All acids were considered to be acetic acid for this analysis, and all gases were considered to be CO_2. The heat of formation for the char product was calculated from its ultimate analysis and from its HHV. The heat of reaction was calculated at three reaction temperatures (200, 230, and 260 °C), and the results showed that the heat of reaction was slightly endothermic, and appeared to become less endothermic with increasing reaction temperature. A significant shortcoming of that analysis was the failure to account for production of water during reaction. As shown in this chapter and elsewhere [61] the amount of water produced can be quite significant. Had water production been included in the analysis of Yan et al. [7], the result would have shown an exothermic reaction, with increasing exothermicity as reaction temperature increases.

Funke and Ziegler [108] reported on a different approach to measure the heat of chemical reaction. They used differential scanning calorimetry (DSC) with glucose, cellulose, and wood, and found an exothermic reaction in all cases, with $\Delta H \approx -1$ MJ/kg at 240 °C for reaction times between 4 and 6 h. Hydrolysis of the carbohydrate fractions of biomass (hemicellulose and cellulose) produces oligosaccharides and simple sugars, among many other produces. With time, oligosaccharides further degrade to simple sugars and subsequently to furfural and 5-HMF. Thus, the long reaction time employed by Funke and Ziegler includes hydrolysis of biomass, as well as that of the products of the initial hydrolysis, especially including aqueous phase reactants.

Neglecting the production of compounds other than water and CO_2, one can compute an upper bound to the amount of heat released during reaction. All carbon lost from the biomass can be assumed to become carbon dioxide, while the oxygen lost is apportioned between water and CO_2. With increasing severity of reaction, more of these two low-energy reaction products are produced, and it follows therefore that the reaction must be more exothermic with increasing reaction temperature and time. Funke and Ziegler reported a minimum exothermic heat of reaction of -2.4 MJ/kg for relatively severe reaction conditions of 310–340 °C and times greater than 64 h.

Clearly, more investigation is required to determine the heat of reaction of biomass HTC. The two published works above reference only woody biomass, and contain very little information on the effects of operating conditions. Nonetheless, it is possible to conclude generally that the heat of reaction is likely to be slightly exothermic. The amount of heat produced by reaction is likely to be insufficient to provide for the heat required to operate a continuous process, since heating water to reaction temperature requires significant amounts of heat. This motivates the need for innovative design to best recover process heat with the goal of minimizing the cost of fuels (biomass, char, or other) needed to operate the process at the designated operating conditions.

12.13 Pelletization

12.13.1 Pelletization Benefits

Since biochar particles are hydrophobic, discrete, friable, and less than 1 mm mesh size, they pose potential respiratory and fire hazards. Pelletization can effectively reduce these hazards. Pelletization compacts the biomass particles and, as a result, the mass density increases. HTC biochar pellets have higher mass and energy density. For instance, Table 12.4 shows that pellets from raw loblolly pine have a mass density of 1102.8 kg/m^3, while raw loblolly wood has a density of 813 kg/m^3. Pellets of loblolly pine pretreated at 260 °C have a mass density of 1462.8 kg/m^3, which is 32.6 % higher than the pellets of loblolly pine pretreated at 200 °C and 80 % higher than that of pellets made from raw loblolly pine wood. Yan et al. [8] reported that the product of hydrothermally carbonized lignocellulosic biomass is more friable with increasing pretreatment temperature and it becomes more hydrophobic.

HHV is almost the same for raw biomass and pellets, as shown in Table 12.4. HTC biochar's HHV is similar whether pelletized or not [6]. That implies that the chemical compositions remain the same through the pelletization process. The materials are compressed, without chemical reaction.

In terms of energy density, since the mass density of pellets increases rapidly and the HHV remains same, the energy density increases rapidly. Table 12.4 shows the energy densities of the pellets of pretreated loblolly pine. Pellets of pretreated loblolly pine at 260 °C have an energy density of 38.79 GJ/m^3 which is 70 % higher than raw loblolly pellets and 142 % higher than raw loblolly wood [6]. Table 12.5 shows the abrasion index for pellets of raw loblolly pine as well as

Table 12.4 Mass and energy density of loblolly pine wood, pellets of raw loblolly pine, pellets pretreated at HTC-200, HTC-230, and HTC-260.

	Mass density (kg/m^3)	HHV (MJ/kg)	Energy density (GJ/m^3)
Wood	813	19.5	15.9
Raw pellet	1102	20.6	22.8
HTC-200 pellet	1125	21.2	24.3
HTC-230 pellet	1331	22.5	30.0
HTC-260 pellet	1468	26.4	38.8

Reproduced from Ref. [6] by permission of John Wiley and Sons Ltd

Table 12.5 Abrasion index and durability of pellets of loblolly pine and HTC biochar pretreated at different temperatures.

Pretreated temperature (°C)	Abrasion index (%)	Durability (%)
Raw	1.03	98.1
200	0.47	99.5
230	0.28	99.7
260	0.18	99.8

Reproduced from Ref. [6] by permission of John Wiley and Sons Ltd

loblolly pine HTC pretreated at 200, 230, and 260 °C for a reaction time of 5 min. Abrasion index is the ratio of mass percentage below 1.56 mm mesh to the initial sample mass after 3000 rotations in a tumbler [109]. The smaller the abrasion index, the better quality is the pellet. Durability is inversely related to the abrasion index. Raw loblolly pine pellets have an abrasion index of 1 %, whereas the pretreated loblolly pine pellets have decreased abrasion index with increasing HTC reaction temperature. (Table 12.5).

In the case of pellets of HTC biochar, the abrasion index decreases with the increase of pretreatment temperature, when all the other variables are the same (Table 12.5). The lower abrasion index and higher durability mean the pellets are mechanically more stable. As lignin is inert in the temperature range of 200–260 °C, the lignin percentage increases in the HTC biochar with pretreatment temperature using the same reaction time [81]. Yan et al. [8] reported that the lignin percentage of HTC biochar pretreated at 260 °C is 35 %, while it is 25 % in the raw biomass.

12.13.2 Lignin as a Binder in Pelletization

12.13.2.1 Glass Transition Behavior of Lignin

Lignin is the only component of biomass that shows glass transition behavior [6]. Glass transition is a property of the amorphous portion of a semi-crystalline solid [110]. The crystalline portion remains crystalline during the glass transition. At a low temperature the amorphous regions of a polymer are in the glassy state, where

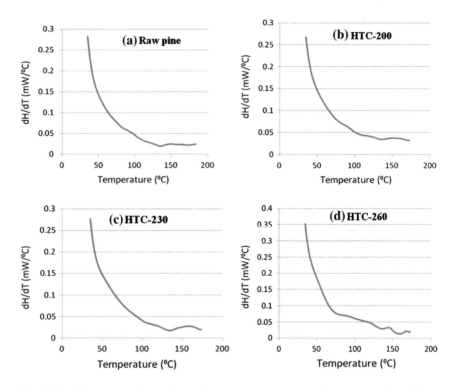

Fig. 12.8 DSC curves (slope of heat flow versus temperature) for determination of glass transition temperature of extracted lignin from HTC biochar and raw loblolly pine. Reproduced from Ref. [6] by permission of John Wiley and Sons Ltd

the molecules are frozen in place. They may be able to vibrate slightly, but do not have any segmental motion in which portions of the molecule rotate. When the amorphous regions are in the glassy state, the polymer generally will be hard, rigid, and brittle [110]. Lignin extracted by fiber analysis showed a glass transition change when undergoing digital scanning calorimetry (DSC), as shown in Fig. 12.8. The lignin extracted from raw loblolly pine and HTC biochar showed a range of glass transition temperature of 135–165 °C in the heat flow curve versus temperature. At a temperature of 135 °C, the slope starts increasing with temperature until 165 °C, and then it follows the same trend again. The change of heat flow pattern with temperature over the range of 135–165 °C indicates the glass transition behavior of lignin [111].

12.13.2.2 SEM Images of HTC Biomass Pellets

At temperatures higher than its glass transition temperature, lignin acts as a binder causing liquid bridges to develop between adjacent particles. In most cases, an immobile thin adsorption layer is produced by the binder which attaches the

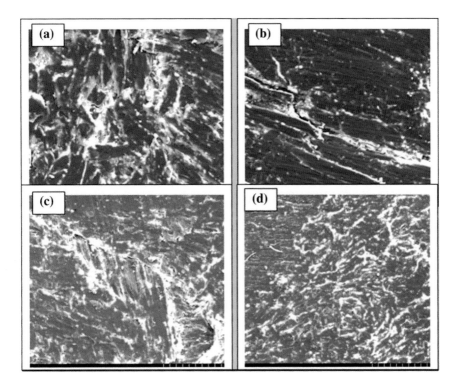

Fig. 12.9 SEM images (200 times magnification) of **a** raw loblolly pine pellet. **b** HTC-200 biochar pellet. **c** HTC-230 biochar pellet. **d** HTC-260 biochar pellet. Reprinted from [56] with permission

particles by effectively reducing surface roughness or decreasing inter-particle distance Attractive forces then come into action resulting in binding and the formation of stable bonds [112]. Lower particle sizes and smaller distances among the particles encourage liquid bridge building [110]. Applying compression reduces interparticle distances and particle size, if the particles are friable. So, pressure is required as well as temperature to activate binding mechanisms among the particles. SEM images can be used to illustrate this binding process.

The raw pine pellet shown in Fig. 12.9a is smooth and seems to have a good solid bridge on the surface. No significant cracks are observed in Fig. 12.9a. But cracks are observed in the HTC biochar pellets. The crack in the HTC-200 biochar pellet is larger than that of the other HTC biochars or even raw pine pellets. HTC biochar has only lignin as a binder as all the water extractives and other natural binders are extracted or destroyed during HTC [109]. The lignin percentage of HTC-200 biochar is probably not enough to fill up the cracks. The crack in the HTC-230 biochar pellet in Fig. 12.9c is smaller than that of HTC-200 biochar pellet, while no significant crack is observed in the HTC-260 biochar pellet in Fig. 12.9d. The percentage of lignin in the HTC-260 biochar pellet is enough to form solid bridges among the particles and make the pellet durable.

12.14 Conclusions

In this chapter, we have discussed the history of HTC of biomass, and reviewed the recent literature. Much has been published recently on HTC of diverse biomass feedstocks, and effects of temperature, time, and pH have been reported, but there is still much left to learn. A picture of the underlying chemistry of HTC is beginning to emerge, but further knowledge is required, especially to learn how to use process parameters to affect desired product characteristics.

HTC pretreatment of lignocellulosic biomass presents unique advantages and unique challenges. In many regard, the process can be viewed as one that mimics earth's natural process of converting biomass to coal, but accelerates the time from millennia to minutes. The solid product has a high energy density, is friable, is easily pelletized, and is very hydrophobic. On the other hand, this is a high pressure process with concomitant capital costs.

A pressing need is development and deployment of pilot scale facilities. Continuous operation of an HTC process presents technical challenges, including: feeding solid and liquid into a high pressure reactor, heat integration, separation and recycling of water without flashing. The experience gained from operating long-term operations will be invaluable in determining the commercial viability of HTC. In particular, capital costs are currently unavailable, and erosion rates and maintenance requirements are entirely unknown.

Further work is required to characterize the aqueous byproducts, which include many useful components, such as organic acids, sugars, and furfurals. A low-cost process to convert these compounds to chemicals or fuels would significantly enhance the commercial prospects of HTC.

References

1. Saidur R, Abdelaziz EA, Demirbas A, Hossain MS, Mekhilef S (2011) A review on biomass as a fuel for boilers. Renew Sust Energy Rev 15(5):2262–2289. doi:10.1016/j.rser.2011.02.015
2. Telmo C, Lousada J (2011) Heating values of wood pellets from different species. Biomass Bioenergy 35(7):2634–2639. doi:10.1016/j.biombioe.2011.02.043
3. Acharjee TC, Coronella CJ, Vasquez VR (2011) Effect of thermal pretreatment on equilibrium moisture content of lignocellulosic biomass. Bioresour Technol 102(7):4849–4854. doi:10.1016/j.biortech.2011.01.018
4. Lynam JG, Coronella CJ, Yan W, Reza MT, Vasquez VR (2011) Acetic acid and lithium chloride effects on hydrothermal carbonization of lignocellulosic biomass. Bioresour Technol 102(10):6192–6199. doi:10.1016/j.biortech.2011.02.035
5. Lynam JG, Reza MT, Vasquez VR, Coronella CJ (2012) Effect of salt addition on hydrothermal carbonization of lignocellulosic biomass. Fuel 99:271–273. doi:10.1016/j.fuel.2012.04.035
6. Reza MT, Lynam JG, Vasquez VR, Coronella CJ (2012) Pelletization of biochar from hydrothermally carbonized wood. Environ Prog Sustain Energy 31(2):225–234. doi:10.1002/ep.11615

7. Yan W, Hastings JT, Acharjee TC, Coronella CJ, Vasquez VR (2010) Mass and energy balances of wet torrefaction of lignocellulosic biomass. Energy Fuels 24:4738–4742. doi:10.1021/ef901273n
8. Yan W, Acharjee TC, Coronella CJ, Vasquez VR (2009) Thermal pretreatment of lignocellulosic biomass. Environ Prog Sustain Energy 28(3):435–440. doi:10.1002/ep.10385
9. Funke A, Ziegler F (2010) Hydrothermal carbonization of biomass: A summary and discussion of chemical mechanisms for process engineering. Biofuels Bioprod Biorefin 4(2):160–177. doi:10.1002/bbb.198
10. Bergius F (1913) Die Anwendung hoher Drücke bei chemischen Vorgängenund eine Nachbil dung des Entstehungsprozesses der Steinkohle. In: Wilhelm Knapp, pp 41–58
11. Bergius F, Erasmus P (1928) Naturwissenschaften 16:1
12. Berl E, Schmidt A (1932) Liebigs Ann 493:97
13. Bobleter O, Niesner R, Rohr M (1976) Hydrothermal degradation of cellulosic matter to sugars and their fermentative conversion to protein. J Appl Polym Sci 20(8):2083–2093. doi:10.1002/app.1976.070200805
14. Fischer F (1921) Schrader H BrennstChemie 2:37
15. Schuhmacher JP (1960) Chemical structure and properties of coal XXVI-studies on artificial coalification. Fuel 39(3):223–234
16. Smith RC, Howard HC (1937) J Amer Chem Soc 59:234
17. Tropsch H, Von Philippovich A (1925) Abb Kennt Kohle 7:84–105
18. Guiotoku M, Rambo CR, Hansel FA, Magalhaes WLE, Hotza D (2009) Microwave-assisted hydrothermal carbonization of lignocellulosic materials. Mater Lett 63(30):2707–2709. doi:10.1016/j.matlet.2009.09.049
19. Hoekman SK, Broch A, Robbins C (2011) Hydrothermal carbonization (HTC) of lignocellulosic biomass. Energy Fuels 25(4):1802–1810. doi:10.1021/ef101745n
20. van Krevelen DW (1950) Graphical-statistical method for the study of structure and reaction processes of coal. Fuel 29:269–284
21. Niesner R, Bruller W, Bobleter O (1978) Carbohydrate analysis in hydrothermally degraded plant material by high-pressure liquid-chromatography (hplc). Chromatographia 11(7):400–402. doi:10.1007/bf02312653
22. Davis HG (1983) Direct liquefaction of biomass final report and summary of effort 1977–1983. LBL- 16243, Lawrence Berkeley Laboratory, University of California, Berkeley
23. Ergun S (1981) Review of biomass liquefaction efforts. LBL-13957, Lawrence Berkeley Laboratory, University of California, Berkeley
24. Ruyter HP (1982) Coalification model. Fuel 61(12):1182–1187. doi:10.1016/0016-2361(82)90017-5
25. Bonn G, Concin R, Bobleter O (1983) Hydrothermolysis—a new process for the utilization of biomass. Wood Sci Technol 17(3):195–202
26. Schwald W, Concin R, Bonn G, Bobleter O (1985) Analysis of oligomeric and monomeric carbohydrates from hydrothermal degradation of cotton-waste materials using hplc and gpc. Chromatographia 20(1):35–40. doi:10.1007/bf02260484
27. Schwald W, Bobleter O (1989) Hydrothermolysis of cellulose under static and dynamic conditions at high-temperatures. J Carbohydr Chem 8(4):565–578. doi:10.1080/07328308908048017
28. Adschiri T, Hirose S, Malaluan R, Arai K (1993) Noncatalytic conversion of cellulose in supercritical and subcritical water. J Chem Eng Jpn 26(6):676–680. doi:10.1252/jcej.26.676
29. Bobleter O (1994) Hydrothermal degradation of polymers derived from plants. Prog Polym Sci 19(5):797–841. doi:10.1016/0079-6700(94)90033-7
30. Yu Y, Wu HW (2010) Significant differences in the hydrolysis behavior of amorphous and crystalline portions within microcrystalline cellulose in hot-compressed water. Ind Eng Chem Res 49(8):3902–3909. doi:10.1021/ie901925g

31. Huber GW, Iborra S, Corma A (2006) Synthesis of transportation fuels from biomass: chemistry, catalysts, and engineering. Chem Rev 106(9):4044–4098. doi:10.1021/cr068360d
32. Nimz H (1973) Chemistry of potential chromophoric groups in beech lignin. Tappi 56(5):124–126
33. Paakkari T, Serimaa R, Fink HP (1989) Structure of amorphous cellulose. Acta Polym 40(12):731–734. doi:10.1002/actp.1989.010401205
34. Fink H-P, Philipp B, Paul D, Serimaa R, Paakkari T (1987) The structure of amorphous cellulose as revealed by wide-angle xray scattering. Polymer 28(8):1265–1270
35. Yu Y, Wu HW (2010) Evolution of primary liquid products and evidence of in situ structural changes in cellulose with conversion during hydrolysis in hot-compressed water. Ind Eng Chem Res 49(8):3919–3925. doi:10.1021/ie902020t
36. Matsumura Y, Yanachi S, Yoshida T (2006) Glucose decomposition kinetics in water at 25 MPa in the temperature range of 448–673 K. Ind Eng Chem Res 45(6):1875–1879. doi:10.1021/ie050830r
37. Kabyemela BM, Adschiri T, Malaluan RM, Arai K (1999) Glucose and fructose decomposition in subcritical and supercritical water: Detailed reaction pathway, mechanisms, and kinetics. Ind Eng Chem Res 38(8):2888–2895. doi:10.1021/ie9806390
38. Yu Y, Lou X, Wu HW (2008) Some recent advances in hydrolysis of biomass in hot-compressed, water and its comparisons with other hydrolysis methods. Energy Fuels 22(1):46–60. doi:10.1021/ef700292p
39. Sevilla M, Fuertes A (2009) The production of carbon materials by hydrothermal carbonization of cellulose. Carbon 47(9):2281–2289
40. Guiotoku M, Hansel FA, Novotny EH, Maia C (2012) Molecular and morphological characterization of hydrochar produced by microwave-assisted hydrothermal carbonization of cellulose. Pesqui Agropecu Bras 47(5):687–692
41. Yu Y, Wu HW (2011) Effect of ball milling on the hydrolysis of microcrystalline cellulose in hot-compressed water. AIChE J 57(3):793–800. doi:10.1002/aic.12288
42. Saha BC (2003) Hemicellulose bioconversion. J Ind Microbiol Biotechnol 30(5):279–291. doi:10.1007/s10295-003-0049-x
43. Lynam JG (2012) Pretreatment of lignocellulosic biomass with acetic acid, salts, and ionic liquids in Reno, Nevada. Master's Thesis, University of Nevada, Reno
44. Hu F, Ragauskas A (2012) Pretreatment and Lignocellulosic Chemistry. BioEnergy Res 5(4):1043–1066. doi:10.1007/s12155-012-9208-0
45. Garrote G, Dominguez H, Parajo JC (1999) Hydrothermal processing of lignocellulosic materials. Holz Als Roh-und Werkst 57(3):191–202. doi:10.1007/s001070050039
46. Pinkowska H, Wolak P, Zlocinska A (2011) Hydrothermal decomposition of xylan as a model substance for plant biomass waste—hydrothermolysis in subcritical water. Biomass Bioenergy 35(9):3902–3912. doi:10.1016/j.biombioe.2011.06.015
47. Bobleter O, Binder H (1980) Dynamic hydrothermal degradation of wood. Holzforschung 34(2):48–51. doi:10.1515/hfsg.1980.34.2.48
48. Jacobsen SE, Wyman CE (2002) Xylose monomer and oligomer yields for uncatalyzed hydrolysis of sugarcane bagasse hemicellulose at varying solids concentration. Ind Eng Chem Res 41(6):1454–1461. doi:10.1021/ie001025+
49. Kabel MA, Carvalheiro F, Garrote G, Avgerinos E, Koukios E, Parajo JC, Girio FM, Schols HA, Voragen AGJ (2002) Hydrothermally treated xylan rich by-products yield different classes of xylo-oligosaccharides. Carbohydr Polym 50(1):47–56. doi:10.1016/s0144-8617(02)00045-0
50. Zhang XL, Yang WH, Blasiak W (2011) Modeling study of woody biomass: interactions of cellulose, hemicellulose, and lignin. Energy Fuels 25(10):4786–4795. doi:10.1021/ef201097d
51. Kang SM, Li XH, Fan J, Chang J (2012) Characterization of hydrochars produced by hydrothermal carbonization of lignin, cellulose, D-xylose, and wood meal. Ind Eng Chem Res 51(26):9023–9031. doi:10.1021/ie300565d

52. Binder JB, Gray MJ, White JF, Zhang ZC, Holladay JE (2009) Reactions of lignin model compounds in ionic liquids. Biomass Bioenergy 33(9):1122–1130. doi:10.1016/j.biombioe. 2009.03.006
53. Hemmingson JA, Leary G (1980) The self-condensation reactions of the lignin model compounds, vanillyl and veratryl alcohol. Aust J Chem 33(4):917–925
54. Lynam JG, Reza MT, Vasquez VR, Coronella CJ (2012) Pretreatment of rice hulls by ionic liquid dissolution. Bioresour Technol 114:629–636. doi:10.1016/j.biortech.2012.03.004
55. Pinkowska H, Wolak P, Zlocinska A (2012) Hydrothermal decomposition of alkali lignin in sub- and supercritical water. Chem Eng J 187:410–414. doi:10.1016/j.cej.2012.01.092
56. Reza MT (2013) Upgrading biomass by hydrothermal and chemical conditioning. Ph.D Dissertation, University of Nevada, Reno
57. Fuertes AB, Arbestain MC, Sevilla M, Macia-Agullo JA, Fiol S, Lopez R, Smernik RJ, Aitkenhead WP, Arce F, Macias F (2010) Chemical and structural properties of carbonaceous products obtained by pyrolysis and hydrothermal carbonisation of corn stover. Aust J Soil Res 48(6–7):618–626. doi:10.1071/sr10010
58. Petersen MO, Larsen J, Thomsen MH (2009) Optimization of hydrothermal pretreatment of wheat straw for production of bioethanol at low water consumption without addition of chemicals. Biomass Bioenergy 33(5):834–840. doi:10.1016/j.biombioe.2009.01.004
59. Chen SF, Mowery RA, Scarlata CJ, Chambliss CK (2007) Compositional analysis of water-soluble materials in corn stover. J Agric Food Chem 55(15):5912–5918. doi:10.1021/jf0700327
60. Reza MT, Lynam JG, Uddin MH, Yan W, Vasquez VR, Hoekman K, Coronella CJ (2013) Reaction kinetics and particle size effect on hydrothermal carbonization of loblolly pine. Bioresour Tech 139:161–169. doi:10.1016/j.biortech.2013.04.028
61. Uddin MH (2013) Master's Thesis, University of Nevada Reno, USA
62. Khajavi SH, Kimura Y, Oomori T, Matsuno R, Adachi S (2005) Kinetics on sucrose decomposition in subcritical water. LWT-Food Sci Technol 38(3):297–302. doi:10.1016/j.lwt.2004.06.005
63. Khajavi SH, Kimura Y, Oomori R, Matsuno R, Adachi S (2005) Degradation kinetics of monosaccharides in subcritical water. J Food Eng 68(3):309–313. doi:10.1016/j.jfoodeng.2004.06.004
64. Oomori T, Khajavi SH, Kimura Y, Adachi S, Matsuno R (2004) Hydrolysis of disaccharides containing glucose residue in subcritical water. Biochem Eng J 18(2):143–147. doi:10.1016/j.bej.2003.08.002
65. Sevilla M, Macia-Agullo JA, Fuertes AB (2011) Hydrothermal carbonization of biomass as a route for the sequestration of CO_2: chemical and structural properties of the carbonized products. Biomass Bioenergy 35(7):3152–3159. doi:10.1016/j.biombioe.2011.04.032
66. Dinjus E, Kruse A, Troger N (2011) Hydrothermal carbonization-1. influence of lignin in lignocelluloses. Chem Eng Technol 34(12):2037–2043. doi:10.1002/ceat.201100487
67. Heilmann SM, Jader LR, Sadowsky MJ, Schendel FJ, von Keitz MG, Valentas KJ (2011) Hydrothermal carbonization of distiller's grains. Biomass Bioenergy 35(7):2526–2533. doi:10.1016/j.biombioe.2011.02.022
68. Stemann J, Ziegler F (2012). Hydrothermal carbonization (HTC): recycling of process water. 19th European biomass conference and exhibition, Berlin, pp 1894–1899
69. Inoue S, Hanaoka T, Minowa T (2002) Hot compressed water treatment for production of charcoal from wood. J Chem Eng Jpn 35(10):1020–1023. doi:10.1252/jcej.35.1020
70. Saeman JF (1945) Ind Eng Chem 37:42–52
71. Kobayashi T, Sakai Y (1956) Bull Agr Chem Soc Jpn 20:1–7
72. Grant GA, Han YW, Anderson AW, Frey KL (1977) Kinetics of straw hydrolysis. Dev Ind Microbiol 18:599–611
73. Tellez-Luis SJ, Ramırez JA, Vazquez M (2002) Mathematical modeling of hemicellulosic sugar production from sorghum straw. J Food Eng 52:285–291
74. Sarkar N, Aikat K (2013) Kinetic study of acid hydrolysis of rice straw. ISRN Biotechnol, vol 2013:5. Article ID 170615. doi:10.5402/2013/170615

75. Guerra-Rodriguez E, Portilla-Rivera OM, Jarquin-Enriquez L, Ramirez JA, Vazquez M (2012) Acid hydrolysis of wheat straw: a kinetic study. Biomass Bioenergy 36:346–355. doi:10.1016/j.biombioe.2011.11.005
76. Conner AH, Wood BF, Hill CG, Harris JF (1986) In: Young RA, Rowell RM (eds) Cellulose: structure, modification and hydrolysis. Wiley, New York, pp 281–296
77. Mok WSL, Antal MJ (1992) Uncatalyzed solvolysis of whole biomass hemicellulose by hot compressed liquid water. Ind Eng Chem Res 31(4):1157–1161. doi:10.1021/ie00004a026
78. Bobleter O, Concin R (1979) Cellul Chem Technol 13:583–593
79. Kim SB, Lee YY (1987) Biotechnol Bioeng Symp 17:71–84
80. Kubikova J, Zemann A, Krkoska P, Bobleter O (1996) Hydrothermal pretreatment of wheat straw for the production of pulp and paper. Tappi J 79(7):163–169
81. Zhang B, Huang HJ, Ramaswamy S (2008) Reaction kinetics of the hydrothermal treatment of lignin. Appl Biochem Biotechnol 147(1–3):119–131. doi:10.1007/s12010-007-8070-6
82. Peterson AA, Vogel F, Lachance RP, Froling M, Antal MJ, Tester JW (2008) Thermochemical biofuel production in hydrothermal media: a review of sub- and supercritical water technologies. Energy Environ Sci 1(1):32–65. doi:10.1039/b810100k
83. Grenman H, Ramirez F, Eranen K, Warna J, Salmi T, Murzin DY (2008) Dissolution of mineral fiber in a formic acid solution: kinetics, modeling, and gelation of the resulting sol. Ind Eng Chem Res 47(24):9834–9841. doi:10.1021/ie800267a
84. Prins MJ, Ptasinski KJ, Janssen F (2006) Torrefaction of wood—Part 1. Weight loss kinetics. J Anal Appl Pyrolysis 77(1):28–34. doi:10.1016/j.jaap.2006.01.002
85. Yan W, Islam S, Coronella CJ, Vasquez VR (2012) Pyrolysis kinetics of raw/hydrothermally carbonized lignocellulosic biomass. Environ Prog Sustain Energy 31(2):200–204. doi:10.1002/ep.11601
86. Sjöström E (1993) Wood chemistry, fundamentals and applications, 2nd edn. Academic Press, New York
87. Xu CL, Leppanen AS, Eklund P, Holmlund P, Sjoholm R, Sundberg K, Willfor S (2010) Acetylation and characterization of spruce (Picea abies) galactoglucomannans. Carbohydr Res 345(6):810–816. doi:10.1016/j.carres.2010.01.007
88. Stoklosa RJ, Hodge DB (2012) Extraction, recovery, and characterization of hardwood and grass hemicelluloses for integration into biorefining processes. Ind Eng Chem Res 51(34):11045–11053. doi:10.1021/ie301260w
89. Ando H, Sakaki T, Kokusho T, Shibata M, Uemura Y, Hatate Y (2000) Decomposition behavior of plant biomass in hot-compressed water. Ind Eng Chem Res 39(10):3688–3693. doi:10.1021/ie0000257
90. Nonaka M, Hirajima T, Sasaki K (2011) Upgrading of low rank coal and woody biomass mixture by hydrothermal treatment. Fuel 90(8):2578–2584. doi:10.1016/j.fuel.2011.03.028
91. Phaiboonsilpa N, Yamauchi K, Lu X, Saka S (2010) Two-step hydrolysis of Japanese cedar as treated by semi-flow hot-compressed water. J Wood Sci 56(4):331–338. doi:10.1007/s10086-009-1099-0
92. Lynam JG, Reza MT, Yan W, Vasquez VR, Coronella CJ (2014) Hydrothermal carbonization of various lignocellulosic biomass. Biomass Conversion and Biorefinery (Submitted March 15, 2014)
93. Chiaramonti D, Prussi M, Ferrero S, Oriani L, Ottonello P, Torre P, Cherchi F (2012) Review of pretreatment processes for lignocellulosic ethanol production, and development of an innovative method. Biomass Bioenergy 46:25–35. doi:10.1016/j.biombioe.2012.04.020
94. Rogalinski T, Liu K, Albrecht T, Brunner G (2008) Hydrolysis kinetics of biopolymers in subcritical water. J Supercrit Fluids 46(3):335–341. doi:10.1016/j.supflu.2007.09.037
95. Sun Y, Cheng JJ (2005) Dilute acid pretreatment of rye straw and bermudagrass for ethanol production. Bioresour Technol 96(14):1599–1606. doi:10.1016/j.biortech.2004.12.022
96. Mosier N, Hendrickson R, Ho N, Sedlak M, Ladisch MR (2005) Optimization of pH controlled liquid hot water pretreatment of corn stover. Bioresour Technol 96(18):1986–1993. doi:10.1016/j.biortech.2005.01.013

97. Chen WH, Ye SC, Sheen HK (2012) Hydrothermal carbonization of sugarcane bagasse via wet torrefaction in association with microwave heating. Bioresour Technol 118:195–203. doi:10.1016/j.biortech.2012.04.101
98. Sasaki M, Adschiri T, Arai K (2003) Fractionation of sugarcane bagasse by hydrothermal treatment. Bioresour Technol 86(3):301–304. doi:10.1016/s0960-8524(02)00173-6
99. Liu ZG, Quek A, Hoekman SK, Balasubramanian R (2013) Production of solid biochar fuel from waste biomass by hydrothermal carbonization. Fuel 103:943–949. doi:10.1016/j.fuel.2012.07.069
100. Roman S, Nabais JMV, Laginhas C, Ledesma B, Gonzalez JF (2012) Hydrothermal carbonization as an effective way of densifying the energy content of biomass. Fuel Process Technol 103:78–83. doi:10.1016/j.fuproc.2011.11.009
101. Saddawi A, Jones JM, Williams A, Le Coeur C (2012) Commodity Fuels from biomass through pretreatment and torrefaction: effects of mineral content on torrefied fuel characteristics and quality. Energy Fuels 26(11):6466–6474. doi:10.1021/ef2016649
102. Cuiping L, Chuanzhi W, Yanyonjie Haitao H (2004) Chemical elemental characteristics of biomass in China. Biomass Bioenergy 27:119–130
103. Masia AAT, Buhre BJP, Gupta RP, Wall TF (2007) Characterising ash of biomass and waste. Fuel Process Technol 88(11–12):1071–1081. doi:10.1016/j.fuproc.2007.06.011
104. Jenkins BM, Baxter LL, Miles TR (1998) Combustion properties of biomass. Fuel Process Technol 54(1–3):17–46. doi:10.1016/s0378-3820(97)00059-3
105. Fahmi R, Bridgwater AV, Darvell LI, Jones JM, Yates N, Thain S, Donnison IS (2007) The effect of alkali metals on combustion and pyrolysis of Lolium and Festuca grasses, switchgrass and willow. Fuel 86(10–11):1560–1569. doi:10.1016/j.fuel.2006.11.030
106. Jung BJ, Schobert HH (1992) Improved prediction of coal ash slag viscosity by thermodynamic modeling of liquid-phase composition. Energy Fuels 6(4):387–398. doi:10.1021/ef00034a007
107. Reza MT, Lynam JG, Uddin MH, Coronella CJ (2013) Hydrothermal carbonization: fate of inorganics. Biomass Bioenergy 49:86–94
108. Funke A, Ziegler F (2011) Heat of reaction measurements for hydrothermal carbonization of biomass. Bioresour Technol 102(16):7595–7598. doi:10.1016/j.biortech.2011.05.016
109. Gil MV, Oulego P, Casal MD, Pevida C, Pis JJ, Rubiera F (2010) Mechanical durability and combustion characteristics of pellets from biomass blends. Bioresour Technol 101(22):8859–8867. doi:10.1016/j.biortech.2010.06.062
110. Kaliyan N, Morey RV (2010) Natural binders and solid bridge type binding mechanisms in briquettes and pellets made from corn stover and switchgrass. Bioresour Technol 101(3):1082–1090. doi:10.1016/j.biortech.2009.08.064
111. Reza MT, Uddin MH, Lynam JG, Coronella CJ (2014) Engineered pellets from dry torrefied and HTC biochar blends. Biomass Bioenergy (in press). doi:10.1016/j.biombioe.2014.01.038
112. Gilbert P, Ryu C, Sharifi V, Swithenbank J (2009) Effect of process parameters on pelletisation of herbaceous crops. Fuel 88(8):1491–1497. doi:10.1016/j.fuel.2009.03.015

Part V
Hydrothermal Conversion of Biomass Waste into Fuels

Chapter 13
Organic Waste Gasification in Near- and Super-Critical Water

Liejin Guo, Yunan Chen and Jiarong Yin

Abstract The treatment and utilization of organic wastes is important. The governments have invested huge funds and made great efforts on the research. Among the various options, the near- and supercritical water gasification (NSCWG) is a most promising method. The main advantage is that organic wastes, containing a high water content of 80 wt% or more, could be converted to other substances without drying. This chapter reviews the current status of NSCWG of organic wastes. The reaction systems are introduced first. Then, the theoretical values of gas yields are predicted by thermodynamics analysis and the experimental results without catalysts are investigated extensively. In order to better understand and improve the reactions, the reaction processes, the application of catalysts, and the analysis of kinetics are also discussed.

Nomenclature

GE: Gasification efficiency, the mass of product gas/the mass of feedstock, %
CE: Carbon gasification efficiency, carbon in product gas/carbon in feedstock, %
CODr: COD removal efficiency, 1-COD of aqueous residue/COD of feedstock, %
ER: Oxidant equivalent ratio, amount of oxidant added/the required amount for complete oxidation by stoichiometry calculation, %

L. Guo (✉) · Y. Chen · J. Yin
International Research Center for Renewable Energy, State Key Laboratory of Multiphase Flow in Power Engineering (SKLMF), Xi'an Jiaotong University, Xi'an, Shaanxi 710049, China
e-mail: lj-guo@mail.xjtu.edu.cn

13.1 Introduction

Due to the increasing price and decreasing amount of fossil fuels, it is urgent to find new renewable and substitute energy resources. In recent years, energy obtained from biomass and organic wastes has gained much attention. Thermochemical and biochemical/biological processes are two of the main technologies in the conversion of biomass and organic wastes, and thermochemical gasification is likely to become a cost-effective method to produce gas fuel. However, a large portion of biomass and organic wastes are wet biomass containing much water. They should be dried first so as to be used in the process of conventional thermochemical gasification. Thus, the NSCWG of biomass and organic wastes has been developed and it is an advanced technology which can deal with high-moisture biomass and produce more hydrogen. The supercritical water (≥ 374 °C, ≥ 22.1 MPa) also has special physical and chemical properties, such as high diffusion rates, low viscosity, and low dielectric constant [1–3]. These unique features make it a better solvent for the dissolution of various organic matters, so that the reaction could happen in a homogenous condition and could be better promoted. Compared with the process of conventional gasification, the reaction temperature of NSCWG of biomass and organic wastes is much lower. The temperature of conventional steam gasification is always above 1,000 °C, while the complete gasification of glucose could be achieved in supercritical water (650 °C, 35.4 MPa). It does not generate NOx and SOx in the gasification process, and the concentration of CO is very low when the reaction temperature is high or the catalyst is used.

In recent decades, the NSCWG technology has been investigated by some institutions, such as laboratories at University of Hawaii, University of Michigan, Pacific Northwest National Laboratory, Forschungszentrum Carlsruhe (FzK) of Germany, Hiroshima University, SKLMF of Xi'an Jiaotong University, and in the Netherlands (University of Twente, TNO-MEP and BTG) etc. In their studies, various reactor systems, thermodynamics analysis, experimental conditions, related reaction process and pathway, the application of catalysts in the reaction, and kinetics analysis were investigated. Various model compounds and biomass were investigated [4–13], however, the investigation of organic wastes was relatively less.

In organic wastes, their treatment in near- and supercritical water could not only obtain the useful products but also reduce the pollution simultaneously. In this chapter, the NSCWG of organic wastes in recent decades will be reviewed. First, the different reactor systems used in NSCWG of biomass and organic wastes are introduced, which includes batch reactor, tubular continuous reactor, and fluidized bed continuous reactor. Second, the thermodynamics analysis on the gasification of organic wastes is introduced and it will provide the theoretical results of the gasification of organic wastes. Third, the experimental results on the gasification of organic wastes are introduced and the effects of various parameters such as temperature, pressure, concentration, residence time, and oxygen addition are discussed. Fourth, the related reaction process and pathway in the gasification reaction is introduced. Fifth, the application of catalysts in the gasification of organic wastes is introduced, and the catalysts include homogenous catalysts and heterogeneous catalysts. Sixth, the analysis of kinetics on the gasification of organic waste is introduced and the investigation of it is helpful for improving the process of gasification. Finally, the challenges and prospect of the NSCWG technology is discussed.

13.2 Reactor System

The reactor system is important for the gasification. Due to the high reaction temperature and pressure in NSCWG of biomass and organic wastes, the reactors which are used in the condition have some special characteristics. The batch reactor, tubular continuous reactor, and fluidized bed continuous reactor have been developed for NSCWG.

13.2.1 Batch Reactor

Several batch reactors were used in some institutes [4, 8, 14–18]. The steel batch reactors were favored for studying the yield and distribution of products of the gasification of various feedstock materials, and they were also used to investigate the reaction process and compare the different catalytic activities when different catalysts were used. In SKLMF, a Parr 4575A steel batch reactor with a stirring paddle was purchased. The volume of this reactor is 0.5 L, the maximum operating temperature and pressure are 773 K and 35 MPa, respectively. Another steel batch reactor was developed. Figure 13.1 shows the schematic diagram of the batch reactor system which was designed for the temperature up to 650 °C and pressure up to 35 MPa. The volume of the reactor chamber is 140 ml and the reactor is made of 316 L stainless. The reactor is heated by 1.2 kW temperature controlled electric furnace and cooled by water. The temperature is controlled by the PID controller and pressure is monitored by type K thermocouple and pressure gauge, respectively. More detail information about the batch reactor can be found elsewhere [19]. A disadvantage of the steel batch reactor is that it takes a longtime to heat the reactor and feedstock to meet the specified reaction conditions. If the gasification rate is higher than the heating rate, the conversion would proceed at undefined temperatures.

In order to overcome the disadvantage of steel batch reactor, the quartz tube batch reactor was made in SKLMF. The inner diameter of the quartz tube batch reactor is 0.75 mm and the external diameter is 1.5 mm, and the length is 200 mm. The quartz tube batch reactor has the following advantages:

Fig. 13.1 Schematic diagram of the batch reactor system [19]

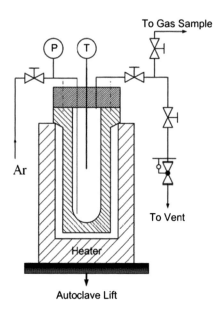

(1) The resistance of high temperature and high pressure. The melting point of quartz glass is 1,730 °C and it could be used for a long time at 1,100 °C. The highest pressure reached 60 MPa when the reaction temperature is 900 °C.
(2) Fast heating rate. The heating rate is very fast and the heating time is only 20 s when the reaction temperature increases to 850 °C from room temperature.
(3) Corrosion resistance. Quartz glass does not react with acids except for hydrofluoric acid and it has excellent chemical stability at high temperature.
(4) Quartz glass does not have any effect (incl. catalysis) on the reaction kinetics.
(5) Due to the transparency of quartz glass, some reaction products such as tar and char could be directly observed.

The defects of quartz tube batch reactor are that the pressure inside a capillary could not be measured and it is derived indirectly from the implied temperature. And it is not resistant to alkali at high temperature.

Although the quartz tube batch reactor has some advantages, the disadvantages also exist. Therefore, the micro-steel tube batch reactor was developed in SKLMF and it overcomes the disadvantages of quartz tube batch reactor. It is designed for the temperature up to 800 °C and pressure up to 40 MPa. The inner diameter of the reactor chamber is 77 mm and the length is 200 mm, and the reactor is made of inconel 625. The reactor is heated by 0.8–1 kW temperature-controlled electric furnace and cooled by air. The temperature is controlled by the PID controller and pressure is monitored by type K thermocouple and pressure gauge, respectively. The heating rate is also fast and the heating time only is 10 min when the reaction temperature increases to 750 °C from room temperature. This reactor could not only solve the problems existing in quartz tube batch reactor but also could investigate more influence factors on the NSCWG of biomass and organic wastes.

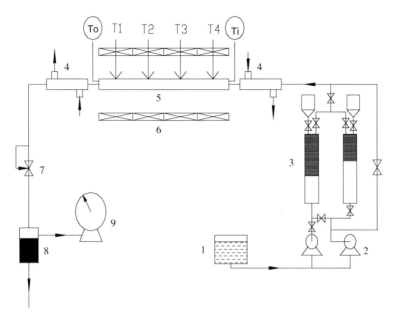

Fig 13.2 Schematic diagram of the miniature scale tubular continuous reactor system: *1* water tank; *2* high pressure pump; *3* feedstock tank; *4* cooler; *5* reactor; *6* heater; *7* back pressure regulator; *8* gas liquid separator; *9* wet test meter [20]

13.2.2 Tubular Continuous Reactor

The advantages of batch reactor system are that the structure is simple and without high-pressure transport system. Therefore, it could be used to study the gasification of almost all biomass and organic wastes. However, the reaction processes are not isothermal. It needs some time to heat and cool the reactor system. In NSCWG of biomass and organic wastes, there are some reactions that occur during the heating stages of the experiments in the batch reactor. Although the yield of gas products is little below 250 °C, feedstock transformation appears to become significant. The formation of gas products becomes significant above 300 °C.

The tubular continuous reactor could make up for the disadvantage of the batch reactor. Figure 13.2 shows the schematic of the miniature scale tubular continuous reactor in SKLMF. The reactor is made of special stainless steel with 6 or 9 mm inner diameter and the corresponding external diameter is 14 or 18 mm, respectively, and the length is 650 mm. it is designed for temperature up to 650 °C and pressure up to 35 MPa. The unique feature of the experimental apparatus is that it could realize the overall continuous reaction by operating the valves. There are two feed tanks in the experimental system. One tank could be kept at high pressure for reaction and the other could be fed with the material at atmospheric pressure simultaneously. Both the feeder and the reactor are pressurized at the beginning of a run. Water is pumped into the reactor tube directly and pressurized in the reactor.

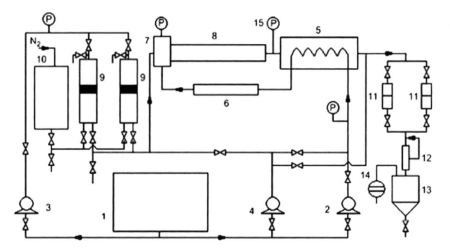

Fig. 13.3 Schematic diagram of the bench scale tubular continuous reactor system: *1* water tank; *2* preheated water pump; *3* feedstock pump; *4* wash pump; *5* cooler and heat exchanger; *6* preheater; *7* mixer; *8* reactor; *9* feeder; *10* feed tank; *11* filter; *12* back pressure regulator; *13* gas–liquid separator; *14* gas flowmeter; *15* pressure gauge [22]

Then, the reactor is heated and the system reaches the set temperature and pressure. When the temperature and pressure meet the reaction condition, the valve between the feeder and the reactor is opened. Then, water flowing in the reactor is replaced by feeding. At the exit of the reactor, product is rapidly cooled down by a water jacket. A back pressure regulator is used to decrease the exit pressure of product fluid to atmospheric pressure. After leaving the back pressure regulator, the products pass through a gas–liquid separator. The more detailed experimental procedures were provided by Hao et al. [20].

In order to further optimize the miniature scale tubular continuous reaction system and enhance the gasification, a novel bench scale continuous flow system was developed. The experimental apparatus of the continuous flow system is shown in Fig. 13.3. The system includes reactor, heat exchanger, high pressure pumps, feeders, separator, cooler, etc. The reactor is made of Hastelloy C276 tubing and the external diameter, inner diameter, and the length of tube are 17.15 mm, 10.85 mm, and 1.24 m, respectively. The design temperature and pressure are 800 °C and 30 MPa, respectively. Fast heating of reactant is realized by mixing with the preheated water. This could suppress the side reactions at the lower temperature which is unfavorable for the NSCWG [21]. The reaction temperature is controlled by six external electric heaters connected to a PID controller. The temperature is measured by a K type thermocouple inserted in the center of reactor. Behind the reactor, the fluid is cooled down by a cooler to realize the fast termination of the reaction. The pressure of the system is controlled by a back pressure regulator. The reaction products are separated into liquid and gas phases in a gas–liquid separator after the back pressure regulator. The flow rate of gaseous

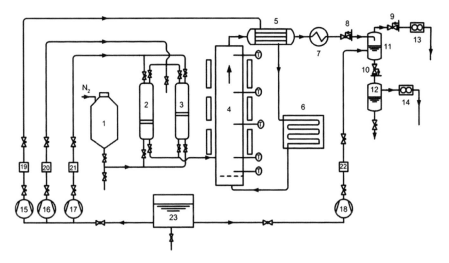

Fig. 13.4 Scheme diagram of sewage sludge gasification in supercritical water with a fluidized bed reactor: *1* feedstock tank; *2, 3* feeder; *4* fluidization bed reactor; *5* heat exchanger; *6* preheater; *7* cooler; *8–10* back pressure regulator; *11* high pressure separator; *12* low pressure separator; *13, 14* wet test meter; *15–18* high pressure metering pump; *19–22* mass flow meter; *23* water tank. (Reprinted with permission from Ref. [23]. Copyright 2008, International journal of hydrogen energy)

product is measured using a wet type gas meter and the liquid is collected as the effluent for further analyses. More detailed description of this system could be found in our previous work [22].

13.2.3 Fluidized Bed Continuous Reactor

Although the NSCWG of biomass and organic wastes has been widely investigated in the continuous tubular reactor, reactor plugging is still a critical problem. In order to solve this problem, a novel supercritical water fluidized bed system which prevents the plugging was developed successfully in SKLMF, and it is designed for the temperature up to 650 °C and the pressure up to 30 MPa. The schematic diagram of supercritical water fluidized bed gasification system is shown in Fig. 13.4. The fluidized bed reactor is made of 316 stainless, and the total length is 915 mm. The bed diameter and freeboard diameter are 30 and 40 mm, respectively. The design temperature and pressure of reactor are 650 °C and 30 MPa, respectively. The reactor is heated by three 2 kW electrical heaters which are coiled around the outer surface of the reactor. The temperature of the reactor wall is measured by some type K thermocouples held on the outer wall of the reactor. The temperature of fluid is measured by five type K thermocouples inserted in the center of reactor. The distributor is located in the bottom of the reactor. In order to prevent the bed material escaping from the reactor, a metal foam filter is installed at the exit of the reactor. The pressure of the system is controlled by a back pressure regulator with a deviation of ±0.5 MPa. More details about the reactor and system could be found elsewhere [23]. This system solved the plugging problem and enhanced the gasification of higher concentration of biomass and organic wastes.

13.3 Thermodynamics Analysis

Thermodynamic analysis is very important and helpful for design, optimization, and operation of gasification system, which could provide the theoretical base. It has been used in NSCWG of model compounds and biomass by some researchers [24–32], and it includes equilibrium calculations and exergy analysis. The theoretical product compositions of gasification of model compounds and biomass could be obtained by equilibrium calculations and the exergy analysis could analyze the degree of thermodynamic perfection of an energy system. The exergy calculation of flow streams of different units in NSCWG needs the results of equilibrium calculation.

Equilibrium models which are used in equilibrium analysis could predict equilibrium compositions and thermodynamic limit of NSCWG of model compounds and biomass. They could be used as a guide to the process design, evaluation, and improvement. There are two general approaches in equilibrium modeling: stoichiometric and nonstoichiometric. The stoichiometric approach requires a clearly defined reaction mechanism, including all the chemical reactions and species involved in the process. However, the nonstoichiometric method does not require reaction mechanism and species involved in the process and the only inputs are reaction temperature, pressure, and elemental compositions of the feedstock. Therefore, this nonstoichiometric method is particularly suitable for unclear reaction mechanisms and chemical compositions of feeding such as biomass and organic wastes.

The nonstoichiometric equilibrium based on Gibbs free energy minimization approaches is used by researchers for the studies of NSCWG of model compounds and biomass. Tang and Kitagawa [26] studied the thermodynamic analysis of the gasification of model compounds and biomass in supercritical water with direct Gibbs free energy minimization and the Peng–Robinson EoS formulation being used. In SKLMF, Yan et al. [27] and Lu et al. [24] investigated the thermodynamic analysis in the gasification of model compounds and biomass in supercritical water with Gibbs free energy minimization and the Duan EoS formulation being used. Lu et al. [24] also studied the exergy and energy analysis of the whole system and some advices were provided for the improvement and operation optimization of the experimental system. In NSCWG of organic wastes, the thermodynamic analysis method used by Yan et al. [27] and Lu et al. [24] could also be used to

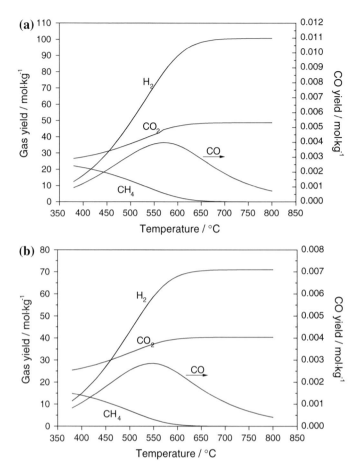

Fig. 13.5 Effect of temperature on equilibrium gas yields of 5 wt% feedstock gasification at 25 MPa: **a** sewage sludge, **b** black liquor

calculate the equilibrium compositions. The equilibrium compositions in NSCWG of sewage sludge and black liquor were studied.

Figure 13.5 shows the effect of temperature on equilibrium gas yields of gasification of sewage sludge and black liquor at 25 MPa. The change trends of equilibrium gas yields of gasification of sewage sludge and black liquor were similar. The equilibrium yields of H_2 and CO_2 increased with the increase of temperature, but the yield of CH_4 decreased sharply. The maximal equilibrium yields of H_2 reached 100.7 mol/kg dry sewage sludge and 80 mol/kg dry black liquor. The equilibrium yield of CO was very small. The CO yield firstly increased and then decreased when the temperature increased from 380 to 800 °C. The effect of pressure on equilibrium gas yield at 600 °C with 5 wt% feedstock content is shown in Fig. 13.6. As can be seen, the equilibrium gas yields had no significant

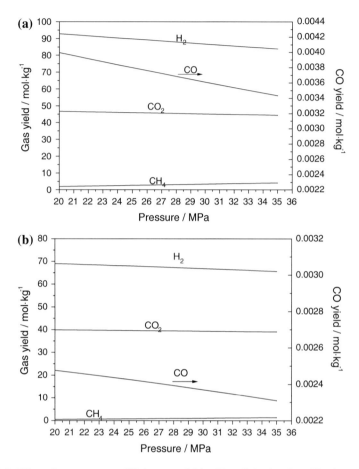

Fig. 13.6 Effect of pressure on equilibrium gas yields of 5 wt% feedstock gasification at 600 °C: **a** sewage sludge, **b** black liquor

change with the increase of pressure. It indicated that the pressure had no significant effect on gasification. Figure 13.7 shows the effect of feedstock concentration on equilibrium gas yield at 600 °C and 25 MPa. With the increase of feedstock concentration, the equilibrium gas yields of H_2, CO_2 decreased and the yield of CO, CH_4 increased. The effect of oxidant addition on equilibrium gas yield was predicted and the results are shown in Fig. 13.8. It showed that the yields of H_2, CO, and CH_4 decreased with the increase of the oxidant addition, and the yield of CO_2 increased with the increase of the oxidant addition.

The results obtained from the thermodynamics analysis indicated that temperature was important and higher temperature was essential to hydrogen production. The pressure had no significant effect on gasification. The concentration and oxidant addition had negative effects on the process of gasification.

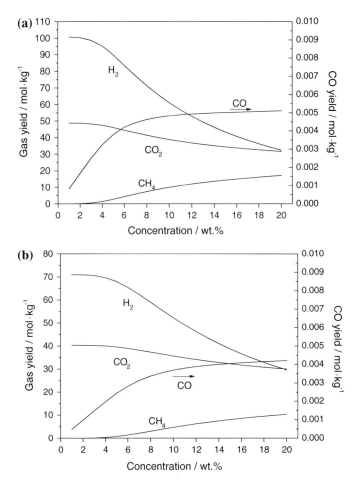

Fig. 13.7 Effect of feedstock concentration on equilibrium gas yields at 600 °C and 25 MPa: **a** sewage sludge, **b** black liquor

13.4 Experiments with Organic Wastes

From the thermodynamics analysis, the different influence factors on NSCWG of organic wastes were investigated. The change trends and equilibrium yields of gas products were obtained. In order to know the actual change trends and yields of gas products, the NSCWG of organic wastes were also investigated by some researchers. However, the number of reports is still small. Jarana et al. [33] gasified the industrial organic wastewaters including oleaginous wastewater and alcohol distillery wastewater in supercritical water with a laboratory-scale continuous flow system. Nakhla et al. [34] investigated the gasification and partial oxidation of hog manure in supercritical water with batch reactor. Onwudili and Williams [35] studied the

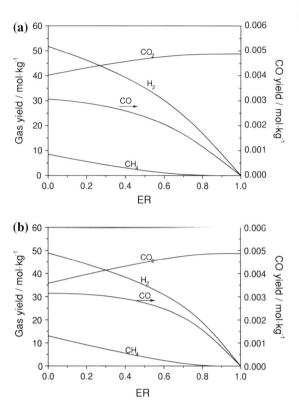

Fig. 13.8 Effect of oxygen addition on equilibrium gas yields of 5 wt% feedstock gasification at 500 °C and 25 MPa: **a** sewage sludge, **b** black liquor

reforming of biodiesel plant waste in sub- and supercritical water. Yanik et al. [36] studied the gasification of leather waste in supercritical water with batch reactor. In sewage sludge gasification, Schmieder et al. [16], Yamamura et al. [37], Zhang et al. [38] and Afif et al. [39] studied the NSCWG of sewage sludge with batch reactor. Vostrikov et al. [40] studied the gasification of sewage sludge in supercritical water with semi-batch reactor and Antal et al. [41, 42] studied the gasification of sewage sludge in supercritical water with microminiature continuous reactor. The NSCWG of sewage sludge was also investigated in SKLMF, and they did not only study it in batch reactor but also study it in the bench scale-fluidized bed reactor for the first time [43, 44]. In black liquor gasification, Sricharoenchaikul [45] investigated the NSCWG of black liquor in a quartz capillary reactor. The NSCWG of black liquor in batch reactor (140 ml volume) was investigated in SKLMF. At present, the NSCWG of it in microminiature continuous reactor and laboratory-scale continuous reactor was only investigated in SKLMF. The gasification of anaerobic organic wastewater which is from the anaerobic fermentation of wheat straw including acids (acetic acid, butyric acid) and ethanol, etc., in supercritical water was also investigated with microminiature continuous reactor system in SKLMF. In the above studies, they both found that H_2, CO_2 and CH_4 were the main gas products and some trace of gas such as CO, C_2H_4, and C_2H_6 also existed. The effects of temperature, pressure, concentration, residence time, the addition of catalyst, and oxygen on the gasification of organic wastes have been investigated widely whenever the reaction happened in the batch or continuous reactors.

13.4.1 Effect of Temperature

Based on the results of thermodynamics analysis, it is known that temperature is the most important influence factor in the theoretical analysis. And the temperature is perhaps the most important influence factor in the experiments of gasification [1]. It had a significant effect on NSCWG of biomass especially in absence of catalyst or less-effective catalyst [10]. In NSCWG of organic wastes, the temperature might be also important.

Jarana et al. [33] studied the effect of reaction temperature on the gasification of oleaginous wastewater in supercritical water with a laboratory-scale continuous flow system. The temperature was 450, 500, and 550 °C respectively, the pressure was maintained at 25 MPa and the initial concentration of the wastes expressed as COD was around 12 g O_2/l. They found that the main gas products are H_2, CO_2, CH_4, and CO. The COD removal efficiency (CODr), yield of H_2 and yield of CO_2 increased markedly with the increase of temperature, and the yield of CH_4 increased and the yield of CO was practically negligible. The maximum CODr reached 80 % at 550 °C. Onwudili and Williams [35] investigated the effect of temperature on reforming of biodiesel plant wastes in sub- and supercritical water. The temperature was from 300 to 450 °C and the pressure was from 8 to 31 MPa. They found that the total yield of gas and the yield of H_2, CO_2, C1–C4 hydrocarbons increased with the increase of temperature. However, the yield of CO decreased with the increase of temperature.

Yamamura et al. [37] studied the gasification of sewage sludge containing 99.1 wt% water in supercritical water with batch reactor, and the reaction temperature was 400, 450, 500 °C. They found that the yield of H_2 increased with the increase of temperature. Zhang et al. [38] studied the gasification of secondary pulp/paper sludge containing 98 wt% water in supercritical water with batch reactor, and the reaction temperature is 400, 450, 500, and 550 °C. They found that the yield of H_2, CO_2, and CH_4 increased continuously with the increase of temperature. As the temperature increased from 500 to 550 °C, the yield of H_2 and CH_4 increased obviously. The yield of H_2 and CH_4 had an almost 10-fold increase in the whole temperature range. Afif et al. [39] studied the NSCWG of activated sludge in batch reactor, and the effect of varying temperature from subcritical (320 °C) to supercritical (410 °C) conditions was reported. They found that the total gas yield increased with the increase of temperature, and the yield of H_2, CH_4 also increased with the increase of temperature. The temperature also had a positive effect on carbon gasification efficiency (CE) and hydrogen gasification efficiency (HE). In SKLMF, the NSCWG of sewage sludge was also studied in batch reactor and fluidized bed reactor. Guo et al. [43] studied the effect of

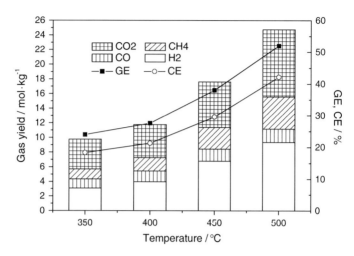

Fig. 13.9 Effect of temperature on gas yield, GE, and CE of NSCWG of 7.8 wt% black liquor in batch reactor (black liquor: 10 g, N_2 initial pressure: 4 MPa, residence time: 10 min)

temperature on municipal sludge gasification with batch reactor. They found that the molar fraction of H_2 and CODr increased with the increase of temperature. The maximum CODr reached 95.3 % without catalysts. Chen et al. [44] studied the effect of reaction temperature on the gasification of activated sludge in supercritical water with fluidized bed reactor, and it was the first time to study the gasification of activated sludge in fluidized bed reactor. The range of temperature was from 480 to 540 °C, the pressure was 25 MPa and the concentration of sludge was 4 wt%. They found that the yield of H_2 and CH_4 increased with the increase of temperature. The GE and CE also increased with the increase of temperature.

Black liquor is wastewater which is produced in pulping process and contains about 90 % COD concentration of the pulping wastes. It mainly contains lignin and its derivatives and high content of alkali wastes. At present, the NSCWG of black liquor was investigated in a quartz capillary reactor by Sricharoenchaikul [45]. They found that the increase of reaction temperature leaded to an increase in the product distribution of gas and a decrease in solid and liquid fractions at all operating conditions. The maximum distribution of gas products reached 75.4 % at 650 °C and 10 wt% concentration. In SKLMF, the gasification of black liquor was also studied in batch reactor. Figure 13.9 shows the effect of temperature on NSCWG of black liquor in batch reactor (140 ml volume). With the increase of temperature, the GE and CE increased. The yield of H_2 and CH_4 also increased with the increase of temperature. So far, the gasification of black liquor in continuous reactor has only been studied in SKLMF. It has been not only studied in microminiature continuous reactor but also in laboratory-scale continuous reactor. In microminiature continuous reactor, the black liquor contained 7.8 wt% solid material was gasified in supercritical water at 22.5 MPa, the flow rate was 0.36 kg/h and the temperature was in the range from 600 to 750 °C. The effect of temperature on gasification is shown in

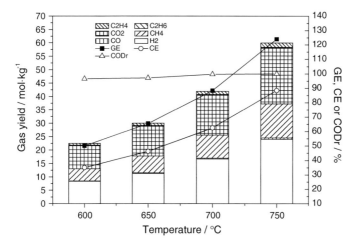

Fig. 13.10 Effect of temperature on gas yield, GE, CE, and CODr of 7.8 wt% black liquor gasification in supercritical water with microminiature continuous reactor (22.5 MPa, flow rate: 6 g/min)

Fig. 13.10. As the temperature increased, the yield of H_2, GE, CE and CODr increased significantly. The maximum yield of H_2, GE, CE, and CODr reached 24 kg/mol, 123, 88, and 100 %, respectively at 750 °C. In laboratory-scale continuous reactor, Cao et al. [46] studied the gasification of black liquor in supercritical water. The black liquor contained 9.5 wt% solid material was gasified in supercritical water at 25 MPa, the flow rate was 5 kg/h and the temperature was in the range of 400–600 °C. With the increase of temperature, the total gas yield almost doubled and the yield of H_2 increased. The maximum yield of H_2 reached 13.67 mol/kg at 600 °C. The GE and CODr also increased with the increase of temperature. The maximum GE and CODr reached 67.89 and 88.69 % at 600 °C, respectively.

The gasification of anaerobic organic wastewater, which was obtained by fermentation of wheat straw, in supercritical water was investigated with a microminiature continuous reactor. The pressure was 25 Mpa, the flow rate was 6 ml/min and the temperature was in the range from 700 to 775 °C. The effect of temperature on the gasification of anaerobic organic wastewater in supercritical water is shown in Fig. 13.11. With the increase of temperature, the GE and CE both increased. The maximum GE almost reached to 100 % at 775 °C. The yield of H_2 and CH_4 also increased with the increase of temperature and the maximum yield of H_2 reached 19.92 mol/kg at 775 °C.

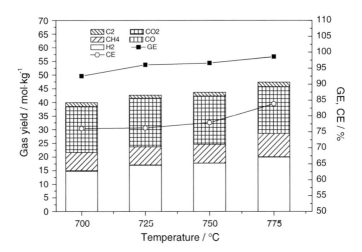

Fig. 13.11 Effect of temperature on gas yield, GE, and CE of anaerobic organic wastewater gasification in supercritical condition with microminiature continuous reactor (25 MPa, flow rate: 6 ml/min, 6 wt% anaerobic organic wastewater +2 wt% CMC)

13.4.2 Effect of Pressure

The effect of pressure on the gasification of biomass in supercritical water is complex and could be explained by roles of water for chemical reaction in supercritical water [47]. Hao et al. [20] and Lu et al. [48] studied the gasification of glucose and biomass in supercritical water respectively, and they found that pressure had no significant effect on gasification. The results obtained from thermodynamics analysis indicated that pressure had no significant effect on NSCWG of organic wastes. Some experimental results were also reported. Figure 13.12 shows the effect of pressure on NSCWG of black liquor in microminiature continuous reactor. The reaction temperature was 750 °C, the flow rate was 6 g/min and the pressure was in the range from 15 to 27.5 MPa. It was found that the increase of pressure could enhance the increase of GE, CE, and the yield of H_2. However, the pressure had no significant effect when the reaction only happened in subcritical condition or supercritical condition. The effect of pressure on NSCWG of anaerobic organic wastewater in microminiature continuous reactor is shown in Fig. 13.13. The reaction temperature was 775 °C, the flow rate was 6 ml/min, and the pressure was in the range from 20 to 27.5 MPa. With the increase of pressure, the change trend of GE, CE, and the yield of H_2 are similar to the change trend of black liquor gasification. The maximum GE, CE, and yield of H_2 reached 109.69 %, 87.12 %, and 5.29 mol/kg, respectively at 27.5 Mpa.

The above analysis results indicate that pressure has no significant effect when the gasification of organic wastes only happened in subcritical water or supercritical water, however, it has more significant effect when the reaction condition transfers from subcritical condition to supercritical condition.

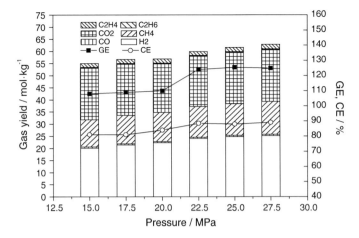

Fig. 13.12 Effect of pressure on gas yield, GE, and CE of NSCWG of 7.8 wt% black liquor in microminiature continuous reactor (750 °C, flow rate: 6 g/min)

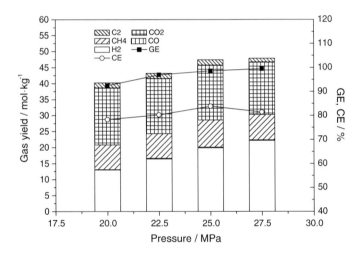

Fig. 13.13 Effect of pressure on gas yield, GE, and CE of NSCWG of anaerobic organic wastewater in microminiature continuous reactor (775 °C, flow rate: 6 ml/min, 6 wt% anaerobic organic wastewater +2 wt% CMC)

13.4.3 Effect of Concentration

The effect of concentration on the gasification of sewage sludge in supercritical water with fluidized bed reactor was investigated by Chen et al. [44]. The range of concentration was from 4 to 12 wt%, the pressure was 25 MPa and the temperature was 540 °C. They found that the yield of H_2 and CH_4 increased with the increase of concentration. In the whole concentration range, the GE and CE

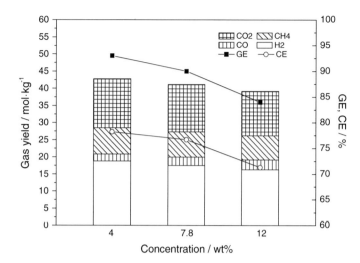

Fig. 13.14 Effect of concentration on gas yield, GE, and CE of 7.8 wt% black liquor gasification in supercritical water with batch reactor (500 °C, black liquor: 10 g, MnO_2: 1 g, N_2 initial pressure: 4 MPa, residence time: 10 min)

decreased with the increase of concentration. Sricharoenchaikul [45] investigated the effect of concentration on black liquor gasification. They found that the yield of H_2 and CH_4 decreased with the increase of concentration. In SKLMF, the effect of concentration on the gasification of black liquor with batch reactor (140 ml volume) was investigated and the results are shown in Fig. 13.14. As can be seen, the GE, CE, the yield of H_2, and CH_4 decreased with the increase of concentration. The effect of concentration on the gasification of black liquor with microminiature continuous reactor was also studied and the results are shown in Fig. 13.15. It also shows that the GE, CE, and yield of H_2 and CH_4 decreased with the increase of concentration. The effect of concentration on the gasification of anaerobic organic wastewater with microminiature continuous reactor is shown in Fig. 13.16. The change trend of GE, CE, and the yield of H_2 is similar to those of black liquor.

The above results indicate that the increase of feedstock concentration has a negative effect on gasification. Gasification of feedstock with high organic waste content is more difficult than that with low organic waste content. The experimental results are in accordance with the results of thermodynamics analysis.

13.4.4 Effect of Residence Time

The residence time also has some effects on the gasification of organic wastes. Most of the works showed that the conversion of biomass increased with the increase of residence time but it had a small change when the residence time increased further [18–20, 49]. In batch reactor, the residence time is the duration of

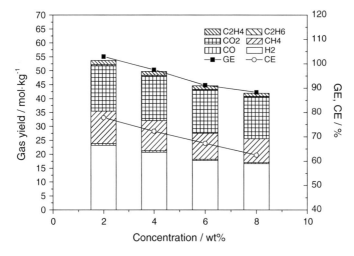

Fig. 13.15 01Effect of concentration on gas yield, GE, and CE of 7.8 wt% black liquor gasification in supercritical water with microminiature continuous reactor (22.5 MPa, 700 °C, flow rate: 6 g/min)

time after the reaction temperature reached the specified condition. In continuous reactor, the residence time is defined as the reactor volume divided by volumetric flow rate of water at the reaction temperature and pressure.

In batch reactor, Nakhla et al. [34] investigated the effect of residence time (30, 60, and 90 min) on gasification of hog manure at 500 °C and 31 MPa when oxygen dose equaled to 80 % of the theoretical COD required to oxidize all the initial COD. They found that there was a little influence on the yield of H_2, CH_4, and CODr. The maximum CODr reached to 85 % at 60 min. In sewage sludge gasification, Yamamura et al. [37] studied the effect of residence time on the gasification of sewage sludge, and they found that the yield of H_2 reached saturation after only 30 min and the residence time had a little influence on H_2 yield. Afif et al. [39] studied the effect of residence time on gasification of activated sludge. They found that the total gas yield increased with the increase of residence time and approached a plateau level after 30 min. As the residence time increased further, there was a small change in the total gas yield. The change trend of GE, CE, the yield of H_2, and CH_4 was similar to the change trend of total gas yield. Zhang et al. [38] studied the effect of residence time (20–120 min) on the gasification of secondary pulp/paper sludge in supercritical water at 500 °C. They found that the residence time had a little influence on the yield of total gas, H_2, and CH_4. Guo et al. [43] also studied the effect of residence time (30–120 min) on the gasification of municipal sludge. They found that the residence time had a little influence on CODr. In black liquor gasification, Sricharoenchaikul [45] investigated the effect of residence time (5, 60, 120 s) on the gasification of black liquor at 650 °C and 30 MPa. They found that the maximum carbon conversion of 84.8 % was achieved at 120 s and 10 wt% concentration. Figure 13.17 showed the

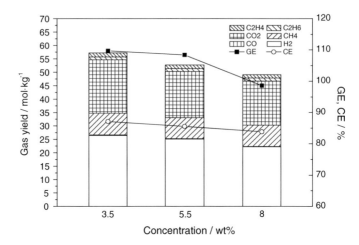

Fig. 13.16 Effect of concentration on gas yield, GE, and CE of anaerobic organic wastewater gasification in supercritical condition with microminiature continuous reactor (27.5 MPa, 775 °C, flow rate: 6 ml/min)

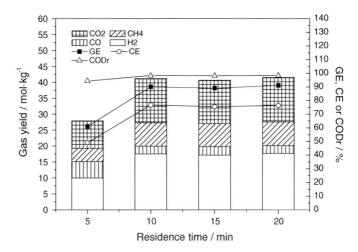

Fig. 13.17 Effect of residence time on gas yield, GE, and CE of 7.8 wt% black liquor gasification in supercritical water with batch reactor (500 °C, black liquor: 10 g, MnO_2: 1 g, N2 initial pressure: 4 MPa)

effect of residence time on the gasification of black liquor at 500 °C with batch reactor (140 ml volume). As the residence time increased from 5 to 10 min, the GE and CE increased from 61, 48 to 90, 76 %, respectively. They had a small change when residence time increased further. The change trend of H_2 yield and

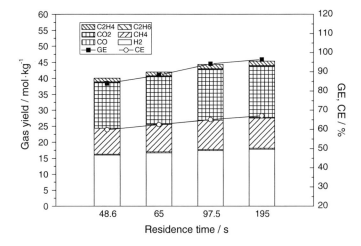

Fig. 13.18 Effect of residence time on gas yield, GE, and CE of 7.8 wt% black liquor gasification in supercritical water with microminiature continuous reactor (700 °C, 22.5 MPa, flow rate: 6 g/min)

CODr was similar to the change trend of GE and CE. The maximum yield of H_2 and CODr reached 17.5 mol/kg and 98.2 % at 10 min.

In continuous reactor, the effect of residence time (48.6–195 s) on the gasification of black liquor with microminiature continuous reactor is shown in Fig. 13.18. As can be seen, the residence time had a little influence on the yield of H_2. However, the increase of residence time enhanced the gasification. As the residence time increased from 48.6 to 195 s, the GE and CE increased from 84 and 59 to 96 and 67 %, respectively. Cao et al. [46] studied the effect of residence time (5.2–12.4 s) on the gasification of black liquor with laboratory-scale continuous reactor. They also found that the residence time also had a little influence on the gas fraction, and the increase of residence time enhanced the gasification. As the residence time increased from 5.2 to 12.4 s, the GE increased from 42.50 to 54.05 % and the CODr increased from 70.84 to 83.56 %.

13.4.5 Effect of Oxygen Addition

The addition of oxygen is also important on the gasification of organic wastes. Jarana et al. [33] investigated the effect of oxygen addition on the gasification of alcohol distillery wastewater and oleaginous wastewater. They found that the addition of H_2O_2 enhanced the increase of CODr, and enhanced the formation of H_2, CH_4, and CO with or without catalyst at 450 °C and 25 MPa. The yield of H_2 increased when the oxygen coefficient increased from 0.07 to 0.19 at 550 °C and CODr is above 80 % in all experiments. Nakhla et al. [50] studied the effect of oxygen dose, equaled to 60 and 80 % of the theoretical COD required to destroy all the initial COD, on the yield of gas products with Pd/AC. They found that the yield of H_2 increased and the yield of CH_4 decreased when the oxygen dose increased from 60 to 80 %, and they indicated that methanation reaction might have occurred and consumed part yield of H_2. The CODr reached 79 and 81 % when an oxygen dose equaled to 60 and 80 %, respectively. Guo et al. [43] studied the effect of oxygen addition on the gasification of municipal sludge. The experiments were conducted at 410 °C, 25 MPa with 30 min reaction time. They found that the molar fraction of H_2 reached to a maximum value at 0.3 oxidant equivalent ratio (OER) and further decreased as OER increased in all experiments with and without activated carbon (AC) catalyst. These results indicated that the addition of appropriate amount of oxidant enhanced the formation of H_2 and CH_4, however, the yield of H_2 and CH_4 decreased when excessive oxidant was used.

13.5 Related Process in Near- and Supercritical Water

Based on the above analysis, the experiments on the gasification of organic wastes were widely investigated and influence factors (reaction temperature, pressure, concentration, residence time, etc.) were widely studied. However, the reactions happened in near- and supercritical water are complex, and they might include pyrolysis, hydrolysis, steam reforming, water–gas shift, etc. In order to know and improve reaction process of the gasification of model compounds and biomass better, some researchers analyzed the gasification products by using different methods. Of course, these methods could also be used to study the reaction process of NSCWG of organic wastes.

13.5.1 Analysis Method

At present, some researchers studied the reaction processes of the gasification of model compounds by analyzing the liquid products. It is known that the studies of liquid products could help to understand the formation process of gas, tar, and char well, and the gasification of tar and char is difficult. Therefore, the understanding of the formation process of tar and char is important and could inhibit the formation of unwanted products. In their studies, the key intermediate compounds were analyzed by qualitative and quantitative analysis. Based on the analysis results and some literatures, the reaction processes were summarized by them. Some other researchers investigated reaction processes of NSCWG of biomass by visual observation and the reaction processes were clear. In their studies, a microreactor, Bassett-type 37 hydrothermal diamond anvil cell (DAC), was used for visual observations. In SKLMF, two methods were used to investigate the reaction process of gasification. The first method was to analyze the liquid products and the solid phase extraction (SPE) was used to pretreat the liquid products first. The SPE was made in a Mediwax 24-ports Vacuum SPE manifold with Agilent SampliQ C18 SPE columns and methanol was used as the elution solvent. The SPE column was first eluted with 3 ml methanol so that it could be activated, and then balanced with 5 ml high-purity water. Following, a certain amount of liquid products was injected into the SPE column in succession when a little water existed in the extraction column. In the whole extraction process, the flow rate was controlled at 5–10 ml/min. The SPE column was drained until no water existed in it. Finally, the SPE column was eluted with 3 ml ethyl acetate and 3 ml carbon dichloride in sequence, and then 6 ml liquid was collected and flushed to 1 ml by nitrogen. The second method was that multiple analysis methods which included Gas chromatographic (GC), Scanning electron microscope (SEM), Fourier transform infrared spectroscopy (FTIR), Specific surface area (BET), Total organic carbon (TOC), NH3-N analysis, and Gas chromatography–Mass spectrometry (GC–MS) were used to analyze the gas, solid, and liquid products, and then the related process was investigated. This method was more helpful for analyzing the macroscopic change process of NSCWG of biomass and organic wastes.

13.5.2 Model Compound

Because the composition of organic wastes is complex, the reaction processes of NSCWG of model compounds were widely investigated. Cellulose, hemicellulose, and lignin which exist in different kinds of organic wastes are typical model compounds. The reaction processes of gasification of them were widely investigated by some researchers. The glucose which is the hydrolysis product of cellulose is also the typical model compound and it was also widely investigated.

The method by analyzing the liquid products is widely used. The reaction processes of glucose conversion in near- and supercritical water were investigated. For example, Kabyemela et al. [51] systematically investigated glucose conversion in near- and supercritical water, and the temperatures were 300, 350, and 400 °C, pressures between 25 and 40 MPa, residence times between 0.02 and 2 s. The liquid products were analyzed and the reaction pathway was proposed. Then, the kinetic based on this pathway was proposed. Williams AND Onwudili [18] investigated the products of NSCWG of glucose. The reaction temperatures were between 330 and 380 °C, and the reaction time was between 0 and 120 min. Based on the analysis of products, the reaction process was proposed by them. Kruse et al. [52] also studied the NSCWG of glucose and they investigated the influence of salts on the gasification of glucose, and the reaction process was proposed by analyzing the liquid products. They built a bridge between model compounds like glucose and real biomass by considering inorganic compounds. The reaction process of cellulose conversion was also investigated. For example, Sasaki et al. [53] studied cellulose hydrolysis in near- and supercritical water and the decomposition experiments were conducted with a flow type reactor in the range of temperature from 290 to 400 °C at 25 MPa. The liquid products were analyzed and the reaction process was proposed by them. Lignin is a complex phenol polymer which consists of syringyl propane, guaiacyl propane, and p-hydroxyphenyl. Saisu et al. [54] and Okuda et al. [55] investigated the lignin conversion in supercritical water and the detailed scheme was proposed. They found that lignin first decomposed by hydrolysis and dealkylation. The formaldehyde and low-molecular weight compounds such as syringols, guaiacols, catechols, and phenols were formed. Then, cross-linking between formaldehyde and these fragments occured, and the residual lignin gave higher molecular weight fragments. Okuda et al. [56] also indicated that no char formation was found in supercritical water when phenol was used. Wahyudiono et al. [57] studied the conversion of lignin in near- and supercritical water at 350 and 400 °C and pressures of 25–40 MPa using batch reactors without catalysts. They found that lignin was successfully degraded and liquified in water at these conditions. Higher temperatures promoted the conversion of lignin into its derivative compounds and the formation of heavier components. The compounds in aqueous phase products were identified as catechol, phenol, m,p-cresol, and o-cresol. With the increase of reaction time, the amount of the lower molecular weight compounds increased. At the same reaction time, disassembly process of ether and carbon–carbon bonds in lignin and its degradation intermediate occurred. Based on the analysis results, a schematic of the reaction mechanism for lignin degradation in near- and supercritical water was proposed. Carrier et al. [58] investigated the degradation pathways of holocellulose, lignin, and a-cellulose from Pteris vittata fronds in sub- and supercritical water. The reaction temperature was 300, 400 °C and the pressure was 25 MPa. By the analysis of the intermediate organic products, the degradation pathways were proposed by them.

The method of visual observation was also used to study the reaction process. Fang et al. [59] investigated the cellulose decomposition in hydrothermal condition which included the condition of near-critical water with a 50-nL microreactor (DAC) coupled with optical and infrared microcopy. The experiments were done without or with catalysts which included Na_2CO_3 and Ni, and the maximum temperature reached 350 °C. They found that solid residues were the main product for noncatalytic decomposition of cellulose, which took place mostly under heterogeneous conditions at slow heating rates (0.18 °C/s). Homogeneous conditions could be achieved at a high heating rate of 2.2 °C/s and solid residues were still the main product. However, there were different structures of solid residues and reaction pathways when reactions happened under the heterogeneous or homogeneous condition. When Na_2CO_3 and Ni were used, oil and gas were the main products and little residues were formed. In both catalytic cases, the reactions seemed to occur in the aqueous phase after cellulose dissolution. Based on the analysis results, different reaction processes for homogeneous and heterogeneous environments were proposed by them. They also studied the catalytic gasification of cellulose and glucose on hydrothermal condition. The different catalysts which included Ni, Ru, and Pt were used and the maximum temperature reached 400 °C. The reaction processes at the slow heating rate (0.18 °C/s) and high heating rate

(9.5, 11.1, 14.4, and 14.5 °C/s) were observed. They found that cellulose completely dissolved in water at 318 °C on the condition of fast heating and Pt was the most active catalyst for glucose reactions [60]. Ogihara et al. [61] investigated the cellulose dissolution in subcritical and supercritical water over a wide range of water densities (550–1,000 kg/m^3) by visual observations. They found that the dissolution temperatures of cellulose depended on water density and decreased from about 350 °C at a water density of 560 kg/m^3 to a minimum of around 320 °C at a water density of 850 kg/m^3. When the water densities was greater than 850 kg/m^3, the dissolution temperatures increased and reached 347 °C at 980 kg/m^3. At low water densities, particles seemed to dissolve into the aqueous phase from the surface. From 670 to 850 kg/m^3, the cellulose particles visibly swelled just before completely collapsing and dissolving into the aqueous phase. From 850 to 1,000 kg/m^3, the particles required longer time to dissolve and many fine brown-like particles were generated as the particles dissolved. The decomposition of organosolve lignin in water/phenol solutions was also studied by them [62]. They found that a homogenous phase was formed when phenol was used, which seemed to promote the decomposition of lignin into phenolic fragments by hydrolysis and pyrolysis. Phenol along with the homogeneous reaction conditions also inhibited re-polymerization of the phenolics and promoted the formation of oil. In the absence of phenol, lignin remained as a heterogeneous phase with water over the range of conditions studied. The homogeneous conditions and conditions for inhibiting char formation by phenol were elucidated by them and it was found that mixtures of phenol and lignin became homogeneous at 400–600 °C and high water densities of 428–683 kg/m^3, corresponding to maximum pressures of 93 MPa. Based on the analysis results, the reaction paths were proposed by them.

In SKLMF, the model compounds were also studied and the reaction processes were investigated by analyzing the liquid products. The NSCWG of lignin was investigated in batch reactor (Parr 4575A). In order to reach the condition of supercritical water, an amount of argon was first added into the reactor to obtain initial pressure. Then, the reactor was heated. In order to control the final pressures were between 24 and 26 MPa, the different initial pressure was applied. The influence of reaction temperature (350–475 °C) on the liquid products when the residence time was 0 min is shown in Fig. 13.19. Eight kinds of typical compounds were analyzed by quantitative analysis. The results showed that the contents of phenol, 3-methyl-phenol, and 4-methyl-phenol increased with the increase of temperature. The maximum yields of them were reached at 450 °C, and then decreased. The contents of 2-methoxy-phenol decreased with the increase of temperature, which might indicate that the compound was the primary product of lignin decomposition and the others were formed successively with the further reaction. In order to research the decomposition of lignin more specifically, more experiments were done (Fig. 13.20). The lignin was investigated in the batch reactor from 225 to 450 °C with the interval of 25 °C, the residence time was 0 min and the initial pressure was 2.8 MPa. The initial reaction conditions were the same. Nine kinds of compounds in the liquid products were analyzed by quantitative analysis. They are acetone, cyclopentanone, 2,3-dimethyl-2-

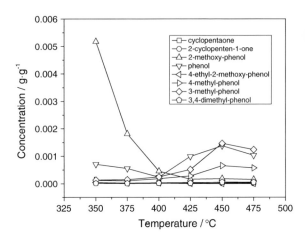

Fig. 13.19 Effect of temperature on liquid products of NSCWG of lignin in batch reactor (water: 45 g, lignin: 5 g, final pressure: 24–26 MPa)

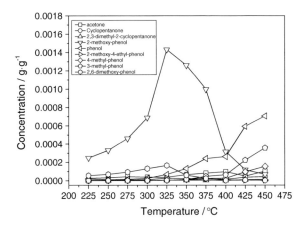

Fig. 13.20 Effect of temperature on liquid products of lignin gasification in sub- and supercritical water with batch reactor (water: 54 g, lignin: 6 g, initial pressure: 2.8 MPa)

cyclopentan-1-one, 2-methoxy-phenol, phenol, 2-methoxy-4-ethyl-phenol, 4-methyl-phenol, 3-methyl-phenol and 2,6-dimethoxy-phenol. Figure 13.20 showed that the contents of 2-methoxy-phenol and 2,6-dimethoxy-phenol increased when the temperature increased from 225 to 325 °C, and then decreased with the temperature increasing further. The change trend of 2-methoxy-phenol was consistent with that in Fig. 13.19, which indicated that it was the primary product of lignin decomposition and it could decompose when the reaction proceeded further. The change trend of 2,6-dimethoxy-phenol was similar to that of 2-methoxy-phenol, which indicated that 2,6-dimethoxy-phenol was also the primary product. The contents of other compounds increased with the increase of temperature, which indicated that they were the successive products and the decomposition of them needed higher temperatures.

13.5.3 Organic Wastes

Although the reaction processes of NSCWG of model compounds are widely investigated, the investigations of reaction process of NSCWG of organic wastes are little. Bocanegra et al. [63] investigated the decompositions of effluent and winegrape slurry in supercritical conditions by visual observation. They found that the decomposition of the two samples began when the supercritical temperature was reached and the dissolution process of two different samples was similar. They also found that the decomposition of samples was almost complete at low concentration levels. The reaction processes of NSCWG of black liquor and sewage sludge were only investigated in SKLMF. The first method (analyzing liquid products) was used to study the gasification of black liquor and the key intermediate compounds were analyzed. Because lignin was the main component of black liquor and was also the main component of wood, the components of liquid products was predictable and mainly contained phenols and its decomposition products [54, 55, 64, 65]. The results of qualitative analysis of products of the gasification of black liquor showed that almost 60 kinds of compounds dissolved in organic solvent and demonstrated the predictable. In these compounds, most of them were phenol derivants which mainly included 2-methoxy-phenol, 4-ethyl-2-methoxy-phenol, and 4-methyl-2-methoxy-phenol, 1,2,3-trimethoxy-5-methyl-benzene etc., and some cyclic ketones, indanones, other compounds also existed. The effect of temperature and different material flows on the main liquid products is shown in Fig. 13.21. It showed that the concentration of 2-methoxy-phenol, 2,6-dimethoxy-phenol, and 4-ethyl-2-methoxy-phenol decreased and the concentration of phenol increased with the increase of temperature. Because the concentration of phenol which existed in original liquid was tiny, the increase of concentration of phenol indicated that phenol was the decomposition products of other matters.

The second method (multiple analysis methods) was used to study the gasification of sewage sludge. The principle diagram of the analysis method is shown in Fig. 13.22. They first analyzed the change of products (gas, liquid, and solid) distribution and gas yields with the increase of temperature. Then, the solid products which were obtained at the same conditions were analyzed. The changes of the morphology structures, BET surface area, total pore volume, average pore diameter, and FITR spectra of sewage sludge and solid products were analyzed. They found that the changes of characteristic of sewage sludge and solid products were in accordance with the changes of products distribution and gas products, and the formation process of gas products could also be explained by the changes of solid products. Finally, the liquid products were also analyzed and it could not only demonstrate the results obtained from the gas and solid analysis but also understand the reaction process more deeply. The liquid products which were obtained at the same conditions were analyzed by TOC and GC/MS. The changes of TOC and liquid compositions with the increase of temperature were investigated. Based on the analysis results, they concluded that the formation of gaseous products could be intensively affected by temperature. Organic matters in sewage sludge are

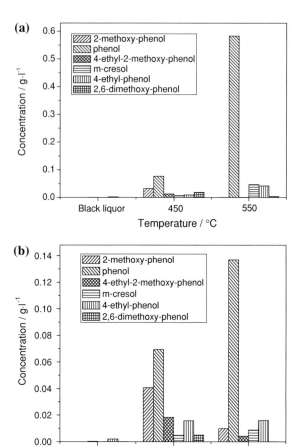

Fig. 13.21 Effect of temperature on liquid products of 9.5 wt% black liquor gasification in supercritical water with microminiature continuous reactor (25 MPa, flow rate: **a** 5 kg/h, **b** 7 kg/h)

almost completely dissolved and hydrolyzed in water at 425 °C. The dissolution and hydrolysis products decompose to form H_2, CH_4, CO_2, and CO via the steam reforming and water–gas shift, dealkylation, decarboxylation, and decarbonylation. The dehydrogenation and polymerization also occurred in dissolution and hydrolysis products, and the Diels–Alder reaction mechanism could be used to explain the phenomenon. Specific details of experimental results will be shown in the future.

13.6 Catalytic Gasification of Organic Wastes

In earlier studies, NSCWG of model compounds and biomass without catalyst were widely investigated by some researchers. They indicated that high reaction temperature was required so that it could get more amount of gas fuel and achieve

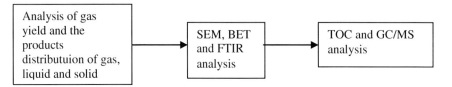

Fig. 13.22 Schematic diagram of the analysis process

the nearly complete gasification. However, the problems of high reactor cost and corrosion, etc., were caused by the high reaction temperature. Therefore, low reaction temperature was required. Osada et al. [66] identified three temperature regions in NSCWG: (1) Region I (500–700 °C supercritical water) biomass decomposed and AC catalyst was used to avoid char formation or alkali catalyst facilitated the water–gas shift reaction. (2) Region II (374–500 °C, supercritical water) biomass hydrolyzed and metal catalysts facilitated gasification. (3) Region III (below 374 °C, subcritical water) biomass hydrolysis was slow and catalysts were required for gas formation. In the low temperature region of Region II and Region III, the use of catalyst was necessary. It could not only reduce reaction temperature but also enhance the formation of gas products. So, the use of catalyst was very important for NSCWG of model compounds and biomass, and it might be also important for organic wastes gasification. The catalysts used in NSCWG could be mainly categorized into two types: One is the homogenous catalyst and the other is the heterogeneous catalyst [67].

13.6.1 Homogenous Catalysts

The alkali catalysts including NaOH, KOH, Na_2CO_3, and K_2CO_3 [15, 16, 20, 52, 68] are the main homogenous catalysts, and other catalysts such as trona (NaHCO$_3$·Na$_2$CO$_3$·2H$_2$O), nickel(II) acetylacetonate (Ni(acac)$_2$), cobalt(II) acetylacetonate (Co(acac)$_2$), and iron(III) acetylacetonate (Fe(acac)$_3$) are also used [69]. In organic waste gasification, the homogenous catalysts are widely used.

Jarana et al. [33] investigated the effect of KOH catalyst on oleaginous wastewater and alcohol distillery wastewater gasification in supercritical water. They found that the addition of KOH catalyst increased the CODr and enhanced the formation of H_2 and CH_4 at the same reaction condition. Nakhla et al. [34] studied the effect of NaOH catalyst on hog manure gasification. They found that the NaOH catalyst could enhance the reduction of COD better than other catalysts, and the CODr achieved 81 %. Onwudili and Williams [35] investigated the effect of NaOH concentration on bio-diesel plant waste reforming at 380 °C and they found that the molar fraction of H_2 increased with the increase of concentration. The maximum molar fraction of H_2 reached 90 % at 3 M NaOH concentration.

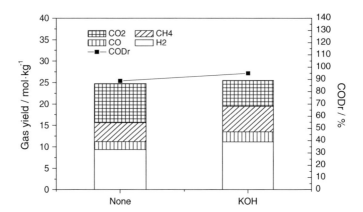

Fig. 13.23 Effect of KOH catalyst on gas yield and CODr of 7.8 wt% black liquor gasification in supercritical water with batch reactor (500 °C, black liquor: 10 g, catalyst amount: 1 g, N_2 initial pressure: 4 MPa, residence time: 10 min)

Schmieder et al. [16] studied the effect of K_2CO_3 catalyst on sewage sludge gasification at 450 °C with batch reactor. They found that the addition of K_2CO_3 catalyst enhanced the decrease of TOC and the TOC destruction efficiency reached 85.3 %. Yanik et al. [36] investigated the effect of K_2CO_3 and trona catalysts on leather waste gasification in supercritical water. They found that the yield of H_2 increased significantly when the two catalysts were used. The K_2CO_3 catalyst was the best catalyst on the leather waste gasification. In SKLMF, Chen et al. [44] studied the effect of NaOH, KOH, Na_2CO_3, and K_2CO_3 on sewage sludge gasification in fluidized bed reactor. They found that the addition of alkali catalysts was suitable for sewage sludge gasification in supercritical water with fluidized bed reactor and the addition of alkali catalysts enhanced the formation of H_2 better. The maximum yield of H_2 reached 15.49 mol/kg by using KOH. The maximum value of GE and CE reached 53.22 and 46.39 %, respectively when K_2CO_3 was used.

The effect of KOH on the gasification of black liquor with batch reactor (140 ml volume) was also studied in SKLMF. As shown in Fig. 13.23, the yield of H_2 increased when KOH catalyst was used. The maximum yield of H_2 reached 11.15 mol/kg and the CODr reached 95 %.

13.6.2 Heterogeneous Catalysts

Compared with the homogenous catalysts, the heterogeneous catalysts have the advantages of high selectivity, recyclability, environment-friendly, etc. They mainly include AC, metal, and metal oxide catalyst. Metal catalysts, including noble metals such as Pt, Pd, Ru, Rh [70–73] and low-cost metal such as Ni [69, 74–76], are widely used in NSCWG. Ca(OH)$_2$, AC [41, 42] and metal oxide [77, 78] were also used as catalysts in these experiments. In organic waste gasification, the heterogeneous catalysts are widely used.

Nakhla et al. [50] studied the effect of Pd/AC, Ru/Al$_2$O$_3$, Ru/AC, and AC catalysts on hog manure gasification. The addition of catalysts enhanced the formation of H$_2$ and the Pd/AC catalyst produced the highest H$_2$ yield followed by Ru/Al$_2$O$_3$. The negligible CO yield during the Pd/AC, Ru/Al$_2$O$_3$, Ru/AC, and AC catalytic experiments was found. The use of them could get the similar CODr and the maximum value reached to 71 % when Ru/AC catalyst was used. Yanik et al. [36] studied the effect of red mud (Fe-oxide containing residue from Al-production) on leather waste gasification. Although the catalytic efficiency of red mud is lower than other alkali catalysts, it also could enhance the formation of H$_2$ better and the yield of H$_2$ was almost doubled compared to that without catalysts.

Yamamura et al. [37] investigated the effect of RuO$_2$ on paper sludge and sewage sludge gasification at 450 °C and 120 min. They found that RuO$_2$ could enhance the paper sludge gasification better and the yield of H$_2$ was higher. The increase amount of RuO$_2$ resulted in a decrease in the yield of H$_2$ and an increase in the yield of CH$_4$, and they attributed this result to effect of RuO$_2$ catalysis on both the gasification reaction and the methanation reaction. Afif et al. [39] investigated the effect of Raney nickel on activated sludge gasification in near-critical water. They found that the addition of Raney nickel had a positive effect on the total gas yield and it increased almost linearly from 3.5 to 33 mol/kg at a catalyst loading of 1.5 g/g after which the total yield reached a plateau level at 380 °C and 15 min. The yield of H$_2$ and CE also increased with the increase of catalyst loading and the maximum yield of H$_2$ reached 14 mol/kg at a catalyst loading of 1.5 g/g and the maximum value of CE reached 70 % at a catalyst loading of 1.8 g/g. Xu et al. [41, 42] studied the effect of AC on sewage sludge gasification in supercritical water with continuous reactor. They achieved nearly complete gasification of sewage sludge with a feed concentration of 2.8 wt% at 650 °C by using carbon catalyst and also achieved 77 % CE when the feed concentration of sewage sludge is 2.8 wt% at 600 °C by using carbon catalyst and the maximum yield of H$_2$ reached 11 mol/kg. However, catalyst loading is about two-thirds of the heated zone volume of the reactor in their experiments. In SKLMF, the effect of Ru/C and various Raney catalysts on sewage sludge gasification in near- and supercritical water was investigated in batch reactor. The Raney catalysts included Raney-copper (RTH6110), Raney-cobalt (RTH7110), Raney-nickel 3110 (RTH3110 including Mo), Raney-nickel 3140 (RTH3140 including Mo), Raney-nickel 4110 (RTH4110), and Raney-nickel 5110 (RTH5110 including Fe). As shown in Fig. 13.24, the maximum yield of H$_2$ was obtained when the RTH3140 was used whenever the reaction happened in near- or supercritical water. The catalytic activity for H$_2$ production in supercritical water was in the following order: RTH3140 > RTH3110 > Ru/C > RTH5110 > RTH4110 > RTH6110 > RTH7110, and the catalytic activity for H$_2$ production in near-critical water was in the following order: RTH3140 > RTH5110 > Ru/C > RTH3110 > RTH4110 > RTH6110 > RTH7110. They indicated that Raney-nickel catalysts

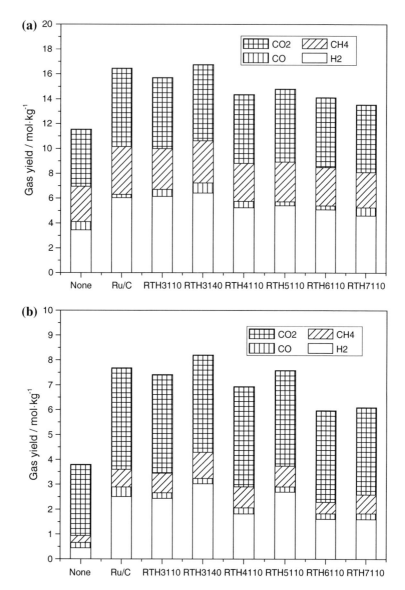

Fig. 13.24 Effect of catalyst type on sewage sludge gasification in near- and supercritical water: feed concentration: 8.9 wt%, feed amount: 10 g, residence time: 20 min, catalyst loading (g catalyst /g dry sludge): 0.23 g **a** 450 °C, **b** 350 °C

with metal components (Fe or Mo) were more helpful for the formation of H_2 in near- or supercritical water than other Raney catalysts. The maximum yield of H_2 reached 20.38 mol/kg dry sludge in the presence of 1.57 g RTH3140 per gram dry sludge. In this case, the maximum GE and CE reached 91.31 and 88.20 %, respectively.

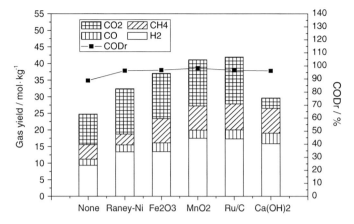

Fig. 13.25 Effect of catalysts on gas yield and CODr of 7.8 wt% black liquor gasification in supercritical water with batch reactor (500 °C, black liquor: 10 g, catalyst amount: 1 g, N_2 initial pressure: 4 MPa, residence time: 10 min)

The effect of various heterogeneous catalysts on the gasification of black liquor with batch reactor was also studied in SKLMF. These heterogeneous catalysts included Raney-nickel, Fe_2O_3, MnO_2, Ru/C, and $Ca(OH)_2$. As shown in Fig. 13.25, the yield of H_2 increased obviously when the five catalysts are used. The maximum yield of H_2 reached 17.5 mol/kg when MnO_2 is used, and the following yield of H_2 reached 17 mol/kg when Ru/C was used. The CODr reached 96.2, 97, 96.7, 98.2, and 96.2 % when Raney-nickel, Fe_2O_3, MnO_2, Ru/C and $Ca(OH)_2$ were used, respectively.

13.7 Kinetics Analysis

For a better rational design of reactor and improving the reaction rate, the kinetic studies on NSCWG of organic wastes are necessary. At present, most kinetic studies are mainly about the biomass and model compounds conversation in near- and supercritical water. Liu et al. [79] investigated the kinetic of epoxy resin flow reactor. Alenezi et al. [80] studied the hydrolysis kinetics of sunflower oil under subcritical water conditions with a continuous flow reactor. They all only studied the kinetics of biomass and model compounds conversion, and they did not study the kinetics on the conversion of biomass and model compounds to gas. In gasification, Lee et al. [81] investigated the kinetics of glucose conversion and COD degradation at temperature range from 480 to 750 °C and 28 MPa with a residence time of 10–50 s, and the kinetic parameters were obtained. Guo et al. [9] investigated the kinetics of glycerol supercritical water gasification at a temperature range from 487 to 600 °C and 25 MPa with a residence time of 3.9–9 s. The detailed kinetic of the gasification of glycerol in supercritical water was also studied by them [82] and the kinetic model was proposed. Savage and Huelsman [83] studied the kinetic of phenol supercritical water gasification at a temperature range from 500 to 700 °C with a residence time of 8–60 min and they also proposed the kinetic models for algae, cellulose, and lignin gasification in supercritical water [84, 85], and they predicted the gas composition and the process of gas formation. These kinetic studies of biomass and model compounds conversion or gasification might be helpful for the kinetic studies of organic wastes gasification.

There are little reports for the kinetic studies on NSCWG of organic wastes. Vostrikov et al. [40] investigated the kinetic of sewage sludge supercritical water gasification in semi-batch reactor and the temperatures range from 600 to 750 °C. The kinetic parameters on the conversion of sewage sludge carbon and conversion of the organic matter of sewage sludge were determined. Blasi et al. [86] studied the supercritical gasification of wastewater from updraft wood gasifiers in a laboratory-scale reactor with a plug-flow behavior and TOC conversion was chosen as the parameter to measure the organic matter decomposition rate, and the kinetic parameters were obtained when temperature ranges from 450 to 548 °C and residence time of 46–114 s.

In SKLMF, Cao et al. [46] studied the kinetic of black liquor supercritical water gasification at a temperature range from 450 to 550 °C and 25 MPa with a reactor residence time of 4.9–13.7 s. CODr was chosen as the parameter to measure the organic matter decomposition rate and the tubular reactor was idealized as plug-flow reactor when the Peclet number satisfied the following criteria:

$$Pe = ud/D \to \infty \quad (13.1)$$

When the length/diameter ratio of the reactor was larger than 100, the axial dispersion coefficient tended to be zero, therefore, the Pe get infinite. The ratio of the reactor in their study was larger than 114, so it could be assumed to be an ideal plug-flow reactor.

They assumed this reaction to be pseudo-first-order reaction and the reaction rate could be expressed as below:

$$-dx/dt = kx \quad (13.2)$$

where x = COD concentration of the liquid effluent/initial COD concentration, k is the reaction rate constant (s^{-1}). The plots of $-\ln x$ versus t at different temperature were reported and they found that the related coefficients was high ($R_2 > 0.97$). So they indicated that the pseudo-first-order reaction assumption was relatively reliable. The reaction rate constants at different temperatures were obtained from linear regression, which were in the range of 0.04–0.23 s^{-1}.

The temperature dependency of the reaction rate could be expressed in Arrhenius equation as below:

$$k = A\exp(-Ea/RT) \quad (13.3)$$

where A is the pre-exponential factor (s^{-1}), Ea is the activation energy (kJ/mol), R is the universal gas constant (8.314×10^{-3} kJ/mol·K) and T is temperature (K).

The activation energy Ea (74.38 kJ/mol) and pre-exponential factor A (104.05 s^{-1}) are obtained respectively through linear regression.

13.8 Challenges and Prospect

From the above works, it is known that NSCWG of organic wastes could not only get the useful energy resource but also could eliminate the pollution. Although many works were done on NSCWG of organic wastes, many challenges still exist and are needed to be overcome.

Because the reactor wall is exposed in severe conditions and the constitutes of organic wastes and intermediate products are complex, the corrosion is an inevitable problem and it was observed in the experiments of NSCWG. However, it is a less serious problem than that happened in supercritical water oxidation. Some methods were used to solve the severe corrosion problem in supercritical water oxidation, and so it would also be possible to solve the corrosion problem which happens in NSCWG. Plugging is also a main problem in NSCWG with continuous flow tube reactor. The solubility of salts in biomass and organic wastes are low under the condition of supercritical water. The salts precipitation and the formation of char are the main reasons for plugging, especially when the catalyst is added to the reactor. The development of reactor is important to solve the problem and the supercritical water fluidized bed reactor is developed. It could resolve the problem of plugging when the catalysts are not used. However, the use of catalysts in the fluidized bed has not been reported. It is also important to investigate whether the plugging occurs when catalysts are used.

The catalysts are also important to the reaction. However, the high pressure in the gasification process and the S in the salt are also a challenge for the durability and life time of the catalyst. The high pressure results in the decrease of the durability and lifetime of the catalyst and the S results in the poisoning of catalyst. Therefore, the development of a high pressure resistant, long-life and cheap catalyst is important, which could increase economical efficiency by improving the gasification and lowering the gasification temperature.

The kinetics is important for the organic wastes gasification and the design of gasification system. Although the total kinetics on organic wastes gasification has been developed, the detailed kinetics on the gasification mechanism and reaction path have not been reported. So the detailed gasification mechanism should be explored by using various methods.

Although some challenges are existing in NSCWG of organic wastes, there are also some good prospects. Especially for the treatment of organic wastes in near- and supercritical water, the purposes of energy recovery and decontamination could be achieved.

Acknowledgements This work was financially supported by the National Key Basic Research Program 973 Project funded by MOST of China (Project No.2009CB220000 and 2012CB215303), the National Natural Science Foundation of China (Project No. 51121092) and the engineering technology research center of renewable energy in Shaanxi (Project No. 2008ZDGC-07).

References

1. Basu P, Mettanant V (2009) Biomass gasification in supercritical water—a review. Int J Chem React Eng 7(1). doi:10.2202/1542-6580.1919
2. Savage PE (1999) Organic chemical reactions in supercritical water. Chem Rev 99(2):603–621
3. Kruse A (2008) Supercritical water gasification. Biofuel Bioprod Biorefin 2(5):415–437. doi:10.1002/Bbb.93
4. Kruse A, Gawlik A (2003) Biomass conversion in water at 330–410 degrees C and 30–50 MPa. Identification of key compounds for indicating different chemical reaction pathways. Ind Eng Chem Res 42(2):267–279. doi:10.1021/Ie0202773
5. Kruse A, Henningsen T, Sinag A, Pfeiffer J (2003) Biomass gasification in supercritical water: influence of the dry matter content and the formation of phenols. Ind Eng Chem Res 42(16):3711–3717. doi:10.1021/ie0209430
6. Antal MJ, Allen SG, Schulman D, Xu XD, Divilio RJ (2000) Biomass gasification in supercritical water. Ind Eng Chem Res 39(11):4040–4053. doi:10.1021/ie0003436
7. Elliott DC (2008) Catalytic hydrothermal gasification of biomass. Biofuel Bioprod Biorefin 2(3):254–265. doi:10.1002/Bbb.74
8. Yoshida T, Matsumura Y (2001) Gasification of cellulose, xylan, and lignin mixtures in supercritical water. Ind Eng Chem Res 40(23):5469–5474
9. Guo S, Guo L, Cao C, Yin J, Lu Y, Zhang X (2012) Hydrogen production from glycerol by supercritical water gasification in a continuous flow tubular reactor. Int J Hydrog Energy 37(7):5559–5568
10. Guo LJ, Lu YJ, Zhang XM, Ji CM, Guan Y, Pei AX (2007) Hydrogen production by biomass gasification in supercritical water: a systematic experimental and analytical study. Catal Today 129(3–4):275–286. doi:10.1016/j.cattod.2007.05.027
11. Savage PE, Resende FLP, Fraley SA, Berger MJ (2008) Noncatalytic gasification of lignin in supercritical water. Energy Fuel 22(2):1328–1334. doi:10.1021/ef700574k
12. Savage PE, Resende FLP, Neff ME (2007) Noncatalytic gasification of cellulose in supercritical water. Energy Fuel 21(6):3637–3643. doi:10.1021/ef7002206
13. Kersten SRA, Potic B, Prins W, Van Swaaij WPM (2006) Gasification of model compounds and wood in hot compressed water. Ind Eng Chem Res 45(12):4169–4177. doi:10.1021/Ie0509490
14. Elliott DC, Sealock LJ Jr, Baker EG (1994) Chemical processing in high-pressure aqueous environments. 3. Batch reactor process development experiments for organics destruction. Ind Eng Chem Res 33(3):558–565. doi:10.1021/ie00027a012
15. Minowa T, Zhen F, Ogi T (1998) Cellulose decomposition in hot-compressed water with alkali or nickel catalyst. J Supercrit Fluids 13(1–3):253–259
16. Schmieder H, Abeln J, Boukis N, Dinjus E, Kruse A, Kluth M, Petrich G, Sadri E, Schacht M (2000) Hydrothermal gasification of biomass and organic wastes. J Supercrit Fluid 17(2):145–153
17. Potic B, Kersten SRA, Prins W, van Swaaij WPM (2004) A high-throughput screening technique for conversion in hot compressed water. Ind Eng Chem Res 43(16):4580–4584. doi:10.1021/ie030732a

18. Williams PT, Onwudili J (2005) Composition of products from the supercritical water gasification of glucose: a model biomass compound. Ind Eng Chem Res 44(23):8739–8749. doi:10.1021/Ie050733y
19. Hao XH, Guo LJ, Zhang XM, Guan Y (2005) Hydrogen production from catalytic gasification of cellulose in supercritical water. Chem Eng J 110(1–3):57–65. doi:10.1016/j.cej.2005.05.002
20. Hao XH, Guo LJ, Mao X, Zhang XM, Chen XJ (2003) Hydrogen production from glucose used as a model compound of biomass gasified in supercritical water. Int J Hydrog Energy 28(1):55–64
21. Sinag A, Kruse A, Rathert J (2004) Influence of the heating rate and the type of catalyst on the formation of key intermediates and on the generation of gases during hydropyrolysis of glucose in supercritical water in a batch reactor. Ind Eng Chem Res 43(2):502–508. doi:10.1021/Ie030475+
22. Guo LJ, Li YL, Zhang XM, Jin H, Lu YJ (2010) Hydrogen production from coal gasification in supercritical water with a continuous flowing system. Int J Hydrog Energy 35(7):3036–3045. doi:10.1016/j.ijhydene.2009.07.023
23. Lu YJ, Jin H, Guo LJ, Zhang XM, Cao CQ, Guo X (2008) Hydrogen production by biomass gasification in supercritical water with a fluidized bed reactor. Int J Hydrog Energy 33(21):6066–6075. doi:10.1016/j.ijhydene.2008.07.082
24. Lu YJ, Guo LJ, Zhang XM, Yan QH (2007) Thermodynamic modeling and analysis of biomass gasification for hydrogen production in supercritical water. Chem Eng J 131(1–3):233–244. doi:10.1016/j.cej.2006.11.016
25. Feng W, van der Kooi HJ, Arons JDS (2004) Biomass conversions in subcritical and supercritical water: driving force, phase equilibria, and thermodynamic analysis. Chem Eng Process 43(12):1459–1467. doi:10.1016/j.cep.2004.01.004
26. Tang HQ, Kitagawa K (2005) Supercritical water gasification of biomass: thermodynamic analysis with direct Gibbs free energy minimization. Chem Eng J 106(3):261–267. doi:10.1016/j.cej.2004.12.021
27. Yan QH, Guo LJ, Lu YJ (2006) Thermodynamic analysis of hydrogen production from biomass gasification in supercritical water. Energy Convers Manag 47(11–12):1515–1528. doi:10.1016/j.enconman.2005.08.004
28. Calzavara Y, Joussot-Dubien C, Boissonnet G, Sarrade S (2005) Evaluation of biomass gasification in supercritical water process for hydrogen production. Energy Convers Manag 46(4):615–631. doi:10.1016/j.enconman.2004.04.003
29. Matsumura Y, Minowa T (2004) Fundamental design of a continuous biomass gasification process using a supercritical water fluidized bed. Int J Hydrog Energy 29(7):701–707. doi:10.1016/j.ijhydene.2003.09.005
30. Voll FAP, Rossi CCRS, Silva C, Guirardello R, Souza ROMA, Cabral VF, Cardozo L (2009) Thermodynamic analysis of supercritical water gasification of methanol, ethanol, glycerol, glucose and cellulose. Int J Hydrog Energy 34(24):9737–9744. doi:10.1016/j.ijhydene.2009.10.017
31. Castello D, Fiori L (2011) Supercritical water gasification of biomass: thermodynamic constraints. Bioresour Technol 102(16):7574–7582. doi:http://dx.doi.org/10.1016/j.biortech.2011.05.017
32. Marias F, Letellier S, Cezac P, Serin JP (2011) Energetic analysis of gasification of aqueous biomass in supercritical water. Biomass Bioenerg 35(1):59–73. doi:10.1016/j.biombioe.2010.08.030
33. Jarana MBG, Saanchez-Oneto J, Portela JR, Sanz EN, de la Ossa EJM (2008) Supercritical water gasification of industrial organic wastes. J Supercrit Fluid 46(3):329–334. doi:10.1016/j.supflu.2008.03.002
34. Nakhla G, Youssef EA, Elbeshbishy E, Hafez H, Charpentier P (2010) Sequential supercritical water gasification and partial oxidation of hog manure. Int J Hydrog Energy 35(21):11756–11767. doi:10.1016/j.ijhydene.2010.08.097

35. Onwudili JA, Williams PT (2010) Hydrothermal reforming of bio-diesel plant waste: products distribution and characterization. Fuel 89(2):501–509. doi:10.1016/j.fuel.2009.06.033
36. Yanik J, Ebale S, Kruse A, Saglam M, Yuksel M (2008) Biomass gasification in supercritical water: II. Effect of catalyst. Int J Hydrogen Energ 33(17):4520–4526. doi:10.1016/j.ijhydene.2008.06.024
37. Yamamura T, Mori T, Park KC, Fujii Y, Tomiyasu H (2009) Ruthenium(IV) dioxide-catalyzed reductive gasification of intractable biomass including cellulose, heterocyclic compounds, and sludge in supercritical water. J Supercrit Fluid 51(1):43–49. doi:10.1016/j.supflu.2009.07.007
38. Zhang LH, Xu CB, Champagne P (2010) Energy recovery from secondary pulp/paper-mill sludge and sewage sludge with supercritical water treatment. Bioresour Technol 101(8):2713–2721. doi:10.1016/j.biortech.2009.11.106
39. Afif E, Azadi P, Farnood R (2011) Catalytic hydrothermal gasification of activated sludge. Appl Catal B-Environ 105(1–2):136–143. doi:10.1016/j.apcatb.2011.04.003
40. Vostrikov AA, Fedyaeva ON, Shishkin AV, Dubov DY, Sokol MY (2008) Conversion of municipal sewage sludge in supercritical water. Solid Fuel Chem 42(6):384–393. doi:10.3103/S0361521908060116
41. Xu XD, Matsumura Y, Stenberg J, Antal MJ (1996) Carbon-catalyzed gasification of organic feedstocks in supercritical water. Ind Eng Chem Res 35(8):2522–2530
42. Xu XD, Antal MJ (1998) Gasification of sewage sludge and other biomass for hydrogen production in supercritical water. Environ Prog 17(4):215–220
43. Guo Y, Wang SZ, Gong YM, Xu DH, Tang XY, Ma HH (2010) Partial oxidation of municipal sludge with activated carbon catalyst in supercritical water. J Hazard Mater 180(1–3):137–144. doi:10.1016/j.jhazmat.2010.04.005
44. Chen Y, Guo L, Cao W, Jin H, Guo S, Zhang X Hydrogen production by sewage sludge gasification in supercritical water with a fluidized bed reactor. Int J Hydrog Energy (0). doi:http://dx.doi.org/10.1016/j.ijhydene.2013.03.165
45. Sricharoenchaikul V (2009) Assessment of black liquor gasification in supercritical water. Bioresour Technol 100(2):638–643. doi:10.1016/j.biortech.2008.07.011
46. Cao CQ, Guo LJ, Chen YA, Guo SM, Lu YJ (2011) Hydrogen production from supercritical water gasification of alkaline wheat straw pulping black liquor in continuous flow system. Int J Hydrog Energy 36(21):13528–13535. doi:10.1016/j.ijhydene.2011.07.101
47. Akiya N, Savage PE (2002) Roles of water for chemical reactions in high-temperature water. Chem Rev 102(8):2725–2750. doi:10.1021/cr000668w
48. Lu YJ, Guo LJ, Ji CM, Zhang XM, Hao XH, Yan QH (2006) Hydrogen production by biomass gasification in supercritical water: a parametric study. Int J Hydrog Energy 31(7):822–831. doi:10.1016/j.ijhydene.2005.08.011
49. Chuntanapum A, Yong TLK, Miyake S, Matsumura Y (2008) Behavior of 5-HMF in subcritical and supercritical water. Ind Eng Chem Res 47(9):2956–2962. doi:10.1021/Ie0715658
50. Youssef EA, Elbeshbishy E, Hafez H, Nakhla G, Charpentier P (2010) Sequential supercritical water gasification and partial oxidation of hog manure. Int J Hydrog Energy 35(21):11756–11767. doi:10.1016/j.ijhydene.2010.08.097
51. Kabyemela BM, Adschiri T, Malaluan RM, Arai K (1997) Kinetics of glucose epimerization and decomposition in subcritical and supercritical water. Ind Eng Chem Res 36(5):1552–1558
52. Kruse A, Sinag A, Schwarzkopf V (2003) Key compounds of the hydropyrolysis of glucose in supercritical water in the presence of K2CO3. Ind Eng Chem Res 42(15):3516–3521. doi:10.1021/ie030079r
53. Sasaki M, Kabyemela B, Malaluan R, Hirose S, Takeda N, Adschiri T, Arai K (1998) Cellulose hydrolysis in subcritical and supercritical water. J Supercrit Fluid 13(1–3):261–268
54. Saisu M, Sato T, Watanabe M, Adschiri T, Arai K (2003) Conversion of lignin with supercritical water–phenol mixtures. Energy Fuel 17(4):922–928. doi:10.1021/ef0202844

55. Okuda K, Umetsu M, Takami S, Adschiri T (2004) Disassembly of lignin and chemical recovery - rapid depolymerization of lignin without char formation in water-phenol mixtures. Fuel Process Technol 85(8–10):803–813. doi:10.1016/j.fuproc.2003.11.027
56. Okuda K, Umetsu M, Takami S, Adschiri T (2004) Disassembly of lignin and chemical recovery—rapid depolymerization of lignin without char formation in water–phenol mixtures. Fuel Process Technol 85(8–10):803–813. doi:http://dx.doi.org/10.1016/j.fuproc.2003.11.027
57. Wahyudiono, Sasaki M, Goto M (2008) Recovery of phenolic compounds through the decomposition of lignin in near and supercritical water. Chem Eng Process: Process Intensif 47(9–10):1609–1619. doi:http://dx.doi.org/10.1016/j.cep.2007.09.001
58. Carrier M, Loppinet-Serani A, Absalon C, Aymonier C, Mench M (2012) Degradation pathways of holocellulose, lignin and α-cellulose from Pteris vittata fronds in sub- and super critical conditions. Biomass Bioenerg 43(0):65–71. doi:http://dx.doi.org/10.1016/j.biombioe.2012.03.035
59. Fang Z, Minowa T, Smith RL, Ogi T, Kozinski JA (2004) Liquefaction and gasification of cellulose with Na2CO3 and Ni in subcritical water at 350 degrees C. Ind Eng Chem Res 43(10):2454–2463. doi:10.1021/Ie034146t
60. Fang Z, Minowa T, Fang C, Smith RL, Inomata H, Kozinski JA (2008) Catalytic hydrothermal gasification of cellulose and glucose. Int J Hydrog Energy 33(3):981–990. doi:10.1016/j.ijhydene.2007.11.023
61. Ogihara Y, Smith RL, Inomata H, Arai K (2005) Direct observation of cellulose dissolution in subcritical and supercritical water over a wide range of water densities (550–1000 kg/m(3)). Cellulose 12(6):595–606. doi:10.1007/s10570-005-9008-1
62. Fang Z, Sato T, Smith RL, Inomata H, Arai K, Kozinski JA (2008) Reaction chemistry and phase behavior of lignin in high-temperature and supercritical water. Bioresour Technol 99(9):3424–3430. doi:10.1016/j.biortech.2007.08.008
63. Bocanegra PE, Reverte C, Aymonier C, Loppinet-Serani A, Barsan MM, Butler IS, Kozinski JA, Gokalp I (2010) Gasification study of winery waste using a hydrothermal diamond anvil cell. J Supercrit Fluid 53(1–3):72–81. doi:10.1016/j.supflu.2010.02.015
64. Yong TL-K, Matsumura Y (2012) Reaction kinetics of the lignin conversion in supercritical water. Ind Eng Chem Res 51(37):11975–11988. doi:10.1021/ie300921d
65. Wahyudiono Sasaki M, Goto M (2008) Recovery of phenolic compounds through the decomposition of lignin in near and supercritical water. Chem Eng Process 47(9–10):1609–1619. doi:10.1016/j.cep.2007.09.001
66. Osada M, Sato T, Watanabe M, Shirai M, Arai K (2006) Catalytic gasification of wood biomass in subcritical and supercritical water. Combust Sci Technol 178(1–3):537–552. doi:10.1080/00102200500290807
67. Guo Y, Wang SZ, Xu DH, Gong YM, Ma HH, Tang XY (2010) Review of catalytic supercritical water gasification for hydrogen production from biomass. Renew Sustain Energy Rev 14(1):334–343. doi:10.1016/j.rser.2009.08.012
68. Kruse A, Meier D, Rimbrecht P, Schacht M (2000) Gasification of pyrocatechol in supercritical water in the presence of potassium hydroxide. Ind Eng Chem Res 39(12):4842–4848. doi:10.1021/Ie0001570
69. Azadi P, Khodadadi AA, Mortazavi Y, Farnood R (2009) Hydrothermal gasification of glucose using Raney nickel and homogeneous organometallic catalysts. Fuel Process Technol 90(1):145–151. doi:10.1016/j.fuproc.2008.08.009
70. Elliott DC, Sealock LJ, Baker EG (1993) Chemical processing in high-pressure aqueous environments. 2. Development of catalysts for gasification. Ind Eng Chem Res 32(8):1542–1548. doi:10.1021/ie00020a002
71. Minowa T, Inoue S (1999) Hydrogen production from biomass by catalytic gasification in hot compressed water. Renew Energy 16(1–4):1114–1117 http://dx.doi.org/10.1016/S0960-1481(98)00436-4

72. Yamaguchi A, Hiyoshi N, Sato O, Bando KK, Osada M, Shirai M (2009) Hydrogen production from woody biomass over supported metal catalysts in supercritical water. Catal Today 146(1–2):192–195. doi:10.1016/j.cattod.2008.11.008
73. Osada M, Sato T, Watanabe M, Adschiri T, Arai K (2004) Low-temperature catalytic gasification of lignin and cellulose with a ruthenium catalyst in supercritical water. Energy Fuel 18(2):327–333. doi:10.1021/Ef034026y
74. Huber GW, Shabaker JW, Dumesic JA (2003) Raney Ni–Sn catalyst for H2 production from biomass-derived hydrocarbons. Science 300(5628):2075–2077. doi:10.1126/science.1085597
75. Sato T, Furusawa T, Ishiyama Y, Sugito H, Miura Y, Sato M, Suzuki N, Itoh N (2006) Effect of water density on the gasification of lignin with magnesium oxide supported nickel catalysts in supercritical water. Ind Eng Chem Res 45(2):615–622. doi:10.1021/Ie0510270
76. DiLeo GJ, Savage PE (2006) Catalysis during methanol gasification in supercritical water. J Supercrit Fluid 39(2):228–232. doi:10.1016/j.supflu.2006.01.004
77. Park KC, Tomiyasu H (2003) Gasification reaction of organic compounds catalyzed by RuO_2 in supercritical water. Chem Commun 6:694–695. doi:10.1039/b211800a
78. Watanabe M, Inomata H, Arai K (2002) Catalytic hydrogen generation from biomass (glucose and cellulose) with ZrO_2 in supercritical water. Biomass Bioenerg 22(5):405–410. doi:Pii S0961-9534(02)00017-X
79. Liu YY, Wei H, Wu S, Guo Z (2012) Kinetic study of epoxy resin decomposition in near-critical water. Chem Eng Technol 35(4):713–719. doi:10.1002/ceat.201100494
80. Alenezi R, Leeke GA, Santos RCD, Khan AR (2009) Hydrolysis kinetics of sunflower oil under subcritical water conditions. Chemical Engineering Research and Design 87(6):867–873. doi:http://dx.doi.org/10.1016/j.cherd.2008.12.009
81. Lee IG, Kim MS, Ihm SK (2002) Gasification of glucose in supercritical water. Ind Eng Chem Res 41(5):1182–1188. doi:10.1021/Ie010066i
82. Guo S, Guo L, Yin J, Jin H (2013) Supercritical water gasification of glycerol: intermediates and KINETICS. J Supercrit Fluids 78(0):95–102. doi:http://dx.doi.org/10.1016/j.supflu.2013.03.025
83. Huelsman CM, Savage PE (2012) Intermediates and kinetics for phenol gasification in supercritical water. Phys Chem Chem Phys 14(8):2900–2910
84. Guan Q, Wei C, Savage PE (2012) Kinetic model for supercritical water gasification of algae. Phys Chem Chem Phys 14(9):3140–3147
85. Resende FLP, Savage PE (2010) Kinetic model for noncatalytic supercritical water gasification of cellulose and lignin. AIChE J 56(9):2412–2420. doi:10.1002/Aic.12165
86. Di Blasi C, Branca C, Galgano A, Meier D, Brodzinski I, Malmros O (2007) Supercritical gasification of wastewater from updraft wood gasifiers. Biomass Bioenerg 31(11–12):802–811. doi:10.1016/j.biombioe.2007.05-002

Chapter 14
Hydrothermal Treatment of Municipal Solid Waste for Producing Solid Fuel

Kunio Yoshikawa and Pandji Prawisudha

Abstract The usage of municipal solid waste (MSW) is usually hindered by its nonuniformity, high moisture, low energy density, and the occurrence of chlorine in the plastic-impregnated waste. A hydrothermal treatment is developed to convert the MSW into solid fuel by employing a commercial scale system of about 1 ton capacity, applying saturated steam at about 2 MPa for about 60 min holding time. It was shown that the product has better uniformity, higher density, and better drying performance compared to MSW without reducing its heating value. The combustion characteristic of the final product was similar to that of sub-bituminous coal, and capable of reducing the SO_2 and NO emissions during co-combustion with coal. Additionally, the product showed that about 80 % of the organic chlorine was converted into inorganic, water-soluble chlorine, and the total chlorine content in the water-washed product was down to 16 %. It was calculated that the required energy for the hydrothermal treatment was 0.8 MJ/kg MSW, lower than conventional RDF production process which needs 1.35 MJ/kg MSW. It can be concluded that the hydrothermal treatment can be employed to convert MSW into a chlorine-free solid fuel suitable for co-combustion with coal.

14.1 Introduction

Municipal solid waste (MSW) has become a severe problem in many countries, not only in developing countries but also in developed countries due to the limited lifetime of final waste disposal. For example in Japan, although the annual MSW

K. Yoshikawa (✉)
Department of Environmental Science and Technology, Tokyo Institute of Technology, Yokohama, Japan
e-mail: yoshikawa.k.aa@m.titech.ac.jp

P. Prawisudha
Department of Mechanical Engineering, Bandung Institute of Technology, Bandung, Indonesia

discharge was slightly decreasing to 48.11 million tons in the fiscal year 2008, the lack of final disposal facilities is still a major concern because the remaining lifetime of those in use is only 18 years [1]. Current waste treatment technologies are still not able to eliminate the waste while meeting three conditions: environmentally friendly, economically feasible, and high processing capacity. Thus, a method based on the aforementioned conditions should be developed for treating the MSW.

On the other hand, the combination of global population and energy demand in the near future will dramatically decrease the fossil resources, which will result in increasing energy price. Worsen with the increasing price of crude oil, many industries are now using coal; this has resulted in a significant increase in the demand and also in the price of coal, which increased by 150 % in 2008 [2]. From [3], it was reported that the price of one of the main energy sources, coal, has risen to approximately double of its price in 1990.

These two conditions would lead to a big opportunity for alternative solid fuel from MSW to replace or partially substitute coal as the main fuel. The problem due to large quantities of MSW can be solved by treating the MSW, and thus the treated MSW having combustion characteristics similar to that of coal can be supplied as solid fuel needed for the industries. Unfortunately, due to its low energy density and high impurities, many industries are still cautious on the usage of this waste-derived fuel in their furnaces. Additionally, raw MSW is mainly consisted of high moisture content organic waste, a common case for generated waste in developing countries [4]. When high moisture content waste occurs, they have to be dried before usage to minimize the energy loss from water evaporation in the waste. This additional drying process would require more energy and thus reduce the overall added value of such fuel.

Another main challenge of this waste-to-fuel system application is that the MSW still contains plastics. In Japan, approximately 9.98 million tons of plastic waste was generated in 2008 [5], which accounted for more than 20 % of the total MSW. Similarly, in Northern Europe plastic waste accounted for 13 % of the total MSW [6]. South America, Northern Europe, and Eastern Asia recorded high plastic content of more than 10 %, while the lowest plastic content in MSW is recorded in Africa, where the plastic content reaches below 6 % [7]. Similar or even worse condition should be observed for the countries with no waste separating system, where organic and plastic wastes are still mixed in the waste disposal.

This plastic waste contains organic chlorine, mostly from poly vinyl chloride (PVC). When such MSW is converted into solid fuel, its high organic chlorine content becomes a major concern and hinders its usage. Evaporated chlorine in the form of hydrochloric acid (HCl) can promote clogging, corrosion, and acts as the chlorine source for dioxin formation in the furnace. In Europe, required maximum chlorine content in these waste-derived fuels, also known as RDFs, are ranging from 0.1 % in Finland (class 1) to 1 % in Sweden [8]. An additional organic chlorine removal process can overcome this limitation, but since organic chlorine is not water soluble, the removal process usually employs thermal degradation

Fig. 14.1 Hydrothermal treatment commercial plant in Japan

principle which requires high energy such as high temperature leaching or by using granular activated carbon, which focuses on the removal of organic chlorines from water [9–13]. Therefore, it can be seen that the use of other methods are still promising to solve the chlorine problem in plastic-impregnated MSW.

An innovative hydrothermal treatment solely designed for converting high moisture content solid wastes into dried, uniform, pulverized, coal-like solid fuel by using low energy consumption [14–17] has been developed in Yoshikawa laboratory, Tokyo Institute of Technology Japan (Tokyo Tech). The commercial scale plant, as shown in Fig. 14.1, has already been working in Japan. The plant has maximum capacity of 1 ton of waste per batch and generally used for treating local medical wastes. The batch process begins by loading the raw material into a reactor, and then injecting saturated steam of about 200 °C and 2 MPa into the reactor. Mixing process is then conducted by a stirrer in the reactor for about 1 hour while holding the temperature and pressure. After finishing the holding period and discharge of the steam, wet uniform product can be extracted.

In the case of MSW treatment, the process basically enhances the drying performance of MSW and thus its moisture content can be easily removed without much energy consumption compared to traditional moisture removal method. This is due to the fact that the bound water in the cells of MSW became free water by destroying the cells under the hydrolysis reaction with steam during the treatment. Hence in principle this treatment itself is not a drying process, but will enhance the drying process by moving bound water into the surface of the cell, i.e., surface water, so that the moisture removal is made easy and accelerated. This improvement in the drying performance leads to a lower energy consumption for removing moisture from MSW.

Subsequent co-combustion researches on the hydrothermally treated MSW with coal showed that the devolatilization properties of coal were improved, and for low-quality coal, such as high ash Indian coal, the ignition of the coal was enhanced by adding hydrothermally treated MSW [16]. From further research, it was confirmed that the treatment for raw MSW produced a solid product with heating value similar to that of low rank coal [15], and exhibited organic chlorine conversion into inorganic, water-soluble chlorine in the presence of alkali [17], which later can be removed by water washing. These results increased the confidence of using the hydrothermal treatment to produce a chlorine-free solid fuel from plastic-impregnated MSW.

14.2 Developed Hydrothermal Treatment

The hydrothermal treatment (HT) in definition uses the combination of water (hydro) and thermal to convert the waste into usable products. The term "hydrothermal" was originally used in geologic field to represent the natural phenomena which produce oil and coal from the remaining of animals and plants by means of earth layer's pressure in a very extensive period. It is generally known that German chemist Robert Bunsen in 1839 used this principle to grow barium carbonate and strontium carbonate crystals in aqueous solutions at temperatures above 200 °C and at pressures above 100 bars, and the same principle is still being used to grow large amount of crystals nowadays [18].

The hydrothermal process involves the application of heat and pressure in aqueous medium, similar to the natural processing of organic remains from plants or animals to become coal or crude oil we consume today. Different from thermal treatment, it bypasses the sorting and crushing processes to directly disintegrate the waste inside the hydrothermal reactor. Added with the fact that the hydrothermal process uses water as reaction medium, high moisture content waste can be directly processed without the need for drying process, and the hot water can serve as a solvent, reactant and even a catalyst for the raw material [19].

However, it was not until the oil crisis in 1970s that the hydrothermal process was used to convert carbonaceous materials. Cole et al. in 1975 filed a patent for preparation of solid fuel-water slurries as gasifier fuel using hydrothermal principle at 400 to 600 °F (about 200–350 °C) [20]. Same group filed another patent for upgrading the low rank coals using superatmospheric pressure in the presence of hydrogen at 400 to 650 °F [21]. Bobleter et al. in 1976 reported the usage of hydrothermal process in the conversion of cellulosic matter to sugar at temperatures of between 200 and 275 °C [22], a base for liquid biofuel production.

The hydrothermal treatment technology are usually categorized according to its processing temperature, they are:

- Supercritical process (Super Critical Water, SCW) is using supercritical water in the temperature above 374 °C and pressure above 22.1 MPa and able to convert organic materials completely into carbon dioxide, water, and nitrogen [23].
- Subcritical process is using the temperature below its critical temperature and pressure (374 °C and 22.1 MPa). A pilot plant for producing liquid biofuel from biomass has been built in TNO Apeldoorn Netherland employing HTU

(Hydro Thermal Upgrading) process developed by Shell Research, with the capacity of 100 kg/h and 3 weeks of continuous run in April 2004 [24].

- Waste autoclave is a technology using saturated steam at 160 °C [25]. Waste is sterilized with steam in an autoclave, like cooking in a pressure cooker. The "cooked" waste pulp is then separated through a series of screens, trommels, and magnets to remove steel, aluminum, and rigid plastics, continued with washing process to remove sand and glass. The washed pulp is dried in preparation for thermal conversion into synthetic gas.
- Waste converter is a technology using superheated steam at 150 °C [26]. The system is capable of performing the pasteurization of organic waste, sterilization of pathogenic or biohazard waste, grinding and pulverization of refuse. The technology is similar to waste autoclave but operated in the atmospheric pressure range. Due to the usage of superheated steam and vacuum pump, the product from waste converter is dehydrated.

Other researchers also showed that the hydrothermal process is causing rapid dehydrochlorination of PVC from about 250 to 350 °C [27, 28], producing hydrochloric acid (HCl) and polyene. This is a positive side effect of the hydrothermal process compared to conventional waste treatment technologies. The waste autoclave and waste converter, however, do not have this benefit due to their low operating temperature below 250 °C.

These MSW treatments using the hydrothermal principle have been developed, but only some of those are designed to obtain solid fuel and commercially available due to high pressure and temperature requirement or the need of catalyst [29–32]. The developed hydrothermal treatment in Tokyo Tech is meant to convert high moisture content solid wastes into dried uniform pulverized coal-like solid fuel using low energy consumption as shown in Fig. 14.2 [33, 34].

The difference between the developed hydrothermal treatment and other hydrothermal treatments is mainly in its operational condition. As shown in Fig. 14.3, the developed process is approximately in the lower temperature and pressure zone compared to the subcritical and supercritical hydrothermal process, thus making the system more constructible. When compared to the waste autoclave and waste converter, the developed hydrothermal treatment employs higher temperature and pressure but also having the benefit of waste dechlorination, a requirement for the product to be acceptable as alternative solid fuel.

14.2.1 Commercial Scale Application

As mentioned before, the commercial scale plant working in Japan has a maximum capacity of 1 ton of waste per batch and employing saturated steam of about 200 °C into the reactor. A total of 3 hours is required to complete the process, including waste loading and unloading. A schematic diagram of the plant is shown

Fig. 14.2 Raw MSW conversion by employing the hydrothermal treatment

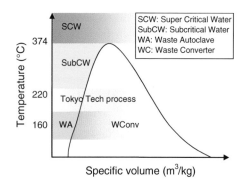

Fig. 14.3 Operational condition summary of various hydrothermal treatments

in Fig. 14.4. It primarily consists of a hydrothermal process reactor, boiler as steam supplier into the reactor, steam condenser and waste water treatment.

The reactor is a cylindrical dual wall pressure vessel of 3 m³ in volume, and high pressure steam is supplied into the reactor for treatment. The reactor consists

Fig. 14.4 Schematic diagram of the hydrothermal treatment process

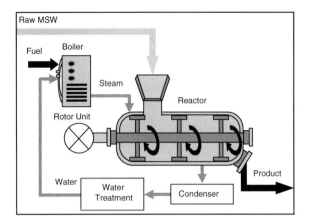

of two ports for supplying MSW into the reactor and discharging treated MSW after the treatment. A screw-type rotor is fitted inside the reactor, which can rotate on either direction by an electric motor at a constant speed of around 20 rpm, ensuring the uniformity of the treatment throughout the MSW sample.

The boiler is a package type, oil fired (kerosene), single fire tube boiler of 1 ton/hr capacity with necessary auxiliary equipment like feed water pump, forced draft fan, etc. The condenser is a direct mixing water cooled ejector type, used to condense the released steam from the reactor after the treatment. It also acts as an ejector to remove steam from the reactor. The water condensed in the condenser is to be sent to the water treatment facility to be fed back to the boiler as feed water for steam generation, constituting a closed-loop of water flow.

In this batch type of treatment, MSW of around 500–1000 kg was supplied into the reactor. Saturated steam at 2 MPa was then gradually supplied into the reactor from the boiler. When the pressure and temperature in the reactor reached the target values, the reactor was held at this condition for a certain time called the holding period. When the holding time finished, steam supply was stopped and the pressure of the reactor was reduced by discharging the residual steam. The released steam was condensed, and after water treatment will be reused for producing steam in the boiler. Once the reactor pressure fell down to atmospheric, the treated product from the reactor was extracted by rotating the stirrer acting as a screw conveyor.

The extracted product is then undergoing natural drying process in about 2 days to obtain the final product. If necessary, after steam extraction from the inner wall, lower pressure steam can be supplied between the inner and outer walls to conduct drying of the product. In the end, glass and metal material in the final product is easily separable by sieving.

Another example of large-scale application of this process is in China as shown in Fig. 14.5. Two units 5 m^3 reactors of the hydrothermal treatment has been working since 2011 for treating municipal solid waste collected from nearby cities.

Fig. 14.5 Large-scale reactor of the hydrothermal treatment in China

It is capable of treating up to 3 tons of waste per batch per reactor, applying saturated steam with middle pressure at around 2 MPa with total required time around 2 hours to complete the overall process.

14.2.2 Treatment Characteristics

Typical hydrothermal process characteristics are shown in Fig. 14.6 to represent short holding period type of 30 min and Fig. 14.7 to represent long holding period type of 90 min. Three main process stages of the heating, holding, and steam release periods as previously mentioned are clearly presented.

It can be seen from both figures that about 60 % of water and fuel were consumed in the heating period to generate steam. In the holding period, small amount of water supply was still required to maintain the steam and reactor temperature, but in the latter part of the holding period, more water was consumed to generate steam, especially in the long holding period type shown in Fig. 14.6.

The increasing consumption of water in the latter part of the holding period can be attributed to the usage of steam boiler as a heat source. Steam boiler senses the reactor pressure rather than the reactor temperature. During the holding period, the density of the steam was increased due to the heat transfer to overcome the heat loss outside, sometimes very high due to the change of phase from vapor to liquid. In the same time, bulk density of MSW was also increased due to the solving of fibrous MSW in the water.

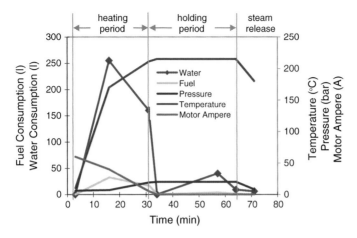

Fig. 14.6 Hydrothermal process characteristics with short holding time

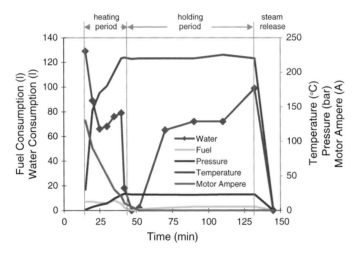

Fig. 14.7 Hydrothermal process characteristics with long holding time

14.3 Characteristics of Hydrothermal Treatment Product

Hydrothermal treatment has resulted in grayish, uniform sized slump products as shown in Fig. 14.8.

When the reaction temperature was lower and the holding period was shorter (center figure), some rubber (shoe sole) materials were still present. At higher reaction temperature and longer holding period (right figure), however, the materials became uniform and more mud-like. It can be concluded that the product is more uniform when treated under higher reaction temperature and longer holding period.

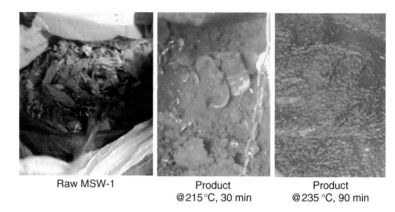

| Raw MSW-1 | Product @215 °C, 30 min | Product @235 °C, 90 min |

Fig. 14.8 Raw MSW and products appearance in various processing conditions

Table 14.1 Typical MSW composition

Organic	77 %
Paper	7 %
Plastic	10 %
Textile	5 %
Rubber	1 %

Table 14.2 Particle size distribution of total samples

Particle size (µm)	Particle size distribution (%)	
	Sample-30	Sample-60
>1180	81.8	69.6
600–1180	7.6	13.4
75–600	9.6	14.1
<75	1.0	3.0

Following analysis on the particle size distribution of the hydrothermally treated product from typical MSW with composition shown in Table 14.1 were conducted using mechanical sieve for larger particle and Shimadzu SALD-3000 laser diffraction method for the measuring range of 0.05–3,000 µm. The treatments were conducted at 2–3 tons capacity reactor at 2 MPa and 200 °C, and in two holding periods of 30 and 60 min.

The larger particle size distribution of the product from the hydrothermal treatment as dry base is shown in Table 14.2. As shown in this table, the particle size of the initial Sample-60 (60 min holding period) was considerably smaller compared to Sample-30 (30 min holding period), suggesting that more severe treatment resulted in smaller product particle size.

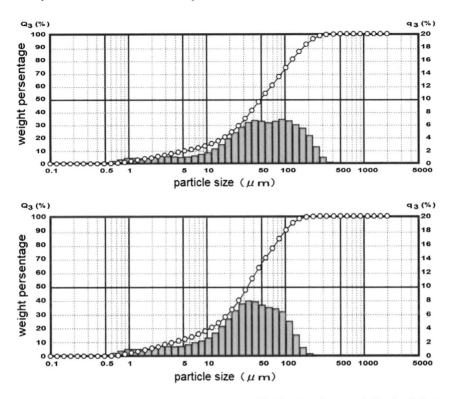

Fig. 14.9 Particle size distribution of the product with 30 min (*above*) and 60 min (*below*) holding time

The smaller particle size distributions of solid product were also confirmed that longer holding time resulted in smaller particle size distribution, as shown in Fig. 14.9. From the figure, it can be seen that the particle size of the solid product after the hydrothermal treatment with 30 min holding time was less than 400 μm, while the particle size of the solid product after the hydrothermal treatment with 60 min holding time was less than 200 μm. Comparing the particle size distributions of the solid product for the two holding periods, it is shown that a longer holding time resulted in more uniform, denser, and smaller particle size products.

The uniformity and smaller particle size can be explained by the fact that the pressure and temperature in the hydrothermal process, such as in coalification of wood into coal by natural hydrothermal process in earth layer, would result in the particle breakage of MSW as shown in Fig. 14.10. This condition, added with the presence of water acts as a solvent would produce smaller particle size of hydrothermally treated product from MSW in higher temperature and longer holding period conditions.

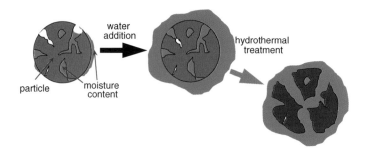

Fig. 14.10 Visualization of particle breakage due to hydrothermal treatment

Table 14.3 Raw MSW characteristics

	Density(g/cm)[a]	Moisture content (wt%)[a]	Combustible content (wt%)[a]	Ash content (wt%)[a]
MSW-1	0.15 ±0.02	33.0 ±10.1	50.0 ±6.5	17.0 ±6.0
MSW-2	0.15 ±0.02	33.2 ±8.9	54.5 ±8.9	12.3 ±9.1

[a] Wet basis

Table 14.4 Varied experimental operating conditions

Pressure (MPa)	2.0	2.0	2.4	2.4	2.4
Temperature (°C)	215	210	225	225	225
MSW mass (kg)	705	670	665	640	599
Holding period (min)	30	90	30	90	90
Experiment ID	A11	A21	B11	B21	B22

14.3.1 Physical and Chemical Characteristics of Product

Two typical MSW characteristics are being used and shown in Table 14.3. It can be seen that both types had low density, high variation in their values, relatively high moisture content and high ash content which would reduce the heating value of MSWs. It also should be noted that both raw MSW visually had high plastic content, which would increase the chlorine content in the MSWs.

After hydrothermal treatment in various conditions as shown in Table 14.4, the product characteristics were showing significant increase of density up to four times of that of MSW in dry basis, shown in Table 14.5. Noting these results, it is predicted that 75–80 % waste volume reduction by the hydrothermal treatment can be achieved. It should be underlined that contrary to raw MSW characteristics, the products showed high repeatability and low variation, which means that the product is more uniform compared to raw MSW.

14 Hydrothermal Treatment of Municipal Solid Waste

Table 14.5 Product characteristics

Experiment ID	Density (g/cm^3)[b]	Moisture content (wt%)[a]	Combustible content (wt%)[a]	Ash content (wt%)[a]
A11	0.61	43.7	46.1	10.2
A21	0.70	43.9	46.0	10.1
B11	0.71	43.4	44.6	12.0
B21	0.75	39.4	45.8	14.8
B22	0.48	55.9	36.3	7.8

[a] Wet basis
[b] Dry basis

Table 14.6 Effect of the hydrothermal treatment on the properties of products

	Paper				Kimchi			
	Raw	@180 °C	@200 °C	@220 °C	Raw	@180 °C	@200 °C	@220 °C
Moisture (%)	2.3	4.0	4.2	6.5	92.4	93.8	93.2	93.3
Proximate analysis (wt%) *(dry basis)*								
Volatile matter	87.0	76.2	58.4	56.6	67.1	60.3	60.9	57.8
Fixed carbon	5.3	14.1	27.0	29.2	22.6	29.8	29.7	31.0
Ash	7.7	9.7	14.6	14.2	10.3	10.0	9.4	11.3
Ultimate analysis (wt%) *((dry basis))*								
C	40.3	45.0	54.5	54.8	33.6	34.4	35.8	37.0
H	5.6	5.4	5.0	4.8	5.3	4.6	4.6	4.5
N	0.2	0.1	0.4	0.2	3.5	3.2	3.2	3.0
O	46.2	39.8	25.5	26.0	47.3	47.8	47.0	44.1

It can be observed that higher reaction temperature and longer holding period would produce denser products. This phenomenon should be attributed to the similar reason as shown in Fig. 14.10, where the hydrothermal process produced smaller particle size and at the same time reduced the product bulk void. As a result, higher bulk density of the product can be achieved.

In the term of composition, all product samples exhibited similar combustible and ash content but with higher moisture content. The high moisture content from hydrothermally treated product reduced the confidence whether they can be used as solid fuel, but one should analyze their heating values, which will be presented afterwards.

Subsequent experiments have been conducted in Tokyo Tech, focusing in hydrothermal treatment on basic material such as paper to represent cellulose, and kimchi (traditional vegetable pickles in Korea) to represent the food waste. Hydrothermal treatment breaks the physical and chemical structure in the materials such as cellulose, hemicellulose, and lignin [35–39], and these biomasses were broken down into smaller and simpler molecules.

Table 14.6 shows the property of raw samples of kimchi and paper and their products after the hydrothermal treatment which were produced at 180, 200, and

Fig. 14.11 SEM microphotographs of kimchi treatment. **a** raw kimchi **b** product at 200 °C **c** product at 220 °C

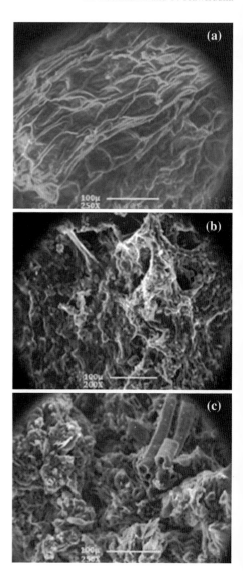

220 °C. It is shown that paper and kimchi had high volatile matter content (87.0 % and 67.1 %) and oxygen content (46.2 % and 47.3 %) like other biomass. Similar to that of MSW treatment, with the increase of the hydrothermal reaction temperature, the volatile matter and oxygen content decreased while the fixed carbon content increased. Comparing between paper and kimchi, the hydrothermal products of paper showed more significant changes than that of kimchi, suggesting that cellulosic materials are easier to degrade compared to typical food waste.

Considering the microphysical structure of the two materials, Figs. 14.11 and 14.12 show the SEM microphotographs of the paper and kimchi before and after

Fig. 14.12 SEM microphotographs of the paper treatment. **a** raw paper **b** product at 200 °C **c** product at 220 °C

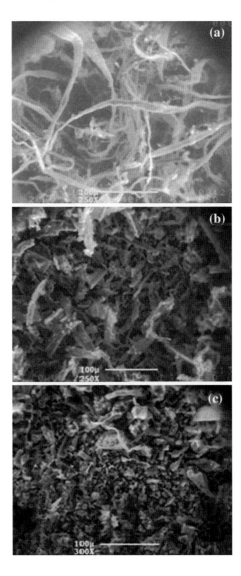

the hydrothermal treatment. These SEM microphotographs reveal the changes between the raw materials and the upgraded solid products, showing disruption of physical structures and formation of individual grains in the products. Apparently the hydrothermal treatment breaks the structure of the paper and kimchi and converts them into smaller, uniform particle products.

The hydrothermal treatment apparently has disrupted the structure of kimchi and paper that contains cellulose, hemicelluloses, and lignin like woody biomass, the structure of biomass was decomposed and disrupted to become small molecular and particle size as shown in Fig. 14.13. Moreover, colloidal bond between

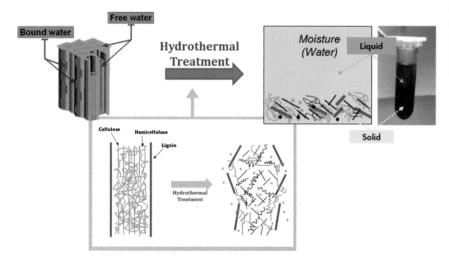

Fig. 14.13 Change of the physical structure of biomass by the hydrothermal treatment

water and material in the sample was broken to separate moisture. The results of changing of physical structure were one of significant causes to improve the dewatering and drying performance of hydrothermal products. Removal of moisture contents in MSW is a major target of the pretreatment, and the moisture content of MSW has a strong influence on the characteristics and treatment method of MSW [40–43].

14.3.2 Drying Characteristics of Product

Since the product needs to be dried before it can be used as a fuel, an improved drying performance is preferable. It was confirmed that the treatment enhances the drying performance of MSW and thus the moisture content of the treated MSW was reduced down to 11.6 from 33 % in 48 h, leading to a lower energy consumption for removing moisture from MSW.

To compare their drying performances, two raw MSW representatives of rice and vegetable were also naturally dried. From Fig. 14.14, it is clearly shown that the drying performances of the hydrothermally treated products were higher compared with two raw MSWs in the term of critical drying curve: that is, the hydrothermally treated products reached stable moisture content faster than raw MSW.

It should be noted that the final equilibrium moisture content of hydrothermally treated products were about 5 %, significantly lower than that of raw MSW which were about 20 %. The phenomenon is related to the fact that the breakage of particle by the hydrothermal treatment suggested that the bound water (bound moisture) inside the raw MSW will be reduced and converted to free water (free moisture),

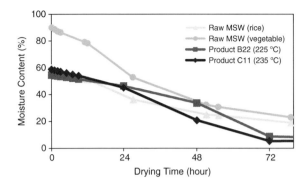

Fig. 14.14 Natural drying characteristics of MSWs and treated products

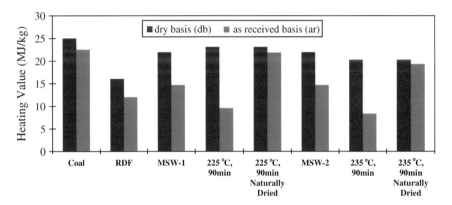

Fig. 14.15 Heating value of hydrothermally treated products compared to conventional solid fuels

which resulted in a lower final equilibrium moisture content. This effect becomes more significant if the reaction temperature and the holding period increased.

14.3.3 Characteristics of Product as Solid Fuel

Observing the heating values of hydrothermally treated products compared to raw MSWs and conventional solid fuel as shown in Fig. 14.15, it can be shown that the heating values of the products were relatively similar regardless of their hydrothermal treatment conditions. The hydrothermally treated products showed the heating value of about 18 MJ/kg to 22 MJ/kg, competitively similar to that of high quality sub-bituminous coal (class A) with the heating value of about 25 MJ/kg [44], and typical RDF with the heating value of 16 MJ/kg [45].

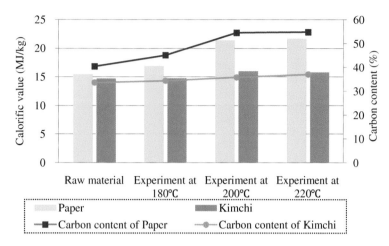

Fig. 14.16 Effect of the hydrothermal treatment on the calorific value and carbon content of paper and kimchi

Table 14.7 Result of HGI measurements		Sample-30	Sample-60
	HGI	30.1	60.3

In the case of kimchi and paper, the calorific values for both materials increased with the increase of the reaction temperature due to the increase of the fixed carbon content as shown in Fig. 14.16. Due to higher carbon content of paper, the calorific value increase of paper is more significant than in the case of kimchi.

Since the hydrothermally treated product will be used in the pulverized coal burner, its grindability, known as HGI (Hardgrove Grindability Index), was also observed for two samples processed in 2 MPa, 200 °C and holding time of 30 min and 60 min, respectively. It is shown from Table 14.7 that the HGI of Sample-60 (60 min holding time) was higher than that of Sample-30 (30 min holding time); therefore, longer holding time is considered to produce easy-to-pulverize solid fuel. As for 60 min holding time, its HGI was higher than 60, which is higher compared to that of coal (normally within 40–60). This means that the product is easier to be pulverized compared to coal.

As a result of good grindability of the product, complete burn out, easy mixing, no blockage or bridging to the feeding system, and no sedimentation can be expected [46]. However, the adhesion of the sample was observed to be high compared to coal, as many of the samples were seen adhered to the ball mill.

Combustion tests have been conducted to confirm the usage of treated MSW as solid fuel. The burning characteristics of the Indian and Australian coals, treated MSW and their various blends have been studied using thermogravimetry analysis (TGA) as shown in Fig. 14.17.

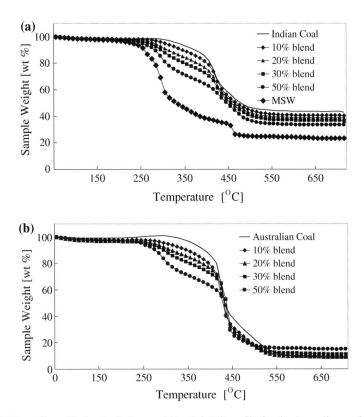

Fig. 14.17 a TG profile for the Indian coal blend. b TG profile for the Australian coal blend

With the rise of the temperature, after a release of moisture, combustion of samples took place with associated weight losses. For the blends, quicker weight losses were observed, mainly due to early emission of volatile matter, which differentiates burning behavior of treated MSW compared with coal. The combustion of coal is mainly due to the combustion of the fixed carbon whereas in the case of treated MSW, this was dominated by combustion of the volatile matters, obviously due to their high volatile content. As for Indian coal, the addition of treated MSW in the blend reduce the ash component in the mixed fuel, resulting in improved combustion performance of Indian coal.

In larger scale experiment, a drop tube furnace (DTF) was used to observe the burnout efficiency of high ash content Indian coal with its blend with treated MSW as shown in Table 14.8.

This shows that the burnout efficiency increased 3.5 % when the Indian coal is blended with treated MSW. This increase in the burnout efficiency can significantly reduce the unburnt loss in a power plant and thus increase in the combustion efficiency. This improvement is attributed to the enhancement of the low reactive char component of the coal due to the addition of treated MSW. It has been

Table 14.8 Burnout efficiency

	Burnout efficiency (%)
Indian coal	89.5
Indian coal + 20 % MSW	93.0

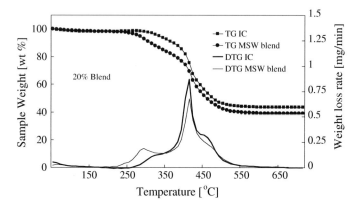

Fig. 14.18 TG and DTG profile for Indian coal (IC) and MSW 20 % blend

demonstrated that MSW blending with Indian coal enhances the reactivity of the low reactive char component [47].

Figure 14.18 shows the mass loss profile for treated MSW blends of 20 % studied by TGA analysis. When comparing their DTG (differential thermogravimetry) profiles, the profile for the MSW blended fuel differs with coal profile in three ways. A clear difference can be found in the low temperature region i.e., around 250 °C, due to the early release of volatile from the MSW fuel which will support the ignition of coal. Second, there is a difference in the height of the DTG profile between coal and treated MSW blend, which was lower. Hence, the weight loss rate for the MSW blended fuel is slightly lower than coal combustion in the char burning stage. It can be concluded that the blending of MSW with coal increases the reactivity of volatile component and reduces the reactivity of char component. The improvement in the reactivity of the volatile component enhances the ignition characteristics and hence the reduction in the ignition temperature [48].

The third and important difference lies in the higher temperature region at around 450 °C. In general char has two regions of reactivity, low reactive component and high reactive component [49]. The low reactive component is responsible for a slow weight loss rate creating a shoulder in the right hand side of the DTG curve, act as the main contributor for the unburnt carbon present in the fly ash of a thermal power plant.

Treated MSW blending with coal can significantly enhance the reactivity of the low reactive component of the coal char; hence the shoulder portion of the DTG

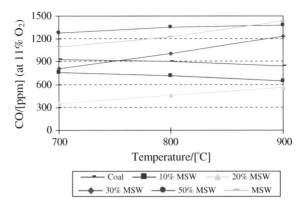

Fig. 14.19 CO emissions as a function of the temperature

profile is completely absent for the MSW blended fuel, and this improvement in the combustibility reduces the unburned char of the coal.

A coal-based bubbling fluidized bed (BFB) reactor was adopted for the co-combustion behavior investigation to verify the feasibility of the replacement of coal with treated MSW without major furnace modification. The BFB was made of a 77 mm i.d. corundum ceramic tube, with a total height of 1100 mm. The operating temperatures were 700, 800, and 900 °C and the excess air of 1.3 with residence time was kept in the range of 1.5–2.0 s. Contents of the flue gas were analyzed by a GASMET™ DX-4010 FT-IR Gas Analyzer.

As shown in Fig. 14.19, when it comes to the co-combustion practice at low blending ratios (10, 20 %), dramatically lower CO concentrations than that of coal were observed, while at high blending ratios (30–50 %) higher CO concentrations were occurred. For coal monocombustion, the higher CO emission could be ascribed to diffusion controlled reactions due to its high ash content (12.01 %) compared to the treated MSW (1.09 %). It is reported that the samples with high ash content is more likely to have the trend to follow a shrinking sphere model which results in an ash layer surrounding the FC making oxygen diffusion difficult [50]. This kind of incombustible ash layer led to the accumulation of unburnt char particles which eventually created a fuel-rich zone in the bed thus resulting in higher CO emission.

When the treated MSW was mixed as a co-combustion fuel, first the ash content was reduced so that the oxygen diffusion could be promoted. Second, burning of the treated MSW during co-combustion was expected to occur at a rapid rate compared to coal and therefore the heat released helped to provide a higher temperature zone which elevated the speed of coal combustion as well as the CO burnout. Third, the high percentage of oxygen content in the treated MSW might increase the amount of O, OH radicals near the char particles which promoted the heterogeneous char oxidation.

Figure 14.20 shows the changes of SO_2 emission as a function of the temperature at different combustion conditions. It is obvious that the SO_2 emitted from the combustion of low sulfur content treated MSW was almost negligible and the

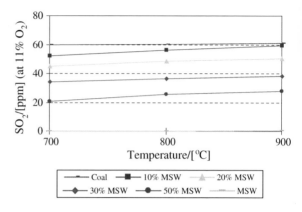

Fig. 14.20 SO$_2$ emissions as a function of the temperature

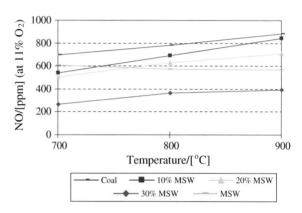

Fig. 14.21 NO emissions as a function of the temperature

SO$_2$ concentrations decreased with the increase of the treated MSW share. Furthermore, there was a monotonic increase trend for SO$_2$ with temperature rising for all the samples, though the levels were rather small due to the low sulfur content in coal (0.44 %).

In the BFB combustion system, fuel NOx is believed to be predominant, which primarily formed via two different pathways: one is through gas-phase oxidation of the nitrogenous group in the volatiles, and the other is through the heterogeneously catalyzed oxidation of the char-bound nitrogen species [51–54]. Moreover, the distribution of the nitrogen between the volatiles and the chars is roughly proportional to the volatile matter in the fuel. Especially, during the devolatilization of treated MSW, as the char content is relatively small, most of the nitrogen is believed to be emitted during the devolatilization phase (66–75 %) [55].

As shown in Fig. 14.21, in the case of co-combustion at low blending ratios (10, 20 %), the blending of the treated MSW contributed to the abatement of NO emitted from coal by providing more available fixed carbon surfaces. In addition to that, the burning of the treated MSW around the coal particles also offered more available NH$_3$ effective for NO reduction. For the co-combustion with a higher

Table 14.9 Raw MSW analysis (dry basis)

	Organic chlorine (wt%)	Inorganic chlorine (wt%)	Total chlorine (wt%)	Higher heating value (MJ/kg)
MSW-1	0.97 ± 0.33	0.47 ± 0.06	1.27 ± 0.25	18.0 ± 2.4
MSW-2	1.38 ± 0.32	0.37 ± 0.08	1.63 ± 0.39	21.9 ± 4.4

MSW blending ratio (30 %), it is noteworthy that the NO emission was even lower than the case of individual MSW. Besides the reasons abovementioned, the rapid heat release from the burning of VM of the HT MSW along with the more available oxygen emitted from high oxygen content HT MSW enhanced the thermal devolatilization of coal thus creating more pore structure of coal char which could be useful for NO reduction in both of the cases of coal and the HT MSW. Additionally, the VM released from the HT MSW also created temporarily the reducing environment with high concentrations of CH_i and HCCO groups near the coal particles, which assisted the reduction of NO strongly as well.

14.4 Beneficial Features of Hydrothermal Treatment

Besides the advantage of no required separation and crushing of MSW in hydrothermal treatment, environmental advantage of organic chlorine reduction is also observed as additional features of the process.

The raw MSW characteristics in dry basis (db) can be observed in Table 14.9. It can be seen that although the heating value of MSW was similar to that of low-grade sub-bituminous rank coal, the chlorine content of MSW was not fulfilling the requirements of low-chlorine solid fuel. MSW-1 has a chlorine content of 1.27 %, or about 12,700 ppm, of which 9,700 ppm was organic or water-insoluble chlorine. MSW-2 exhibited a higher chlorine content of 1.63 %, or 16,300 ppm, with 13,800 ppm organic chlorine.

The total chlorine content of the products obtained from MSW-1 (A11 to B21, see Table 14.4) should range from 1.02 to 1.52 % (1.27 ± 0.25 %), while those from MSW-2 (B22 to C11, see Table 14.4) should have a chlorine content range of 1.24 to 2.02 % (1.63 ± 0.39 %). The chlorine content results in Fig 14.22 show that the chlorine content for all products was in the expected range, suggesting that little or no chlorine was vented out of the system.

As shown in Fig. 14.22, the organic chlorine contents of the products were generally reduced. The case of B21 showed a reduction down to 0.16 % (1,600 ppm), approximately an 83 % reduction of chlorine, even though the treatment temperature of 225 °C at 90 min was lower than conventional dechlorination temperature processes. Since the total chlorine content of the product remained at the same level as raw MSW, consequently the water-soluble, inorganic chlorine contents of the products were increased up to 1.04 %. This phenomenon suggested that in the case of MSW hydrothermal process, the organic

Table 14.10 Energy requirement of hydrothermal and conventional pelletizing

Process	Hydrothermal treatment	Conventional pelletizing
MSW heating energy (MJ/kg MSW)	0.5	0.25
Steam generation energy (MJ/kg MSW)	0.3	0
Drying energy (MJ/kg MSW)	0	0.67
Mechanical energy (MJ/kg MSW)	0.03	0.43
Total required energy (MJ/kg MSW)	0.8	1.35

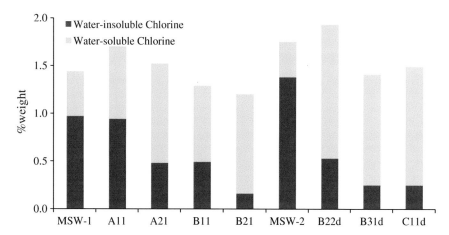

Fig. 14.22 Organic to inorganic chlorine conversion in MSW under various processing conditions

chlorine was converted into inorganic chlorine in the lower temperatures compared to conventional dechlorination process.

It can be seen from Fig. 14.22 that after hydrothermal treatment, the organic chlorine contents of the products were low compared to that of raw MSW. The increase of the reaction temperature (for example, A11 compared with B11, see Table 14.4) and the holding period (for example, B11 compared with B21, see Table 14.4) would further reduce the organic chlorine content.

The water-soluble inorganic chlorine content in the product can be easily removed by water-leaching and dewatering process. It was shown that one-time leaching can effectively reduce 80 % of the water-soluble chlorines [56]. In the case of B21, 80 % reduction equals to 0.83 % inorganic chlorine, reducing total residual chlorine in the product to 0.37 % (3,700 ppm). Therefore, it is suggested

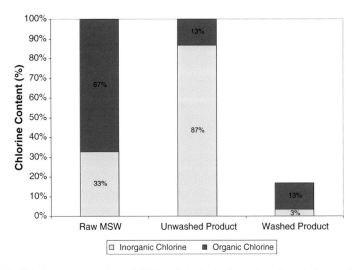

Fig. 14.23 Chlorine contents of raw MSW and the hydrothermal products before and after the washing process

that a combination of the hydrothermal process and multiple water-washing processes could further reduce the chlorine content of hydrothermally treated product from MSW.

The chlorine contents of the raw MSW (Raw MSW), the product after the hydrothermal treatment before the washing process (Unwashed Product) and the product after the hydrothermal treatment followed by the washing process (Washed Product) are presented in Fig. 14.23, where the total chlorine content in the raw MSW is taken as 100 %. The unwashed product exhibited low organic chlorine content because it was converted into inorganic chlorine during the hydrothermal process.

In the term of energy consumption, hydrothermal treatment also showed superior performance compared to conventional waste treatment system, also known as RDF (Refuse Derived Fuel) system. According to Caputo and Pelagagge [57], the power requirements for a hammer mill as a particle shredder are the largest, and usually reach about 360 kJ/kg MSW. Total power requirements for other equipment such as a densifier, pelletizer, magnetic separator, and shredder usually reach about 72 kJ/kg input material. The energy required for drying, however, is usually significant for the overall process, especially as it is obligatory to dry the MSW before inputting it into the shredder or mill. Sikka [58] reported an energy requirement as low as 324 kJ/kg MSW. In Perry [59], the energy requirements for a plate dryer type, to dry foodstuff, was about 420 kJ/kg, while for a turbo-tray dryer it was about 560 kJ/kg.

Those drying numbers are considered to be low, since theoretically the energy required to dry the MSW should be at least equal to the latent heat of vaporization of the water inside the MSW, that is, about 2257 kJ/kg water [60]. Using the typical moisture content of MSW, which is about 30 %, the energy required to dry the MSW should be approximately 677 kJ/kg MSW.

The hydrothermal treatment does not necessarily need a drying system since the particle size reduction is performed in wet conditions inside the reactor. It does, however, need steam generation energy to supply steam to the reactor. Table 14.10 compares the difference between the characteristics and energy requirements of a conventional pelletizing process to produce RDF and the developed hydrothermal treatment. The final product temperature is assumed to be 110 °C for the conventional pelletizing process, and 215 °C for the hydrothermal treatment.

It can be seen that both systems have their own advantages and disadvantages, since the product appearances for both systems were also different; pulp for hydrothermal treatment and pellet for conventional pelletizing. From the table, it can be summarized that the total specific energy requirement for conventional RDF production is about 1.35 MJ/kg MSW, and hydrothermal treatment exhibits a slightly lower energy requirement of about 0.8 MJ/kg MSW. It then can be expected that the total required energy for hydrothermal treatment will be about 40 % less, compared to that of conventional RDF production.

References

1. Ministry of Environment Japan (2010) State of discharge and treatment of municipal solid waste in FY 2008. http://www.env.go.jp/en/headline/headline.php?serial=1333
2. The Asahi Shimbun (2008)Coal prices surging due to global demand. Australia Flooding. Accessed 4 March 2008
3. British Petroleum (BP) (2010) BP statistical review of World energy. http://www.bp.com/statisticalreview/
4. UNEP (2005) Solid waste management, vol. I, chapter X: types of waste-to-energy systems. http://www.unep.or.jp/ietc/publications/spc/solid_waste_management/index.asp
5. Plastic Waste Management Institute Japan (2010) Breakdown of total plastic waste. In: Plastic products, plastic waste and resource recovery [2008], PWMI Newsletter No. 39, (2010). http://www2.pwmi.or.jp/siryo/ei/ei_pdf/ei39.pdf (accessed December 2010)
6. IPCC's Task Force on National Greenhouse Gas Inventories (2010) MSW composition data by percent-regional defaults. In: 2006 IPCC Guidelines for National Greenhouse Gas Inventories, vol 5: Waste, Chapter 2. Waste generation, composition and management data. http://www.ipcc-nggip.iges.or.jp/public/2006gl/vol5.html. Accessed Dec 2010
7. IPCC Guidelines for National Greenhouse Gas Inventories (2006) Vol. 5: Waste, Chapter 2. Waste generation, composition and management data. IPCC's task force on national greenhouse gas inventories, 2010. http://www.ipcc-nggip.iges.or.jp/public/2006gl/vol5.html
8. European Commission—Directorate General Environment (2003) Refuse derived fuel, current practice and perspective: final report. http://ec.europa.eu/environment/waste/studies/pdf/rdf.pdf. Accessed Dec 2010
9. Zevenhoven R, Axelsen EP, Hupa M (2002) Pyrolysis of waste-derived fuel containing PVC. Fuel 81:507–510
10. Ma S, Lu J, Gao J (2002) Study of the low temperature pyrolysis of PVC. Energy Fuels 16:338–342
11. Miranda R, Yang J, Roy C, Vasile C (1999) Vacuum pyrolysis of PVC. Polym Degrad Stab 64:127–144

12. Xiao X, Zeng Z, Xiao S (2008) Behavior and products of mechano-chemical dechlorination of polyvinylchloride and poly (vinylidene chloride). J Hazard Mater 151:118–124
13. Kamo T, Kondo Y, Kodera Y, Sato Y, Kushiyama S (2003) Effects of solvent on degradation of poly(vinyl chloride). Polym Degrad Stab 81:187–196
14. Sato K, Jian Z, Soon JH, Namioka T, Yoshikawa K, Morohashi Y et al (2004) Studies on fuel conversion of high moisture content biomass using middle pressure steam. In: Proceeding of the thermal engineering conference, pp 259–260
15. Yoshikawa K (2005) Fuelization and gasification of wet biomass with middle-pressure steam. Eco Ind 10:29–37 (in Japanese)
16. Muthuraman M, Namioka T, Yoshikawa K (2010) Characteristics of co-combustion and kinetic study on hydrothermally treated municipal solid waste with different rank coals: a thermogravimetric analysis. Appl Energy 87:141–148
17. Prawisudha P, Namioka T, Yoshikawa K (2012) Coal alternative fuel production from municipal solid wastes employing hydrothermal treatment. Appl Energy 90:298–304
18. Roditi International Co. Ltd. (2010) Hydrothermal crystal growth—Quartz. http://www.roditi.com/SingleCrystal/Quartz/Hydrothermal_Growth.html. Accessed Dec 2010
19. Savage PE, Levine RB, Huelsman CM (2010) Hydrothermal processing of biomass. In: Thermochemical conversion of biomass to liquid fuels and chemicals, Ch. 8. RSC Publishing, Cambridge
20. Cole EL, Hess HV, Guptill FE (2010) Preparation of solid fuel-water slurries. United States Patent No. 4,104,035, http://www.google.com/patents?hl=id&lr=&vid=USPAT4104035&id=c9EuAAAAEBAJ&oi=fnd&dq=hydrothermal+fuel&printsec=abstract#v=onepage&q=hydrothermal%20fuel&f=false. Accessed Dec 2010
21. Cole EL, Hess HV, Wong J (2010) Upgrading of solid fuels. United States Patent No. 4,047,898, http://www.google.com/patents?hl=id&lr=&vid=USPAT4047898&id=VJU3AAAAEBAJ&oi=fnd&dq=hydrothermal+fuel&printsec=abstract#v=onepage&q=hydrothermal%20fuel&f=false. Accessed Dec 2010
22. Bobleter O, Niesner R, Röhr M (1976) The hydrothermal degradation of cellulosic matter to sugars and their fermentative conversion to protein. J Appl Polym Sci 20:2083–2093
23. Goto M, Obuchi R, Hirose T, Sakaki T, Shibata M (2004) Hydrothermal conversion of municipal organic waste into resources. Bioresour Technol 93:279–284
24. Goudriaan F, Naber JE (2008) HTU diesel from wet waste streams. In: Symposium new biofuels, Berlin 2008. http://www.fnr-server.de/cms35/fileadmin/allgemein/pdf/veranstaltungen/NeueBiokraftstoffe/5_HTU.pdf. Accessed Dec 2010
25. Brightstar Environmental (2010) Solid Waste & Energy Recycling Facility (SWERF) Technology. http://www.sovereignty.org.uk/features/eco/swerf.html. Accessed Dec 2010
26. Ompeco (2010) Converter MO Series. http://www.ompeco.com/converter/en_serie_mo.html. Accessed Dec 2010
27. Endo K, Emori N (2001) Dechlorination of poly(vinyl chloride) without anomalous units under high pressure and at high temperature in water. Polym Degrad Stab 74:113–117
28. Takeshita Y, Kato K, Takahashi K, Sato Y, Nishi S (2004) Basic study on treatment of waste polyvinyl chloride plastics by hydrothermal decomposition in subcritical and supercritical regions. J Supercrit Fluids 31:185–193
29. Shanableh A (2000) Production of useful organic matter from sludge using hydrothermal treatment. Water Resour 34:945–951
30. Wenzhi H, Guangming L, Lingzhao H, Hua W, Juwen H, Jingcheng X (2008) Application of hydrothermal reaction in resource recovery of organic wastes. Resour Convers Recycl 52:691–699
31. Ishida Y, Kumabe K, Hata K, Tanifuji K, Hasegawa T, Kitagawa K et al (2009) Selective hydrogen generation from real biomass through hydrothermal reaction at relatively low temperatures. Biomass Bioener 33:8–13
32. Jomaa S, Shanableh A, Khalil W, Trebilco B (2003) Hydrothermal decomposition and oxidation of the organic component of municipal and industrial waste products. Adv Environ Resour 7:647–653

33. Sato K, Jian Z, Soon JH, Namioka T, Yoshikawa K, Morohashi Y et al (2004) Studies on fuel conversion of high moisture content biomass using middle pressure steam. In: Proceeding of thermal engineering conference, G132
34. Yoshikawa K (2005) Fuelization and gasification of wet biomass with middle-pressure steam. Eco Ind 10:29–37 (in Japanese)
35. Funke A, Ziegler F (2010) Hydrothermal carbonization of biomass: A summery and discussion of chemical mechanisms for process engineering. Biofuels Bioprod Bioref 4:160–177
36. Krammer P, Vogel H (2000) Hydrolysis of esters in subcritical and supercritical water. J Supercrit Fluids 16(3):189–206
37. Bobleter O (2005) Hydrothermal degradation of polymers derived from plants. Prog Polym Sci 19:797–841
38. Sakaguchi M, Laursen K, Nakagawa H, Miura K (2008) Hydrothermal upgrading of Loy Yang Brown coal-Effect of upgrading conditions on the characteristics of the products. Fuel Prog Technol 89:391–396
39. Yuliansyah T, Jirajima Y, Kumagai S, Sasaki K (2010) Production of solid biofuel from agriculture wastes of the palm oil industry by hydrothermal treatment. Waste Biomass Valor 1:395–405
40. Hammerschimidt N, Boukis E, Hauer U, Galla E, Dinjus B, Hitzmann T, Larsen S, Nygaard D (2011) Catalytic conversion of waste biomass by hydrothermal treatment. Fuel 90:555–562
41. Nonaka M, Hirajima T, Sasaki K (2011) Upgrading of low rank coal and woody biomass mixture by hydrothermal treatment. Fuel 90:2578–2584
42. Luo SY, Xiao B, Ho ZQ (2009) An experimental study on a novel shredder for municipal solid waste (MSW). Int J Hydrogen Energy 34(3):1270–2272
43. Luo S, Xiao B, Xiao L (2010) A novel shredder for municipal solid waste (MSW): influence of feed moisture on breakage performance. Bioresour Technol 101:6256–6258
44. ASTM Standard D388-99 (1999) Standard classification of coals by rank
45. European Commission—Directorate General Environment (2003) Refuse derived fuel, current practice and perspective: final report. http://ec.europa.eu/environment/waste/studies/pdf/rdf.pdf. Accessed Dec 2010
46. Ryu C, Yang YB, Khor A, Yates NE, Sharifi VN, Swithenbank J (2006) Effect of fuel properties on biomass combustion: Part I. Experiments—fuel type, equivalence ratio and particle size. Fuel 85(7–8):1039–1046
47. Muthuraman M, Namioka T, Yoshikawa K (2009) A comparison of co-combustion characteristics of coal with wood and hydrothermally treated municipal solid waste. Bioresour Technol. doi:10.1016/j.biortech.2009.11.060
48. Muthuraman M, Namioka T, Yoshikawa K (2010) Characteristics of co-combustion and kinetic study on hydrothermally treated municipal solid waste with different rank coals: a thermogravimetric analysis. Appl Energy 87:141–148
49. Chen Y, Mori S (1995) Estimating the combustibility of various coals by TG-DTA. Energy Fuels 9:71–74
50. Khan AA, de Jong W, Spliethoff H (2005) Biomass combustion in fluidized bed boiler. Bioenergy for Wood Industry, Jyväskylä, Finland, pp 365–370
51. Werther J, Saenger M, Hartge E-U, Ogada T, Siagi Z (2000) Combustion of agricultural residues. Prog Energy Combust Sci 26(1):1–27
52. Löffler Gerhard, Wargadalam Verina J, Winter Franz (2002) Catalytic effect of biomass ash on CO, CH4 and HCN oxidation under fluidised bed combustor conditions. Fuel 81(6):711–717
53. Lu Y, Hippinen I, Jahkola A (1995) Control of NOx and N_2O in pressurized fluidized-bed combustion. Fuel Energy Abstracts 36(3):216
54. Khan AA, de Jong W, Jansens PJ, Spliethoff H (2009) Biomass combustion in fluidized bed boilers: potential problems and remedies. Fuel Process Technol 90(1):21–50
55. Amand LE, Leckner B (1991) Influence of fuel on the emission of nitrogen-oxides (NO and N2O) from an 8-MWth fluidized-bed boiler. Combust Flame 84(1–2):181–196

56. Hwang IH, Matsuto T, Tanaka N (2006) Water-soluble characteristics of chlorine in char derived from municipal solid wastes. Waste Manag 26:571–579
57. Caputo AC, Pelagagge PM (2002) RDF production plants I: Design and costs. Appl Therm Eng 22:423–437
58. Sikka P Energy from MSW: RDF pelletization—A pilot Indian plant
59. Moyers CG, Baldwin GW (1999) Psychrometry, evaporative cooling, and solids drying. In: Perry's chemical engineer's handbook, 7th edn, Section 12. McGraw-Hill, New York
60. Glasser L (2004) Water, ordinary water substance. J Chem Educ 81:414–418

Chapter 15
Sewage Sludge Treatment by Hydrothermal Process for Producing Solid Fuel

Kunio Yoshikawa and Pandji Prawisudha

Abstract Sludge treatment and disposal is one of the focus points in the waste treatment technology research due to the fact that sewage sludge is a form of pollution. One of the promising methods of sewage sludge treatment, considering its maximum sludge reduction and short processing time, is incineration. However, the usage of sewage sludge as fuel in incinerator is hindered by its high water content and high nitrogen content. A hydrothermal process for producing solid fuel from sewage sludge is developed by using saturated steam at 160–200 °C and about 60 min holding time. It was shown that the product has improved dehydrability, but at the same time exhibiting higher solubility in the water, resulting in slightly lower calorific value due to the loss of dissolved solids, which have significant calorific value; therefore, an optimum operating condition is required to improve the dehydrability of sludge without reducing its solid content and calorific value. In the term of sewage odor, it was shown that after the hydrothermal treatment the sulfur-containing compound concentration was decreased, the same with the total odor intensity in the solid product. On the other hand, the odor intensity in the liquid and gaseous products were increased, suggesting the transfer of these compounds to the liquid and gaseous parts. During the combustion experiment, it was shown that the hydrothermally treated sludge emits lower NO emission compared to the raw sludge, promoting its possibility to be used in incinerator as direct or co-fuel without resulting in secondary pollution.

K. Yoshikawa (✉)
Department of Environmental Science and Technology, Tokyo Institute of Technology, Tokyo, Japan
e-mail: yoshikawa.k.aa@m.titech.ac.jp

P. Prawisudha
Department of Mechanical Engineering, Bandung Institute of Technology, Bandung, Indonesia
e-mail: pandji@termo.pauir.itb.ac.id

15.1 Background

Due to the rapid growth of industrialization and its following urbanization in developing countries, sludge discharge became higher in these countries, and their treatments became more and more difficult. For example, in 2004, the industrial waste discharge in Japan was recorded as high as 417 million tons per year. Compared to the data in [1], from Fig. 15.1 it can be observed that 39 % or 162 million tons of such waste was contributed by animal manure and sewage sludge.

Similar condition was reported in China. In 1980s, China started a large-scale construction of modern sewage treatment plants, and in 2009 the nation's total discharge of wastewater was 590 million tons [2], an increment of 3 % compared to the previous year. After the wastewater treatment process, about 1–3 % sewage sludge with 80 % moisture content was discharged as a by-product, resulted in the sewage sludge production of about 20 million tons (Fig. 15.2).

Sludge treatment and disposal is one of the focus points in the waste treatment technology research due to the fact that sewage sludge is a form of pollution and able to threaten people's lives if not treated properly. It is putrescible, has a foul smell, and in addition to large amounts of pollutant precursors such as proteins, fats, and other organic compounds, it also contains large amount of pathogenic organisms, parasites, salts, and heavy metals, as well as polychlorinated biphenyls, dioxins, radio nuclides, and other components that are difficult to decompose.

Currently, conventional disposal methods can be differentiated as agricultural use (by composting, stabilizing), land/ocean disposal and incineration (by drying, carbonization, oxidation). In developed countries, the sludge treatment and disposal has 100 years of history, and various technology according to the practical situation are already available. For example, in Western Europe indirect drying is mainly used, the United States is mainly applying landfills and agricultural use, while Japan is usually employing incineration process.

When sludge is treated by means of landfilling, its volume reduction is not effective, influencing the operation of landfill due to sedimentation, sinking, and other related problems. On the other hand, when sludge is treated by composting, there are also some issues related with processing time, costs, and the quality of compost.

Incineration is one of the promising methods of sewage sludge treatment. Co-combustion of coal and sludge, or sole combustion of sludge in the incinerator can generate heat and power, reducing conventional fossil fuel consumption and greenhouse gas emissions [3–7]. Incineration also resulted in maximum sludge reduction (up to about 95 %), and during the process all the pathogenic bacteria is eliminated, and toxic organic residues are oxidized and decomposed by heat. At the end of the process, the incineration ash is used as raw material for cement and other construction materials production, so that heavy metals can be consolidated in concrete to avoid their release to the environment.

Another possible obstacle of sludge utilization by incineration is its high nitrogen content. During the incineration, the nitrogen contained in sludge would

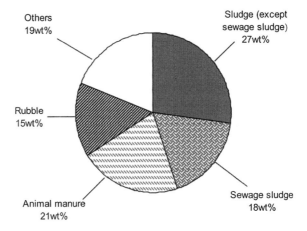

Fig. 15.1 Japanese industrial waste composition in 2004

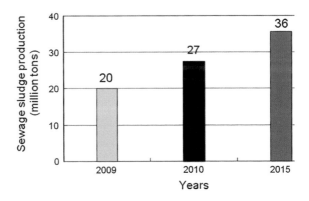

Fig. 15.2 Sewage sludge production of China and its predicted amount in the future

be finally converted into NO and N_2O associated with direct combustion or indirect combustion, resulting in secondary pollution.

Furthermore, since sludge is consisted mainly of water, the main technical challenge of sewage sludge treatment by incineration is to establish the dehydration technology with minimum usage of energy. Current available dehydration technologies are heat drying, sludge carbonization, and hydrothermal drying.

Heat drying uses heat to evaporate the water in the sewage sludge. After drying sludge turns into granules or powder, the volume is reduced to 20–25 % of the original volume. Moreover, due to low moisture content in the dried sludge, the microbial activity is completely inhibited, preventing the bad smell [8]. In 1990s, heat drying treatment of sludge was developing quickly in United States and in 2000 the production of dry sludge reached 10 times higher compared to the amount in 1990 [9]. The dried sludge can be used not only as the energy resource, but also as fertilizer, soil amendment and construction material [10].

Sludge carbonization technology is a certain procedure during which the water and volatile matter contained in sludge are released, and at the same time the carbon content in sludge is maximized. There are three main methods of sludge carbonization: High temperature carbonization [11] is an atmospheric carbonization process in a carbide furnace at 649–982 °C after pre-drying process. The produced carbide granules can be used as low grade fuel, calorific value with heating value of 8,360–12,540 kJ/kg. Medium temperature carbonization [12, 13] is an atmospheric carbonization process in a carbide furnace at 426–537 °C, while during low temperature carbonization process [14–17], pre-drying is not required before carbonization and the pressure needs to be set to 6–8 MPa at 315 °C, and after dewatering a low grade fuel with calorific value of 15,048–20,482 kJ/kg will be produced.

Hydrothermal drying treatment, which is the main topic in this chapter, is a new technology reacting sewage sludge and saturated steam at 160–200 °C which will improve the dewatering ability of the sludge. Previous researches on the application of hydrothermal treatment to sewage sludge have been focused on the improvement of the anaerobic digestion ability of the sludge [18]. After the treatment, the sludge turns into liquid, and after drying and granulation it can be used as low grade fuel with calorific value of about 2,000 kcal/kg.

15.2 Hydrothermal Treatment of Sludge

Hydrothermal treatment of sludge is mainly a process during which sludge is heated at certain temperature and pressure, where the hydrolysis of sludge viscous organic matters is taking place, the colloidal structure of sludge is destroyed resulted in lower viscosity, and at the same time improving its dehydration and anaerobic digestion performance [19]. During the rising temperature and pressure, the particle collision rate is increasing and resulting in the destruction of colloidal structure, leading to the separation of bound water and solid particles.

Compared to conventional sludge drying process, hydrothermal treatment requires lower energy due to the avoidance of latent heat requirement from the water evaporation. Theoretically the energy consumption for evaporating 1 kg of water is about 620 kcal, but in hydrothermal process the energy consumption for heating 1 kg of water is merely 155 kcal or about a quarter of that of drying. It is predicted that "hydrothermal process + mechanical dewatering" would be able to reduce 50 % water content of sludge using lower energy consumption compared to conventional sludge drying.

Other researchers have studied the effect of heat treatment on the sludge dewatering performance. In Belgium, Neyens and Baeyens [20] noticed that extracellular polymer content doubled the effect on sludge dewatering performance with optimum reaction temperature of 175 °C. The so-called Cambi technology is using pyrohydrolysis process during which the organic matters of sludge are biodegraded and anaerobically digested, enhancing the biogas production. In China, Tsinghua University and Beijing's Jian Kun Wei Hua New Energy Co., Ltd cooperated in creating a comprehensive 30 ton/day sludge treatment system combining the thickening, hydrothermal, and centrifugal dewatering process [21].

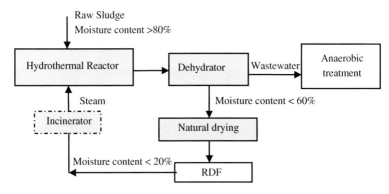

Fig. 15.3 RDF production from sewage sludge employing the hydrothermal process

Tokyo Institute of Technology has been investigating sewage sludge hydrothermal treatment system, resulted in improved natural drying performance and reduced odor [22–24], and recent research confirmed that NO reduction occurred on the hydrothermally treated sludge compared to untreated sludge [25]. The proposed refuse-derived fuel (RDF) production system is shown in Fig. 15.3. In this system, the sludge with 80 % moisture content is inserted into a sealed reactor and heated using 150–300 °C saturated steam. The moisture content of sewage sludge after the hydrothermal treatment and mechanical dewatering can be reduced down to 50 %. After natural drying, the moisture content drops to less than 20 % and the obtained RDF can be directly sent to incinerator. The heat produced from the combustion in the incinerator will be used to produce steam required for the hydrothermal process, while the wastewater produced during the dehydration process will be sent to an anaerobic treatment to fulfill the water standards.

15.2.1 Small-Scale Facility

The drawing and specification of the hydrothermal reactor of the small-scale facility are shown in Fig. 15.4 and Table 15.1, respectively. The 20 L reactor capable of processing 13 kg of sludge is made of stainless steel, containing a horizontal agitator with adjustable speed. The saturated steam generated in the electric boiler is injected into the reactor to provide the heating energy, and after the specified time of agitation and reaction, the steam is discharged into the buffer tank and then to the atmosphere. The product is extruded from the reactor and mechanically dehydrated by a centrifuge before naturally dried to investigate its drying performance.

The dehydration equipment is consisted of a 3 ft centrifuge with the volume of 2.5 L (dia. 200 × 100 mm). The hydrothermal product was supplied into the centrifuge, and mechanical dehydration was conducted in 1,900 rpm speed for 30 min, and afterwards the solid and liquid residue can be weighed and analyzed.

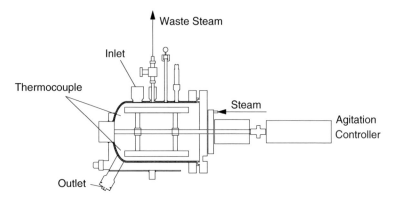

Fig. 15.4 The reactor of hydrothermal treatment of small-scale facility

Table 15.1 Specification of the small-scale reactor

Volume	20 L (dia. 260 × 377 mm)
Rated operating pressure	3.0 MPa
Design temperature	240 °C
Jacket rated operating pressure	1.0 MPa
Jacket temperature	187 °C

15.2.2 Large-Scale Facility

A larger facility consisted of mainly a pressurized vessel and a 500 kg/h high-pressure steam boiler was built in Japan as shown in Fig. 15.5. The approximately 3 m^3 steel vessel is installed horizontally in the frame and covered by heat insulation. It is equipped with a central impeller rotated at 12 rpm and pairs of thermocouples. After sewage sludge insertion, the temperature and pressure gradually increased to about 200 °C. The product was agitated for the following 30 min, and then the steam and other gases inside the vessel were discharged. The solid product was sampled from the outlet port at the bottom of the vessel when the pressure level returned to atmospheric pressure.

In China, a 100 ton/day large-scale commercial plant was built in Hohhot, with the process flow diagram shown in Fig. 15.6. In this system, sewage sludge cake with moisture content of 80 % and saturated steam with the temperature of around 200 °C were supplied into a 8 m^3 batch reactor with a rotor inside to promote the mixing of sludge and steam. The moisture content of the products was around 85 %, but they can be mechanically dehydrated down to 35–60 % moisture content, and 48 h of natural drying was adequate to obtain RDF with the moisture content of less than 10 %.

The drawing and specification of the large-scale facility are shown in Fig. 15.7 and Table 15.2, respectively.

The experimental operating procedure is as follows:

15 Sewage Sludge Treatment by Hydrothermal Process for Producing Solid Fuel 391

Fig. 15.5 Larger-scale hydrothermal reactor in Japan

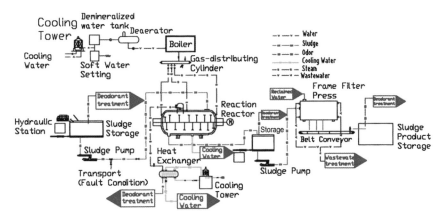

Fig. 15.6 Process flow diagram of the commercial plant in China

1. After closing the steam inlet valve (11) and sludge outlet valve (8) and then opening the pressure relief valve (101), 3 tons of raw sludge was supplied into the reactor, and the rotor starts to rotate with the speed of 22 rpm. The sludge supply inlet valve (13) and pressure relief valve (101) were then closed.
2. By opening the steam inlet valve (11) at the top of the reactor, saturated steam generated in a coal fired boiler was supplied into the reactor. When the sludge temperature reaches the required value, this condition was maintained for certain period, and then the steam supply valve (11) was closed to stop the supplying steam, and reactor shaft (3) stopped rotating.

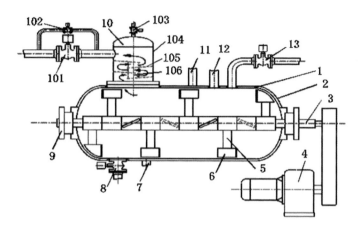

Fig. 15.7 Equipment of the large-scale hydrothermal treatment facility. *1*—inner cylinder, *2*—jacket, *3*—stirrer, *4*—drive unit, *5*—rake arm, *6*—rake, *7*—condensed water port, *8*—sludge discharge valve, *9*—sealing device, *10*—pressure relief device, *11*—steam inlet valve, *12*—jacket steam inlet valve, *13*—sludge inlet valve, *101*—main relief valve, *102*—bypass valve, *103*—purge valve, *104*—relief cylindrical section, *105*—plate block, *106*—strengthening tendon-plate

Table 15.2 Specification of the large-scale reactor

Volume (dia. 1800 × 6000 mm)	7.8 m^3
Rated operating pressure	3.0 MPa
Design temperature	240 °C
Jacket rated operating pressure	1.0 MPa
Jacket temperature	187 °C

3. The pressure bypass valve (102) on the top of the reactor was opened to discharge the steam inside of the reactor. The pressure relief valve (101) was then opened after dropping the pressure of the reactor down to 1.5 MPa.
4. After relieving the steam pressure, the sludge discharge valve (8) was opened to extrude the product, and after extruding the product, the purge valve (103) was opened to clean the pressure relief device (10). The pressure relief and sludge discharge valves were then closed before the next batch.

To further dehydrate the hydrothermally treated sludge, a 5 m^2 frame filter capable of treating 1,000 kg sludge was employed with the squeezing piston with the pressure of 1.5–1.8 MPa in 30–40 min. The sludge flowed through the filter cloth and its solid part was gradually accumulated on the filter cloth to form a filter cake. Another way to dehydrate the product is using a horizontal scroll decanter centrifuge, whose treatment capacity is 1–2 m^3/h and revolution speed of 5,000 rpm for 20 min.

15.3 Sludge Characteristics After Hydrothermal Treatment

To understand the sewage sludge characteristics after hydrothermal treatment, several samples were taken from different municipal sewage treatment plants as shown in Table 15.3 below to be treated in the small-scale facility. The organic content of HS-Sludge2 was the highest due to mixing of untreated industrial wastewater in this sewage treatment plant, resulting in the increasing of the organic content of sludge in the process of treatment.

15.3.1 Effect of Reaction Temperature to Sludge Dehydration

To investigate the effect of reaction temperature on the product, various reaction temperatures were set to 160, 180, 190, 200 and 210 °C, respectively, and the reaction period was fixed to 30 min, followed by dehydration of the products using the centrifuge.

The relationship between the hydrothermal reaction temperature and the moisture content of the solid product after dehydration are shown in Fig. 15.8. It can be seen that with increasing temperature, the moisture content of different samples was decreased; however, the effect was not significant at the reaction temperature of 160 °C. When the hydrothermal reaction temperature was increased to 190 °C, moisture content of the dehydrated product was significantly reduced, showing obvious hydrothermal effect, but the results for higher temperatures were almost similar, indicating that the improvement of the dehydration performance was already saturated.

Due to different sources of sludge, the dehydration performances of sludges were also varied. But it can be observed from the results that the dehydration performance of hydrothermally treated products became more similar regardless of the sludge sources in increasing reaction temperature, and the moisture content of all dehydrated products was reduced to less than 60 % at the reaction temperature higher than 190 °C.

15.3.2 Effect of Reaction Period to Sludge Dehydration

To investigate the effect of reaction period on the product, subsequent experiments were conducted in the small-scale facility at 190 °C for 10, 20, 30, 40, and 50 min, respectively. The relationship between the hydrothermal reaction period and the moisture content of the hydrothermally treated and dehydrated products of the four types of sewage sludge are shown in Fig. 15.9.

Table 15.3 Sewage Sludge Properties

Characteristics	BJ-Sludge (Beijing)	HS-Sludge1 (Hohhot 1)	HS-Sludge2 (Hohhot 2)	WX-Sludge (Wuxi)
HHV (MJ/Kg)	10.53	15.31	19.68	7.10
Water content (%)	85.8	83.7	75.4	82.9
Proximate analysis (db)				
V.M (Wt%)	40.1	47.89	55.42	26.46
F.C (Wt%)	5.4	8.01	11.17	4.18
Ash (Wt%)	54.5	44.1	33.41	69.36
Ultimate analysis (daf)				
C (Wt%)	52.92	57.12	69.19	50.88
H (Wt%)	5.8	7.9	8.37	7.85
N (Wt%)	7.7	9.39	3.58	6.51
O (Wt%)	33.58	25.59	18.86	34.76

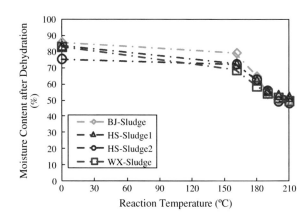

Fig. 15.8 Effect of reaction temperature on the moisture content of dehydrated, hydrothermally treated product

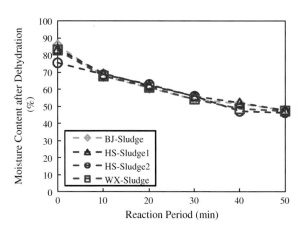

Fig. 15.9 Effect of reaction period on the moisture content of the dehydrated, hydrothermally-treated product

It can be seen from the figure that the reaction period is one of the vital factors affecting the hydrothermal treatment performance under certain reaction temperature. The longer the reaction time, the better the hydrothermal performance, and the lower the moisture content of the dehydrated product. When the reaction period was longer than 30 min, the moisture content of the dehydrated products dropped below 60 %. Moisture contents of the dehydrated products were almost the same, in spite of the different initial moisture contents. Prolonging of the reaction period appeared to promote the sludge colloid structure destruction; however, this phenomenon seemed to reach saturation point when the reaction period reached 50 min.

15.3.3 Effect of Dehydration Equipment to Sludge Dehydration

Subsequent experiment using the large-scale facility was performed to investigate the effect of the dehydration equipment on the sewage sludge after the hydrothermal process. The reaction period was set to 30 min, at the reaction temperature of 160, 180, 190, 200, and 210 °C, respectively. Three types of dehydration equipment of small centrifuge, horizontal decanter(HDC) and frame filter were used. One experiment result from small-scale facility is shown in the result as a comparation.

Like previous result in Fig. 15.8, result in Fig. 15.10 shows that the increase of the hydrothermal reaction temperature and period improves the dehydration performance of the sludge until the temperature exceeds 190 °C. It also shows that the dewaterability of the hydrothermally treated sewage sludge depends on the performance of the dehydration equipment. The order of the dehydration efficiency was: frame filter > HDC > small centrifuge.

Frame filter conducts dehydration by extruding the water portion of sludge through the filter cloth with the pressure force of the plate and frame. The dehydrated solid product is discharged in the form of plate, and still needs to be crushed before subsequent drying process. On the other hand, horizontal decanter is using a rotating-drum which forms a liquid layer due to the centrifugal force, where the solid particles are settling to the rotating-drum wall in the form of power-like, and the separated liquid is discharged through the rotating drum. Even though the dehydration efficiency of frame filter was the highest, in the commercial usage the horizontal decanter can be preferable since it does not need crushing equipment before the subsequent drying process.

In the case of the small centrifuge, its dehydration efficiency was the lowest due to its low speed, and consequently low centrifugal force. The speed of the small centrifuge was merely 1,900 rpm, while in the horizontal decanter, the speed was 5,000 rpm. Thus, the dehydration efficiency of the horizontal decanter is much higher than the small centrifuge.

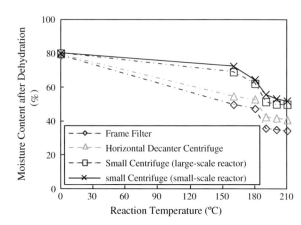

Fig. 15.10 Effect of dehydration equipment on the moisture content of the dehydrated, hydrothermally treated product

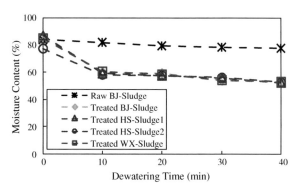

Fig. 15.11 Moisture content changes of raw sludge and hydrothermally-treated sludge as a function of the dehydration time

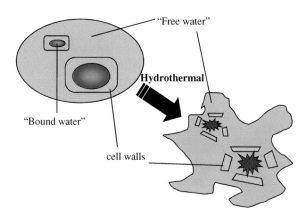

Fig. 15.12 The process of bound water becoming free water

This phenomenon, however, is not occurred in the raw sludge. From Fig. 15.11, it can be seen that the moisture contents of various sewage sludge after hydrothermal treatment at 190 °C and 30 min showed significant decrease by increasing the dehydration time regardless of the raw sludge type, while that of the raw sludge could not be reduced even in longer dehydration time.

15 Sewage Sludge Treatment by Hydrothermal Process for Producing Solid Fuel 397

Table 15.4 Climate in Beijing during the natural drying experiments

Time	Temperature	Wind power	Climate
2010-4-12	3–11 °C	Northerly wind, grade 4–5 (5.5–10.7 m/s)	Sunny
2010-4-13	1–11 °C	Northerly wind, grade 3–4 (5.5–10.7 m/s)	Sunny

It is expected that the hydrothermal process make the bound water in the sludge becomes free water as shown in Fig. 15.12, and promoting its dehydration performance. When the dehydration time was longer, the total centrifugal energy was increased and the water in the sludge molecules was further extricated by the force, lowering the moisture content of the sludge.

15.3.4 Changes in Natural Drying Performance

In order to investigate the natural drying performance of the dehydrated sludge, subsequent natural drying experiment of the raw sewage sludge and the dehydrated sludge after hydrothermal treatment was performed in Beijing. The sample was made into 15 × 15 × 2 cm piece, and dried in natural environment for 48 h. Local climate when the measurements were taken is indicated in Table 15.4 as below.

For the produced sludge at the reaction temperature of 190 °C and reaction time of 30 min, the time change of the moisture content in the raw sludge and the dehydrated sludge during the natural drying is shown in Fig. 15.13. As shown in the figure, all raw sludge could not be dried by natural drying, even though the drying period was prolonged. On the other hand, all dehydrated sludge showed improved natural drying performance (faster drying), and after 48 h all dehydrated, hydrothermally treated products had less than 10 % moisture content.

Based on the result shown in Fig. 15.13, natural drying rate of the dehydrated product can be separated into two stages; constant-rate period in the initial 12 h, continued with a falling-rate period. The constant-rate period represents the evaporation of free water, while the falling-rate period represents the evaporation of bound water. It can be seen that after the hydrothermal treatment and mechanical dehydration, the structure of sewage sludge was altered and most of the water in the sewage sludge was converted into free water, resulted in the occurrence of constant-rate period, which had higher drying rate compared to that of the falling-rate period.

The effect of reaction temperature on the natural drying performance of the hydrothermally treated product after mechanical dehydration was also observed. The raw sludge and solid product samples were tested in the block shape of 15 × 15 × 2 cm and the change of their moisture contents were measured for 48 h in the atmospheric temperature of 8–12 °C. The reaction temperatures were varied from 160 to 210 °C with a fixed reaction period of 30 min. The results are shown in Fig. 15.14.

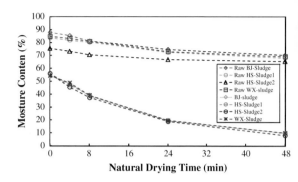

Fig. 15.13 Natural drying performance of the raw sludge and the dehydrated product

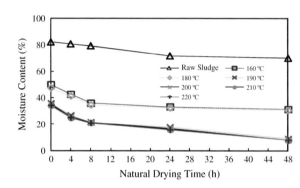

Fig. 15.14 Moisture content changes of raw sludge and various solid products during natural drying

From the figure, we can see that the raw sludge practically could not be dried by natural drying, while the solid products showed good natural drying performance. When the sludge was hydrothermally-treated at the reaction temperature of less than 180 °C, the final moisture content of the products were still above 30 % even after 48 h of natural drying. However, different phenomenon occurred for the products with reaction temperature of 190 °C: the moisture contents of solid products can be reduced to lower than 20 % after 24 h of natural drying, and even lower than 10 % after 48 h of natural drying. Significant difference was not observed for other products with reaction temperature above 190 °C, again showing that the optimum reaction temperature is 190 °C.

15.3.5 Changes on the Product Solubility

The organic matter in the sewage sludge includes mainly microorganisms. Activities of such microorganisms produce extracellular polymers (ECP), for instance high molecular weight matters such as polysaccharides, protein and nucleotides [26], which fill the gap between bacteria and form a flocculated structure of sewage sludge and affect the dehydration ability of sewage sludge [27].

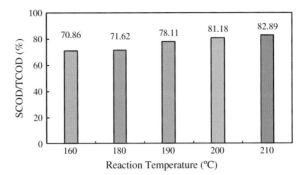

Fig. 15.15 Changes in SCOD/TCOD under various reaction temperatures

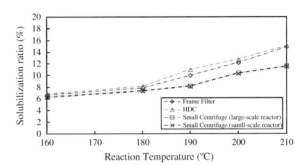

Fig. 15.16 Effect of reaction temperature on solubilization ratio

During the hydrothermal process, the insoluble organic matter and ECPs in the sewage sludge are degraded into smaller molecular weight materials, increasing the solubility of the sludge, and in the end increasing the BOD (Biological Oxygen Demand), COD (Chemical Oxygen Demand) and SS (Suspended Solid) values in the separated water, which requires appropriate wastewater treatment after the hydrothermal process. Therefore, the treatment condition should be optimized so that the breakage of the sludge cell is adequate to increase its dewaterability, but with minimal increase of its solubilization.

Figure 15.15 shows the changes in SCOD/TCOD (soluble COD/total COD) in the separated liquid obtained from the hydrothermal products at various reaction temperatures and reaction period of 30 min. According to the figure, increasing the reaction temperature would increase the SCOD/TCOD in the liquid.

The increasing of SCOD/TCOD was caused by the increasing content of organic matter in the liquid phase due to degradation of organic matter and ECPs. As a consequence, the forms of existent moisture in sewage sludge were affected, i.e., moisture of mechanically inseparable interstitial water, capillary water and intracellular water were reduced, which increased the dehydration performance of sewage sludge.

The solubility of the sewage sludge can be defined as solubilization ratio. It refers to the ratio of the solid-phase material in raw sludge and the solids dissolved into the separated liquid obtained in the mechanical dehydration process.

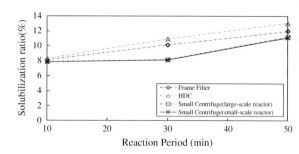

Fig. 15.17 Effect of reaction period on solubilization ratio

Figure 15.16 shows the solubilization ratio for various dehydration equipments as a function of the reaction temperature at reaction period of 30 min, while Fig. 15.17 shows the solubilization ratio as a function of reaction period at the reaction temperature of 190 °C. It can be seen that there was significant increase of the solubilization ratio at the reaction temperature above 190 °C, and at the reaction period of longer than 30 min. These results assure the validity of the optimum reaction temperature of 190 °C and the reaction period of 30 min, where the expected solubilization ratio would be around 10 % using the frame filter or the horizontal decanter.

15.3.6 Changes in Product Heating Value

Table 15.5 shows the properties of sewage sludge after hydrothermal treatment followed by mechanical dehydration. As indicated in the table, the higher heating value (HHV) of the dehydrated products were slightly decreased with increasing reaction temperature. These changes suggesting that the dissolved solids—represented by higher solubilization ratio—have significant calorific value, thus confirming the importance of optimizing the improvement of dewaterability with increasing solubility.

15.3.7 Changes in Product Odor

Because sewage sludge has a specific, intense malodor that is difficult to remove by drying treatment, and there is the threat of spontaneous ignition of dried sewage sludge during transportation and storage, carbonization is often used as the pretreatment for the usage of sewage sludge as fuel [28]. Therefore, additional experiment is needed to confirm the malodor intensity of the hydrothermally-treated product. This experiment was conducted in northern Japan sewage sludge with characteristics shown in Table 15.6 using 3 m^3 horizontal steel vessel shown in Fig. 15.5 at 19.6 MPa and 200 °C for 30 min [24].

Table 15.5 Sewage sludge properties

Sludge sample	HHV (MJ/Kg)	Elemental analysis (wt.%)				Proximate analysis (wt.%)		
		C	H	N	O	VM	FC	Ash
BJ-Sludge (190 °C)	10.29	50.2	5.6	7.4	36.8	38.2	5.2	56.6
BJ-Sludge (200 °C)	9.8	48.2	5.5	7.3	39	35.91	4.89	59.2
HS-Sludge1 (190 °C)	14.14	54.6	7.55	8.54	29.31	43.66	6.86	49.48
HS-Sludge1 (200 °C)	13.9	53.65	7.32	8.36	30.67	42.22	6.5	51.28
HS-Sludge2 (190 °C)	19.34	65.42	8.12	3.41	23.05	50.82	10.37	38.01
HS-Sludge2 (200 °C)	18.9	63.1	8.03	3.34	25.53	50.01	9.5	40.05
WX-Sludge (190 °C)	6.77	48.54	7.62	6.42	37.42	24.84	4.36	70.8
WX-Sludge (200 °C)	6.68	46.23	7.51	6.37	35.76	23.81	4.29	71.9

Table 15.6 Sewage sludge properties used in malodor experiment

Property (wt.%)	Sewage sludge
Moisture[a]	85.3
Ash[b]	19.3
Carbon[c]	50.5
Hydrogen[c]	7.5
Nitrogen[c]	8.6
Sulfur[c]	0.9
Oxygen[c]	32.5

[a] wet base, [b] dry base, [c] dry ash free

After being treated, 150 g of each solid product sample was placed into a 5 L gas sampling bag, and 3 L of fresh air was injected into the bag. Twenty hours later, the gas sampling bag was filled with fresh air and shaken to enhance gas production from the sample. Next, the gas was transferred to a new sampling bag, and the components of the gas were measured by gas chromatography. Specific 21 malodor components based on the Japanese Offensive Odor Control Act was selected and analyzed by Nihon CCL, Japan.

The results are presented in Table 15.7 below. It was found from the experiment that hydrothermal treatment decreased the discomfort associated with the malodor. On the basis of organoleptic evaluation, after this treatment the solid product had somewhat of an acidic and scorched odor rather than a typical fecal odor.

It can be seen from the table that the sulfur-containing compounds mainly characterize the odor from the sewage sludge [29, 30], and after the hydrothermal treatment, the sulfur-containing compound concentrations decreased, whereas the concentrations of aldehydes, light aromatics, and organic acid compounds increased slightly. The perceived acidic odor from the sample after treatment probably came from organic acid compounds produced by hydrolysis of the sewage sludge, and the scorched odor was probably produced by secondary or tertiary reactions, namely, Maillard and/or caramelizing reactions of amino acids and sugars produced by hydrolysis of sewage sludge. Therefore, one possible malodor reduction mechanism might be that the odors from by-products produced

Table 15.7 Malodorous compound concentrations of sewage sludge

Type	Compound	Dried sewage sludge (ppm)	Treated sewage sludge (ppm)
Sulfur-containing compounds	Hydrogen sulfide	0.67	0.39
	Methyl mercaptane	14.2	0.0034
	Dimethyl sulfide	8.1	0.011
	Methyl sulfide	1.65	0.006
Aldehydes	Acetaldehyde	0.19	1.8
	Propionic aldehyde	0.25	0.65
	n-butyraldehyde	–	0.12
	Isobutyraldehyde	–	0.15
	Isovaleraldehyde	–	–
	n-valeraldehyde	–	0.18
Light aromatic compounds	Styrene	–	0.032
	Ethyl acetate	–	0
	4-methyl-2-pentanone	0.055	0.05
	Toluene	0.41	0.072
	Xylene	0.17	0.11
Organic acids	2-methyl-1-propanol	–	0
	Propionic acid	0.042	0
	n-butyric acid	0.027	0.05
	n-valeric acid	–	0
	Isovaleric acid	0.005	0.007

—: under the detection limit

by hydrolysis, thermal decomposition, and Maillard and caramelizing reactions [31, 32] mask the malodor from the treated sewage sludge.

Because it is difficult to sample gas and liquid products from the large-scale reactor, experiments investigating the malodor reduction mechanisms were conducted using 500 ml laboratory-scale stirred pressurized vessel. The vessel was filled with raw sludge and distilled water, then heated at 200 °C for 30 min, after which steam and the gases produced were released via the condenser. Exhausted steam and condensate were sampled at the condensate product sampler, and produced gas was collected in a gas sampling bag.

Odor intensities of each phase of the odor samples were analyzed using the odor-measuring device (ONU-Sn, Futaba Electronics, Japan), equipped with four kinds of metal oxide semi-conductor sensors: for heavy hydrocarbon odor compounds (e.g., benzene and toluene), light hydrocarbon odor compounds (e.g., alcohols, organic acids, and aldehydes), hydrogen sulfide, and ammonia. Odor intensities are quantitatively analyzed by detecting redox reactions of odor substances on the sensors.

The odor intensity distributions of heavy hydrocarbons, light hydrocarbons, ammonia, and hydrogen sulfide in the raw sludge and products (gas, liquid, and solid) are shown in Fig. 15.18a–d, respectively. The heavy hydrocarbon, light hydrocarbon, ammonia, and hydrogen sulfide odor intensities of the solid product

15 Sewage Sludge Treatment by Hydrothermal Process for Producing Solid Fuel

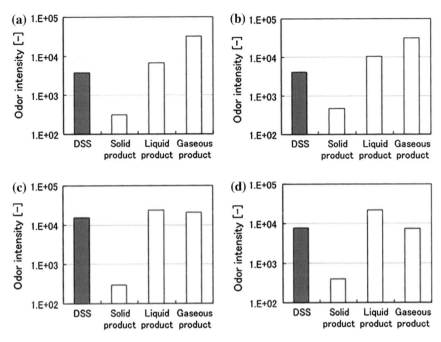

Fig. 15.18 Odor distribution in raw sewage sludge (DSS) and hydrothermally-treated products (solid, liquid and gaseous) for **a** Heavy hydrocarbons, **b** Light hydrocarbons, **c** Ammonia, and **d** Hydrogen sulfide

were one or two orders of magnitude lower than those of raw sewage sludge, whereas those of the gas and liquid products were similar to or one order of magnitude higher than those of raw sludge. It can be concluded that the odor compounds, which included sulfur-containing compounds, were separated from the raw sludge and transferred to the gaseous and liquid products by the thermal effect of hydrothermal treatment.

15.4 Application of Hydrothermally Treated Sludge as a Fuel

15.4.1 Energy Recovery Ratio (ERR)

Since the final purpose of hydrothermal treatment of the sewage sludge is to produce fuel which can be used in coal-fired boilers, the total energy requirement analysis is important. From the previous investigation [33], the moisture content of RDF should be less than 10 % for co-firing with coal in most of the coal-fired boilers. Therefore, if the moisture content of the hydrothermal products cannot

Fig. 15.19 Effects of the reaction temperature and time on the value of ERR

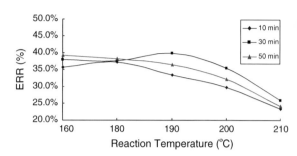

reach 10 % after natural drying, additional drying process should be used to achieve the target moisture content.

A simple parameter to estimate the optimum condition in the term of energy requirement to produce the RDF from sewage sludge was developed, defined as Energy Recovery Ratio (ERR):

$$\mathrm{ERR} = \frac{(\text{Energy in RDF} - \text{Energy for steam generation} - \text{Energy for drying})}{\text{Energy content in sludge}}$$

Figure 15.19 shows the value of ERR as functions of the reaction temperature and period. If the reaction temperature or period is too low, more energy is needed for the dryer due to worse natural drying performance of the dehydrated hydrothermal products. On the other hand, if the reaction temperature or time is too high, more energy is needed for the steam generation to be used in the hydrothermal treatment. The figure shows that the ERR value takes its maximum at the reaction temperature of 190 °C and reaction period of 30 min, suggesting the optimum condition to convert sewage sludge into a solid fuel by hydrothermal treatment.

15.4.2 Combustion Characteristics of Hydrothermally Treated Sludge

To obtain the effect of hydrothermal treatment on the combustion of the product, a laboratory scale combustion experiment was built as shown in Fig. 15.20, including a gas supplying and mixing system, an online flue gas analysis system, and a dual-bed reactor with gas inlet/outlet to simulate various combustion processes. The dual-bed reactor was made from quartz, which is divided into three parts, an outer tube with a gas inlet in the middle, a top cover with a feeding port and also a gas inlet, and an inner tube. A sintered quartz porous plate was fixed in each tube to support the sample or char.

The reactor was surrounded and heated in the temperature range of 870–1173 K by a two-zone electric furnace. Instead of air, the oxidizing gas was replaced with

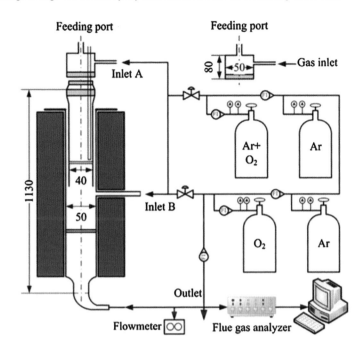

Fig. 15.20 Sewage sludge combustion experimental facility

a mixture gas, which was made from high purity argon (99.99 %) and O2 in order to avoid thermal NOx and prompt NOx generated from N_2. Various combustion mode was achieved by adjusting the gas supplied (Ar, O_2 or pre-mixed gas Ar/O_2) and gas inlets/outlets. The composition of the flue gases was analyzed by an online flue gas analyzer (Testo 350XL, Japan).

In this experiment, five combustion modes were investigated. For all experiments, 10 g of unreacted silica sand was added in the sample to maintain a certain height and to prevent ash agglomeration during the combustion process. The combustion temperature and gas flow rate for all experiments were 1073 K and 2.5 L/min, respectively. In DC mode, 0.5 g of sludge was first pyrolyzed under the oxygen-free atmosphere to produce char with the same temperature as the combustion temperature. The mixture gas of oxygen and argon was then supplied from the inlet B to combust the pyrolysis gas and char.

In CC mode, 0.5 g of sludge was first placed in the upper tube and then the mixture gas of Ar/O_2 was supplied into the reactor to support the reaction, while in SC mode, the combustion of the pyrolysis gas and char was separated intentionally by changing the inlet of the reactant gas. The sample was placed in the inner tube and then argon was injected into the reactor to create an air-free condition. After the reactor was heated up to the predetermined temperature, the inner tube was plugged into the outer one. The mixture gas was supplied from the inlet B to maintain the combustion of the pyrolysis gas. Subsequently, the inlet B was closed and the mixture gas was supplied from the inlet A to burn the char generated in the previous step.

Table 15.8 Proximate and ultimate analysis of raw sludge (RS) and product (HTS)

Parameter	Sample	
	RS	HTS
Ultimate analysis (%, air dried basis)		
C	40.75	41.95
H	5.40	4.99
N	6.60	6.20
S	1.09	0.94
O (by difference)	46.16	46.02
Proximate analysis (%, air dried basis)		
Moisture	1.23	1.08
Volatile matter	76.93	69.07
Fixed carbon	2.53	3.54
Ash	19.31	26.31

In AC mode, the sample was placed in the inner tube, and then the reactant gas was transported to replace the air in the tube. After the reactor was heated up to the predetermined temperature, the inner tube was plugged into the outer one. The mixture gas was supplied from both inlet A and B with a total amount of 2.5 L/min and kept flowing downward. Two gas-supplying modes, defined as AC-A, 1.0 L/min from inlet A and 1.5 L/min from inlet B, and AC-B, 1.5 L/min from inlet A and 1.0 L/min from inlet B, were carried out in the last two combustion modes.

Before the combustion experiment, activated sewage sludge with a moisture content of 85.6 % was taken from a wastewater treatment plant (WWTP) in Japan. The sludge pre-mixed with pure water (Wako Pure Chemical Industries, Ltd. Japan) in the ratio of 10:3 (100 g of sewage sludge and 30 g of pure water) was loaded into a 500 mL reactor filled argon gas. It was then stirred and heated up to 200 °C for 30 min. Finally, the products were mechanically dewatered with a pressing machine and then dried by a convective dryer. All these dried materials were ground and screened to the same size of 0.5–1.0 mm to eliminate the influence of the sample size. Table 15.8 presents the proximate and ultimate analysis of the raw sludge and treated sludge.

Figure 15.21 shows the NO emission from RS and HTS in all five combustion modes. Obviously, the NO emission from HTS was lower than that from the RS no matter which kind of combustion mode was employed. This difference was more outstanding in CC, AC-A and AC-B modes, where the NO emission of HTS was only 49.3, 56.2 and 43.6 %, respectively, of that from RS, while the NO emission from HTS in DC and SC were 96.9 and 92.1 %, respectively, of that from RS. Considering the difference in the initial fuel-N content of RS and HTS, the NO reduction ratio by the HT pretreatment in CC, DC, SC, AC-A and AC-B was 50.7, 3.1, 7.9, 43.8, and 56.4 %, respectively.

Subsequent combustion experiments were conducted in thermogravimetric analyzer (TGA) with three kinds of heating rates. The results are presented in Fig. 15.22, and it clearly indicated that the maximum weight loss rate (combustion rate) of HTS was much higher, around 1.3—2.3 times more than that of RS. At the

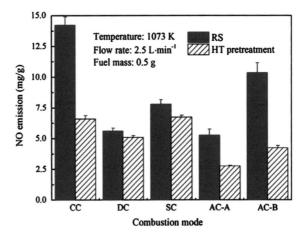

Fig. 15.21 Comparison of NO emission from RS and HTS in five combustion modes

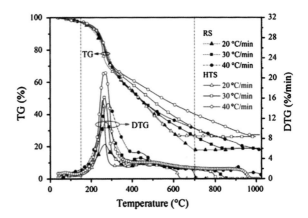

Fig. 15.22 Thermograms of RS and HTS combustion di TGA

heating rate of 20 °C/min, most of the organic matters in HTS have been burnt out over the temperature range of 170–300 °C, whereas the corresponding temperature range was 170–375 °C for RS. These results verified that the hydrothermal treatment has promoted the devolatilization properties of sludge and consequently improved the nitrogen-compound release rate in HTS, reducing the NO emission.

References

1. State of the Discharge and Treatment of Industrial Waste in FY (2004) Ministry of the environment, Japan (in Japanese). http://www.env.go.jp/press/press.php?serial=7928. Accessed in 2013
2. Report on the State Environment in China (2009) Ministry of environmental protection of the People's Republic of China
3. Spliethoff H, Scheurer W, Hein KRG (2000) Effect of co-combustion of sewage sludge and biomass on emissions and heavy metals behavior. Process Saf Environ Prot: Trans Inst Chem Eng, Part B 78:33–39

4. Pronobis M (2005) Evaluation of the influence of biomass co-combustion on boiler furnace slagging by means of fusibility correlations. Biomass Bioenergy 28:375–383
5. Barbosaa R, Lapa N, Boavida D, Lopes H, Gulyurtlu I, Mendes B (2009) Co-combustion of coal and sewage sludge: chemical and ecotoxicological properties of ashes. J Hazard Mater 170:902–909
6. Werther J, Ogada T (1999) Sewage sludge combustion. Prog Energy Combust Sci 25:55–116
7. Folgueras MB, Díaz M, Xiberta J (2004) Sulphur retention during co-combustion of coal and sewage sludge. Fuel 83:1315–1322
8. Shuqin G, Dawei H, Xinhong L et al (2003) Several technologies and equipments of sludge thermal drying [J]. China Water Wastewater 19:5–6
9. Xiaowen Y, Yinghao D (2002) Sludge thermal drying applications in the United States [J]. China Water Wastewater 18:1–3
10. Tarraso'n D, Ojeda G, Ortiz O, Alcan JM (2008) Differences on nitrogen availability in a soil amended with fresh, composted and thermally-dried sewage sludge [J]. Bioresour Technol 99:252–259
11. Hongjiang Y, Jinkai Y (2009) Researches on low-temperature pyrolysis pilot scale experiment of sewage sludge [J]. Information of Water Industry Market in China. 3:55–57
12. Inguanzo M (2002) On the pyrolysis of sewage sludge: the influence of pyrolysis conditions on solid, liquid and gas fractions [J]. J Anal Appl Pyrol 63:209–222
13. Bing L, Qingmei Y, Hua Z, Lei S, Youcai Z (2004) Sludge treatment and disposal methods and resources. Saf Environ Eng 4:52–56
14. Di L, Jianlin C (2008) Researches on low-temperature pyrolysis experiment of sewage sludge. Environ Prot Science 34(4):34–36
15. Longmao L, Jianlin C, Di L, Kai Q, Qiangli G (2009) Experimental study on low-temperature catalytic pyrolysis of sewage sludge. Environ Sci Technol 32(7):156–159
16. Emerging Technologies for Biosolids Management (2006) EPA 832-R-06-005
17. Long-Range Biosolid Management Plan (2003) Orange County Sanitation District
18. Rui X, Wei W, Wei Q et al (2008) Status of urban sludge treatment and hydrothermal reduction technology of enhanced dewatering. Environ Sanit Eng 16(2):28–32
19. Zhijun W, Wei W (2005) Enhancement of sewage sludge anaerobic digestibility by thermal hydrolysis pretreatment [J]. Environ Sci 26(1):68–71
20. Neyens E, Baeyens J (2003) A review of thermal sludge pre-treatment process to improve dewaterability. J Hazard Mater 98:51–67
21. Zhijun W, Wei W (2005) Enhancement of sewage sludge anaerobic digestibility by thermal hydrolysis pretreatment. Environ Sci 26(1):68–71
22. M Yoshiaki (2007) Research of high effect drying process by hydrothermal treatment of sewage sludge. Doctorate Thesis, Tokyo Institute of Technology
23. Jiang ZL, Meng DW, Mu HY, Yoshikawa K (2011) Experimental study on hydrothermal drying of sewage sludge in large-scale commercial plant. J Environ Sci Eng 5:900–909
24. Namioka T, Morohashi Y, Yoshikawa K (2011) Mechanisms of malodor reduction in dewatered sewage sludge by means of the hydrothermal torrefaction. J Environ Eng 6(1):119–130
25. Zhao P, Chen H, Ge S, Yoshikawa K (2013) Effect of the hydrothermal pretreatment for the reduction of NO emission from sewage sludge combustion. Appl Energy 111:199–205
26. Melanie JB, Lester JN (1979) Metal removal in activated sludge, the role of bacterial extracellular polymers. Water Res. 13(1):817–837
27. Bruus JH et al (1992) On the stability of activated sludge flocs with implication to dewatering. Water Res 26:1597–1604
28. Koga Y, Endo Y, Oonuki H, Kakurata K, Amari T, Ose K (2007) Biomass solid fuel production from sewage sludge with pyrolysis and co-firing in coal power plant. Mitsubishi Heavy Ind Tech Rev 44(2):43–47
29. Horta C, Gracy S, Platel V, Moynault L (2009) Evaluation of sewage sludge and yard waste compost as a biofilter media for the removal of ammonia and volatile organic sulfur compounds (VOSCs). Chem Eng J 152:44–53

30. He C, Li X-Z, Sharma VK, Li S-Y (2009) Elimination of sludge odor by oxidizing sulfur-containing compounds with ferrate (VI). Environ Sci Technol 43:5890–5895
31. Fujimaki M (1976) Food Chemistry (in Japanese), 9th edn. Asakura Shoten, Tokyo
32. Food Science Handbook Editorial Committee ed (1978) Food science handbook (in Japanese), 1st edn. Kyoritsu Syuppan, Tokyo
33. Wei LH, Qi D, Li RD, Yang TH (2009) Thermal analysis on ignition characteristics of micro-pulverized sewage sludge with coal. Proceedings of the CSEE 29(35):59–63

Printed by Amazon Italia Logistica S.r.l.
Torrazza Piemonte (TO), Italy